Fundamentals of nuclear physics

Fundamentals of nuclear physics

N. A. Jelley
Department of Nuclear Physics
and Lincoln College
University of Oxford

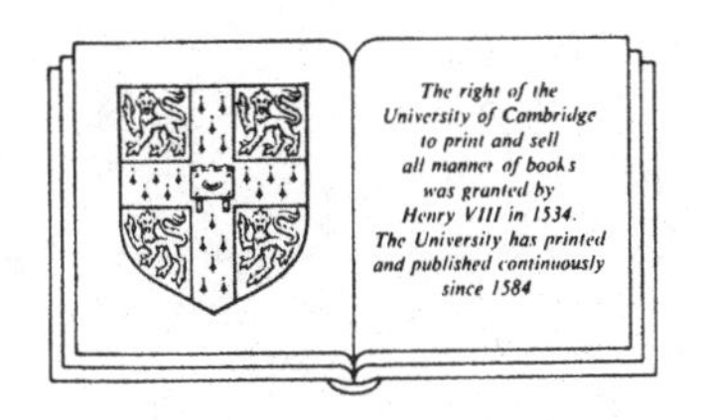

CAMBRIDGE UNIVERSITY PRESS
Cambridge
New York Port Chester
Melbourne Sydney

CAMBRIDGE UNIVERSITY PRESS
Cambridge, New York, Melbourne, Madrid, Cape Town, Singapore, São Paulo

Cambridge University Press
The Edinburgh Building, Cambridge CB2 8RU, UK

Published in the United States of America by Cambridge University Press, New York

www.cambridge.org
Information on this title: www.cambridge.org/9780521264471

First published 1990
Re-issued in this digitally printed version 2007

A catalogue record for this publication is available from the British Library

Library of Congress Cataloguing in Publication data

Jelley, N. A. (Nicholas Alfred), 1946–
Fundamentals of nuclear physics/N. A. Jelley.
p. cm.
Bibliography: p.
Includes index.
ISBN 0-521-26447-2. — ISBN 0-521-26994-6 (pbk.)
1. Nuclear physics. I. Title.
QC776.J45 1990
539.7—dc19 89-565 CIP

ISBN 978-0-521-26447-1 hardback
ISBN 978-0-521-26994-0 paperback

To Jane, John and Tessa

Contents

Preface

The aim of this textbook is to give a student a thorough understanding of the principal features of nuclei, of nuclear decays and of nuclear reactions. The properties of nuclei at low excitation and low angular momentum have been thoroughly studied and are now generally understood, and current research is on nuclei at high interaction energies, high angular momentum and far from the valley of stability. Several models have been developed to explain the observed wide variety of phenomena and I have attempted to describe and justify them, and also to explain the connections between them in some detail. This involves trying to give a microscopic description of nuclei, which is an intriguing many-body problem and forms an important part of this book. Besides this interest, parts of nuclear physics are of importance in the study of elementary particle physics and several nuclear phenomena have particular significance in other fields: for example, fission in nuclear power, fusion in astrophysics and radioactivity in biological tracer techniques. Consequently, nuclear physics is an important part of any physics undergraduate course.

In the first part of the book, several models are described and used to explain nuclear properties with many illustrative examples given. Sections follow on α-, β- and γ-decay, fission, thermonuclear fusion, reactions, nuclear forces and nuclear collective motion. In all of these, many examples are discussed and the student should gain a thorough grounding in our current knowledge of the nucleus. A lot of interesting experimental techniques have been developed to study nuclei and examples of these are also given.

The presentation is quantitative and short derivations are given in full to enable the student to make quantitative predictions about nuclear phenomena. The textbook is aimed at undergraduates in their final year, though the first part will be useful for a more introductory course and the latter chapters contain sections suitable for a first year graduate course. The level of the presentation is aimed at bridging the gap that exists between introductory undergraduate and graduate textbooks.

There are several questions after each chapter; those marked with an asterisk are taken from, or based on, Oxford University Physics Final Honour School questions. Several of the figures are taken from books, articles or papers, and the source is acknowledged in the figure caption.

I would like to thank I. J. R. Aitchison, D. M. Brink, M. Chadwick, P. M. Evans, P. A. Holmes, G. A. Jones, C. N. Pass, W. D. M. Rae, J. R. Rook, A. E. Smith and J. Tigg for kindly reading sections of the book and making many helpful suggestions, Mrs I. M. Smith for her preparation of several of the line drawings and Miss B. A. Roger for her excellent typing of the manuscript. Above all I would like to thank my wife and children for their help and encouragement whilst writing this book.

N. A. Jelley

Physical constants and conversion factors

speed of light in vacuo	$c=(\mu_0\varepsilon_0)^{-1/2}$	2.998×10^8 m s^{-1}
Planck's constant	h	6.626×10^{-34} J s
	$\hbar=h/2\pi$	1.055×10^{-34} J s
		$=6.582\times10^{-16}$ eV s
	$\hbar c$	197.3 MeV fm
elementary charge	e	1.602×10^{-19} C
	$e^2/4\pi\varepsilon_0$	1.440 MeV fm
fine structure constant	$\alpha=e^2/4\pi\varepsilon_0\hbar c$	1/137.0
Bohr radius	$a_0=\hbar/\alpha m_e c$	5.292×10^{-11} m
permittivity of vacuum	ε_0	8.854×10^{-12} F m^{-1}
permeability of vacuum	μ_0	$4\pi\times10^{-7}$ H m^{-1}
Bohr magneton	$\mu_B=eh/2m_e$	9.274×10^{-24} J T^{-1}
nuclear magneton	$\mu_N=eh/2m_p$	5.051×10^{-27} J T^{-1}
Avogadro number	N_A	6.022×10^{23} mol^{-1}
Boltzmann constant	k	1.381×10^{-23} J K^{-1}
		$=8.617\times10^{-5}$ eV K^{-1}
Fermi coupling constant	G_F	1.436×10^{-62} J m^3
Cabbibo angle	θ_c	0.224 rad
nuclear β-decay coupling	$G_\beta=G_F\cos\theta_c$	1.400×10^{-62} J m^3
constant	$G_\beta/(\hbar c)^3$	1.137×10^{-11} MeV^{-2}
gravitational constant	G	6.672×10^{-11} Nm^2 kg^{-2}
atomic mass unit (^{12}C)	u	1.661×10^{-27} kg
		$=931.5$ MeV$/c^2$
neutron-H atom mass difference	m_n-M_H	0.782 MeV$/c^2$

masses (MeV$/c^2$): $m_e=0.511$; $m_\mu=105.7$; $m_{\pi^0}=135.0$; $m_{\pi^\pm}=139.6$; $m_p=938.3$; $m_n=939.6$

1 Å $=10^{-10}$ m; 1 fm $=10^{-15}$ m; 1 b $=10^{-28}$ m^2; 1 eV $=1.602\times10^{-19}$ J;
1 gauss (G) $=10^{-4}$ tesla (T); 0 °C $=273.15$ K

1 Introduction

Nuclear physics is an intriguing subject because of the great variety of phenomena that occur. Nuclei exhibit many different types of behaviour, from the classical, where the nucleus behaves like a liquid drop, to the quantum-mechanical, where nuclei show a shell structure similar to that found in atoms. It is a considerable intellectual challenge to try and understand this behaviour and many models have been devised. These models are described in chapter 2 and used to understand the principal properties of nuclei and their excited states.

The study of beta and gamma decay in nuclei has given considerable information on the structure of nuclei; and from experiments on beta decay on the nature of the weak interaction, for example parity violation and the helicity of the neutrino. These topics are discussed in chapters 3 and 4.

The importance of quantum-mechanical tunnelling in nuclear physics is illustrated by a discussion in chaper 5 of α-decay, fission and thermonuclear fusion. The last two have considerable significance in other fields, for example nuclear power and nuclear astrophysics.

The interactions between nuclei offer a rich variety of phenomena and these are described in chapter 6. The basic types of nuclear reactions: compound nucleus, direct and deep-inelastic, and typical features such as the occurrence of resonances and characteristic angular distributions, are first described before reaction theories are developed.

The forces between nucleons are discussed in chapter 7. The information from neutron–proton and proton–proton scattering is analysed and the characteristics of the nuclear force explained. The connection between the two-nucleon force and the effective interaction between nucleons in nuclei is described and the reason for the success of the independent-particle model of the nucleus is discussed. Finally, the connections between different aspects of nuclear behaviour, in particular collective and single-particle, are drawn together in chapter 8.

In the remainder of the introduction there is a short historical review of nuclear physics followed by a description of the basic features and characteristic dimensions of nuclei. The general technique of scattering particles off nuclei to find out about their size and structure is then explained and examples are given. The spectra of excited states in nuclei with the same number of nucleons are found to have striking similarities which are a consequence of the charge independence of the nuclear force. This is formalised through the introduction of the concept of isospin, which provides an illustration of the important connection between invariance principles and conservation laws.

1.1 Historical review

In 1897 J. J. Thompson discovered the electron and found that most of the mass of an atom was positively charged. He proposed the idea that the atom was rather like a plum pudding with a uniform distribution of positive charge in which the negatively charged electrons were embedded. The previous year Becquerel had established that some atoms gave off ionising radiations, and three types of radioactivity (alpha, beta, gamma) were soon identified. In 1909 Rutherford and Royds identified the alpha rays as ionised helium atoms. Studies of the scattering of alpha particles as they pass through a thin foil led Rutherford to the realisation in 1911 that the positive charge of an atom was concentrated at the centre of the atom, in what is called the nucleus, with the negatively charged electrons surrounding it. From experiments, performed by Geiger and Marsden, on the alpha particle scattering off gold, Rutherford was able to conclude that the radii of the gold and alpha particle nuclei must be of the order of 10 fm (10^{-14} m).

Geiger and Marsden also showed that the positive charge of the nucleus was roughly half of its atomic weight relative to hydrogen. In 1911, Soddy conjectured from studies of radioactivity the existence of isotopes, which are atoms with different nuclear masses but with the same charge and number of atomic electrons and hence the same chemical behaviour. This led to the idea that the nucleus contained protons and electrons, the electrons neutralising some of the protons, but not appreciably altering the mass of the nucleus as the mass of an electron is only $\sim 1/2000$ of the proton's mass. This picture was also thought to account for the origin of electrons in beta radioactivity.

Progress on understanding the atom was made in 1913 when Bohr postulated that atoms only existed in discrete energy levels with the electrons moving around the nucleus with quantised amounts of angular momentum. This model was able to explain the spectrum of the hydrogen atom very well. It was also used by Moseley in 1913 to account for the dependence he had found of the energy of characteristic X-rays on the charge of the atom, which enabled elements to be identified by their characteristic X-rays.

The first nuclear reaction was seen by Rutherford in 1919 when nitrogen was irradiated by alpha particles and protons were produced. But after this, progress on understanding the structure of nuclei was slow, mainly because the only particles available to probe the nucleus were alpha particles from naturally radioactive materials. In the meantime, continued research on beta decay had indicated, by 1928, an apparent violation of energy conservation, which was only resolved by Pauli's hypothesis in 1930 that there was another particle, the neutrino, with which the electron shared the β-decay energy.

After the development of quantum mechanics in 1925 by Schrödinger and Heisenberg, the energetics of alpha decay were explained by Gamow and by Gurney and Condon in 1928 as arising through quantum-mechanical tunnelling of the alpha particle through the nuclear potential barrier. It was also realised that there were considerable difficulties with the proton and electron model of the nucleus. Using the uncertainty principle it was estimated that the energy of an electron within the nucleus was of the order of 50 MeV so it was unclear what was keeping the electron within the nucleus. Furthermore, by 1929 both the spin (integral) and statistics (Bose) of the ^{14}N nucleus were found to be inconsistent with the nucleus containing protons and electrons.

These problems were resolved by the discovery of the neutron by Chadwick in 1932, which led Heisenberg later that year to propose that nuclei consisted of just protons and neutrons (though he envisaged the neutron as made up of a proton and an electron). Such a model was consistent with the observed spin and statistics of ^{14}N and the idea of Iwanenko (1932), made quantitative by Fermi in 1934, that the electron was created in a β-decay, like a photon in a γ-decay, removed the need for electrons to exist in any form within nuclei. The neutron plus proton model of the nucleus was gradually accepted and the subject of modern nuclear physics really dates from this time (1932).

1.2 The scattering of particles by nuclei

The early experiments on α-particle scattering gave an indication of the size of nuclei and with beams of high-energy particles much more detailed information has been obtained. In quantum mechanics the scattering of a particle is described by Fermi's golden rule:

$$w = \frac{2\pi}{\hbar} |M_{if}|^2 \rho_f$$

where w is the scattering probability per unit time, M_{if} is the transition matrix element between initial and final states and ρ_f is the density of final states. This is a general relation and does not require the perturbation V causing the transition to be weak for its validity. However, if V is weak then M_{if} can be evaluated to a good approximation using first-order perturbation theory, also called the Born approximation. In this approximation M_{if} is given by the volume integral:

$$M_{if} = \int \psi_f^* V \psi_i \, d\tau$$

where ψ_i and ψ_f are the initial and final wavefunctions of the scattered particle. For $V = (g^2/4\pi r)e^{-\mu r}$, where g is a measure of the strength of the interaction and μ^{-1} of its range, and a scattering from momentum $\mathbf{p}_i = \hbar \mathbf{k}_i$ to $\mathbf{p}_f = \hbar \mathbf{k}_f$ then $\psi_i = L^{-3/2} \exp(i\mathbf{k}_i \cdot \mathbf{r})$, $\psi_f = L^{-3/2} \exp(i\mathbf{k}_f \cdot \mathbf{r})$ and:

$$\begin{aligned} M_{if} &= \frac{g^2}{4\pi L^3} \int \exp(i\mathbf{q} \cdot \mathbf{r}) \frac{e^{-\mu r}}{r} \, d\tau \qquad (1.1) \\ &= \frac{g^2}{4\pi L^3} \int \exp(iqr \cos\theta - \mu r) r \sin\theta \, d\theta \, dr \, d\phi \\ &= \frac{g^2/L^3}{q^2 + \mu^2} \end{aligned}$$

where $\hbar \mathbf{q} = \mathbf{p}_i - \mathbf{p}_f$ is the momentum transfer in the scattering and L^3 is the normalisation volume.

The matrix element M_{if} is thus the Fourier transform of the potential. For a Coulomb potential $\mu = 0$ and $g = e/\sqrt{\varepsilon_0}$ and in the elastic scattering of a light charged particle by a heavy nucleus the momentum transfer $\mathbf{q}$ is given by $(\hbar q)^2 = 4p_0^2 \sin^2 \frac{1}{2}\theta$ (see figure 1.1) where $|\mathbf{p}_i| = |\mathbf{p}_f| = p_0$ so:

$$M_{if} \propto \frac{1}{p_0^2 \sin^2 \frac{1}{2}\theta} \qquad (1.2)$$

The scattering probability, which is proportional to $|M_{if}|^2$, is therefore proportional to $\sin^{-4} \frac{1}{2}\theta$, which is the angular dependence of the Rutherford scattering formula.

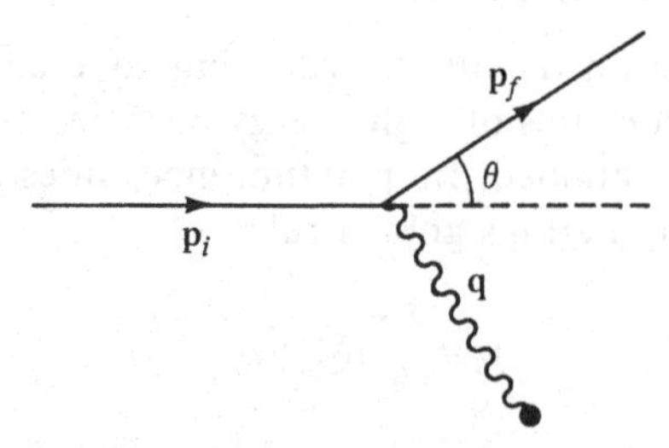

1.1. The relation between the momenta $\mathbf{p}_i$ and $\mathbf{p}_f$, and the momentum transfer $\mathbf{q}$.

1.2.1 Form factor

Generally the source of the perturbation has a spatial extent and this modifies the matrix element M_{if}. For example in the Coulomb scattering by a spherical charge distribution $Z\rho(\mathbf{R})$ the perturbation $V_\rho(\mathbf{r})$ is given by:

$$V_\rho(\mathbf{r}) = \int \rho(\mathbf{R}) V(\mathbf{r}-\mathbf{R})\, \mathrm{d}^3 R$$

where V is the Coulomb potential energy due to a point charge Z.

The matrix element M_{if} is given by the Fourier transform $\bar{V}_\rho(q^2)$ of $V_\rho(\mathbf{r})$. The perturbation $V_\rho(\mathbf{r})$ is a convolution of $\rho(\mathbf{R})$ and $V(\mathbf{r}-\mathbf{R})$ so by the convolution theorem:

$$\bar{V}_\rho(q^2) = F(q^2)\bar{V}(q^2)$$

where

$$\bar{V}(q^2) = \frac{1}{L^3}\int \exp(\mathrm{i}\mathbf{q}\cdot\mathbf{r})\, V(\mathbf{r})\, \mathrm{d}^3 r$$

and

$$F(q^2) = \int \exp(\mathrm{i}\mathbf{q}\cdot\mathbf{R})\rho(\mathbf{R})\, \mathrm{d}^3 R \tag{1.3}$$

So the effect of an extended source is to modify the matrix element M_{if} by the factor $F(q^2)$, which is called the form factor. Note that $F(0) = 1$ for all distributions. Expanding the exponential gives the following expression for $F(q^2)$:

$$F(q^2) = 1 - \frac{q^2\langle R^2\rangle}{6} + \cdots \tag{1.4}$$

so for values of q such that $q^2\langle R^2\rangle \ll 1$ then $F(q^2) = 1$, which is the value for a point charge distribution. The above expression (1.3) for $F(q^2)$ is equivalent to:

$$F(q^2) = \int \rho(R)\frac{\sin qR}{qR} 4\pi R^2\, \mathrm{d}R \tag{1.5}$$

1.2.2 The scattering of electrons by nuclei

If the cross-section for electron scattering off a nucleus is measured over a wide range of momentum transfer q then the charge distribution can be measured from the inverse transform relation:

$$\rho(R) = \frac{1}{2\pi}\int F(q^2)\frac{\sin qR}{qR} 4\pi q^2\, \mathrm{d}q$$

This technique has been used to measure the charge distribution for many nuclei.

A particularly striking example of the information that can be obtained is seen in a comparison of the elastic electron scattering cross-section for ^{205}Tl and ^{206}Pb at an incident energy of 502 MeV. The charge density for each nucleus was deduced using the inverse transform relation together with information from measurements of muonic X-rays. The energies of these X-rays are sensitive to details of the nuclear charge distribution because of the much smaller size of the muon orbitals relative to the electron orbitals. (The radius is proportional to $1/m$ and since $m_\mu = 207 m_e$ their radii are much smaller.) The resulting charge-density difference is shown in figure 1.2. The shape determined by experiment is strikingly close to the one expected for a $3s_{1/2}$ wavefunction. In the simple shell model, using the notation $\pi \equiv$ proton and $\nu \equiv$ neutron, the ground state of ^{205}Tl is described as a pure $(\pi 3s_{1/2})^{-1}(\nu 3p_{1/2})^{-2}$ hole state in ^{208}Pb, i.e. two neutrons and one proton less than the doubly closed shell nucleus ^{208}Pb. The nucleus ^{206}Pb is described as a pure $(\nu 3p_{1/2})^{-2}$ hole state in ^{208}Pb. The difference is therefore one proton in a $3s_{1/2}$ state. The lack of exact agreement reflects the fact that the ground state wavefunctions are more complicated, and, if

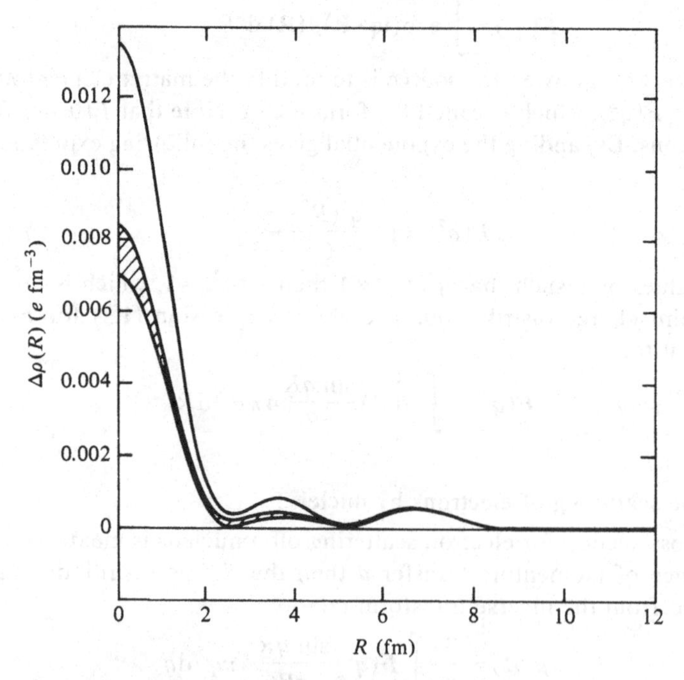

1.2. Experimental charge-density difference between ^{206}Pb and ^{205}Tl together with the shape (solid line) expected for a $3s_{1/2}$ wavefunction. (From Cavedon, J. M. *et al.*, *Phys. Rev. Lett.* **49** (1982) 978.)

this is allowed for, much better agreement is obtained. The similarity in shape, though, provides clear evidence for independent-particle motion within a nucleus.

1.3 Nuclear spectra

Besides giving detailed information on the size and charge distribution of nuclei, scattering experiments enable the excited states of nuclei to be studied. The basic technique is to bombard a thin foil, containing the nuclei under study, with a high-energy beam of particles (e.g. protons, ^{16}O nuclei) provided by an accelerator. The nuclei can be excited, just like atoms or molecules are excited when bombarded by electrons, and the scattered particles, or the gamma radiation following the de-excitation of the excited nuclei (or both), are detected. Their energies give information on the spectra of excited states (see figure 1.3).

The excited states arise when neutrons and protons are excited to higher quantum levels. A typical level scheme is shown in figure 1.4, with each state characterised by its excitation energy, its angular momentum (J) and parity (π). For comparison the excited atomic states of sodium are shown alongside. While a typical excitation energy in an atom is of the order of 1 eV, in nuclei it is of the order of 1 MeV. These magnitudes are what one would expect from applying the uncertainty principle to electrons confined in an atom (size: ~ 1 Å $= 10^5$ fm) and to nucleons (protons or neutrons) confined in a nucleus (size: ~ 10 fm). The natural units of energy and length in nuclear physics are MeV (1 MeV $= 10^6$ eV, 1 eV $= 1.6 \times 10^{-19}$ joules) and fermis (1 fm $= 10^{-15}$ metres). In these units $\hbar c = 197$ MeV fm ($= 1970$ eV Å), which is a useful number when estimating magnitudes in nuclear physics.

Also shown on the level scheme is the excitation energy when the nucleus is unbound to neutron emission, corresponding to the ionisation energy in an atom. Bound excited states generally decay by gamma decay, which is the emission of a photon, just as do atomic excited states. A typical lifetime for such an electromagnetic decay emitting a 1 MeV photon is $\sim 10^{-12}$ seconds in a nucleus, while in atoms for photons of 1 eV it is $\sim 10^{-8}$ seconds. If the excited state is unbound in a nucleus then particle decay usually occurs with a typical lifetime of 10^{-18} seconds. Such a short lifetime is equivalent to an energy width of approximately 1 keV. The connection between mean lifetime, τ, and energy width, Γ, is $\Gamma\tau = \hbar$. Noting that $c = 3 \times 10^{23}$ fm s^{-1} and $\hbar c \approx 200$ MeV fm gives $\Gamma = 0.7$ keV if $\tau = 10^{-18}$ seconds. Nuclear ground states which are beta radioactive have lifetimes which are very strongly dependent on the energy release in the decay but are always greater than 10^{-3} seconds. This much longer timescale

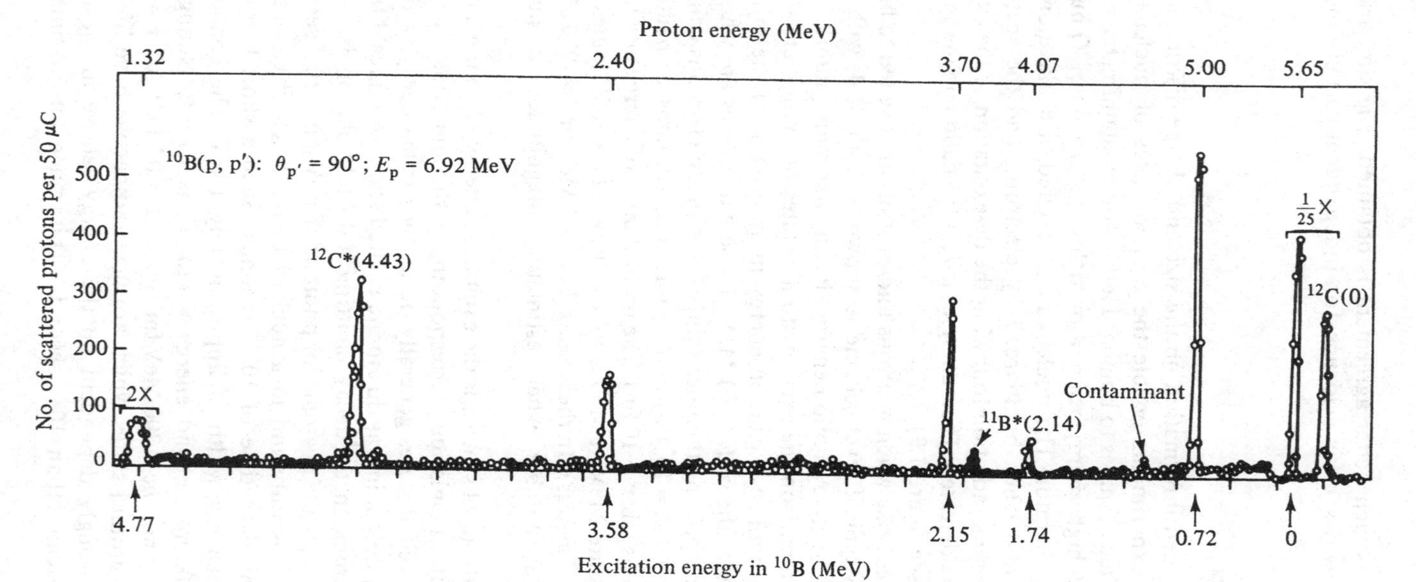

1.3. Spectrum of protons scattered from ^{10}B nuclei. In ^{10}B(p, p′) p refers to the incident and p′ to the scattered proton. Other peaks identified (∗ denotes an excited state) are from ^{12}C, ^{11}B and a contaminant. (Note the non-linear energy scale.) (Data from Bockelman, C. *et al.*, *Phys. Rev.* **92** (1953) 665.)

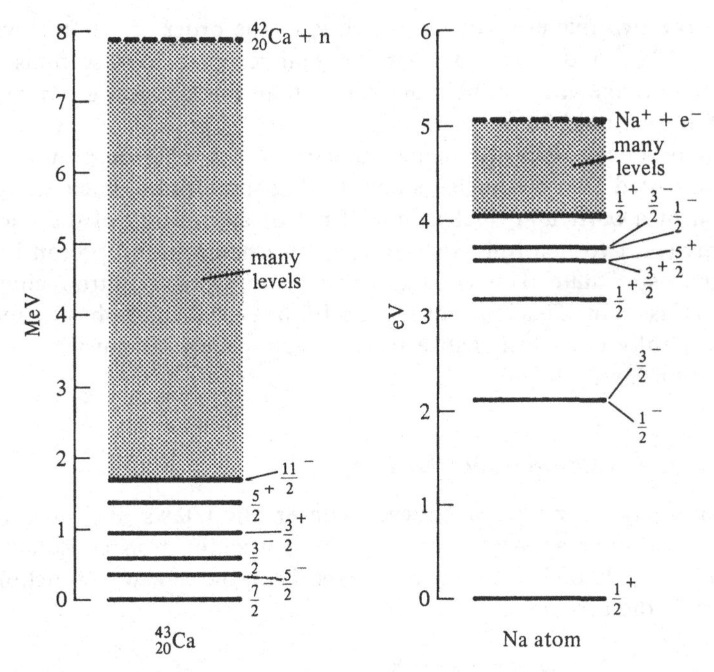

1.4. Diagram comparing the nuclear energy levels of ^{43}Ca with the atomic energy levels of Na. The ionisation energy of Na is ~5 eV and the neutron separation energy of ^{43}Ca is ~8 MeV.

for beta decay is caused by the decay taking place via the weak interaction.

Nucleons interact via the strong, electromagnetic and weak interactions which have characteristic strengths called coupling constants of $f = 1$ (strong), $\alpha = 1/137$ (electromagnetic) and $g \approx 10^{-6}$ (weak). It is now understood that the weak and electromagnetic interactions are part of an electroweak force while the strong force as it is known in nuclear physics is now realised to be a residual (but strong) effect of a more fundamental interaction between the quark constituents of a nucleon, rather like the Van der Waals force between molecules being a residual effect of the electrostatic interactions of the molecular constituents. In nuclear physics a good description of the strong nuclear force is obtained in terms of pion exchange, first proposed by Yukawa in 1935. Its range is short (~1.4 fm).

In the interaction between two nuclei a very important concept is that of cross-section (σ), which is the effective cross-sectional area that the pair of nuclei possess for a particular reaction process. For any interaction to take place the classical estimate of σ would be the sum of the cross-sectional areas of the two nuclei involved, which is of the order of $(10\,\text{fm})^2$ which equals $10^{-28}\,\text{m}^2$ and is called 1 barn. Actual reaction cross-sections vary over a huge range and can be much larger than the classical estimate, but one millibarn would be quite typical.

Levels in nuclei with the same number $(A = Z + N)$ of protons and neutrons, called isobaric nuclei, show striking similarities which are not a reflection of a particular nuclear model but of an equality of the nuclear force between two neutrons, two protons or a neutron and proton in the same space-spin state. This equality can be formalised by introducing the concept of isospin. This concept is a useful one in nuclear physics and to develop it fully one must realise the connection between invariance and conservation laws.

1.4 Invariance and conservation laws

To illustrate the connection between conservation laws and invariance principles consider a system described by a wavefunction ψ. Rotate the system by a small angle ε about the z-axis. Then the new wavefunction ψ' is related to the old one by:

$$\psi' = \psi + \varepsilon \frac{\partial \psi}{\partial \phi} + \cdots$$
$$\equiv \left(1 + \frac{\mathrm{i}\varepsilon}{\hbar} L_z\right)\psi \qquad \left(L_z \equiv -\mathrm{i}\hbar \frac{\partial}{\partial \phi}\right)$$

to first order in ε where L_z is the z component of the angular momentum operator.

If the energy of the system E after rotation by ε is the same as before then:

$$H\psi = E\psi \quad \text{and} \quad H\psi' = E\psi'$$

where H is the Hamiltonian of the system:

$$\therefore \quad \int \psi'^* H \psi' \,\mathrm{d}\tau = \int \psi^* H \psi \,\mathrm{d}\tau$$

$$\therefore \quad \int \psi^* \left(1 - \frac{\mathrm{i}\varepsilon}{\hbar} L_z\right) H \left(1 + \frac{\mathrm{i}\varepsilon}{\hbar} L_z\right) \psi \,\mathrm{d}\tau = \int \psi^* H \psi \,\mathrm{d}\tau \quad (L_z \text{ is Hermitian})$$

$$\therefore \quad \int \psi^* (HL_z - L_z H) \psi \,\mathrm{d}\tau = 0$$

$$\therefore \quad [HL_z] = 0$$

i.e. H must commute with L_z. Heisenberg's equation of motion for L_z is:

$$\frac{\mathrm{d}}{\mathrm{d}t}\langle L_z \rangle = \frac{\mathrm{i}}{\hbar}\langle [HL_z] \rangle$$

so the expectation value of L_z is conserved if H commutes with L_z. Therefore, invariance under rotation about the z-axis is equivalent to conservation of the z component of angular momentum. Similarly it can be shown that invariance under a linear displacement is equivalent to conservation of linear momentum, under a displacement in time to energy conservation and under a reflection to conservation of parity.

Experimental tests of these invariance principles have provided valuable information on the fundamental interactions, e.g. parity violation in nuclear β-decay.

1.4.1 Charge symmetry and charge independence

Examples of the similarities between the level schemes of nuclei of the same atomic number are shown in figure 1.5. In this comparison the neutron-proton mass and Coulomb energy differences have been allowed for and a correspondence between sets of levels of the same J^π can be observed. The nuclei ($^{23}_{11}$Na, $^{23}_{12}$Mg) and ($^{22}_{10}$Ne, $^{22}_{12}$Mg) are pairs of mirror nuclei: that is, the number of protons in one is equal to the number of neutrons in the other, and similarities in their level schemes reflect the equality of the neutron-neutron and proton-proton force when the nucleons are in the same space-spin state. This is called charge symmetry. This similarity though does not imply that the neutron-proton force is the same when the neutrons and protons are in the same state, as the number of neutron-proton pairs is the same in both nuclei. The similarity between certain levels in $^{22}_{11}$Na and levels in $^{22}_{10}$Ne and $^{22}_{12}$Mg though does imply this, as these nuclei are not mirror nuclei and so have different numbers of neutron-proton pairs. This equality is called charge independence.

1.4.2 Isospin and isospin multiplets

This property of charge independence can be formalised by assigning to each nucleon a spin (called isospin) of a half and distinguishing a proton and a neutron by the value of the third component (t_z) of the spin. For a collection of nucleons the total isospin is denoted T and its third component T_z. In nuclear physics it is conventional for a proton to have $t_z = -\frac{1}{2}$ and a neutron to have $t_z = +\frac{1}{2}$. Two protons will therefore have $T_z = -1$; two neutrons $T_z = +1$; and a proton and a neutron $T_z = 0$. The total isospin (T) of two nucleons is $T = 1$ when the nucleons are the same, as $T \geq T_z$ and $T = 0$ or 1 for the proton-neutron pair. The four lowest energy states are shown in figure 1.6. Of these four states only the ground state of the deuteron, which has $J^\pi = 1^+$, is bound. This shows that the nuclear force is spin dependent. All four states are s-states, i.e. $l = 0$, and the Pauli exclusion principle forbids ^{2}n and ^{2}H having $J^\pi = 1^+$. The first excited state of the

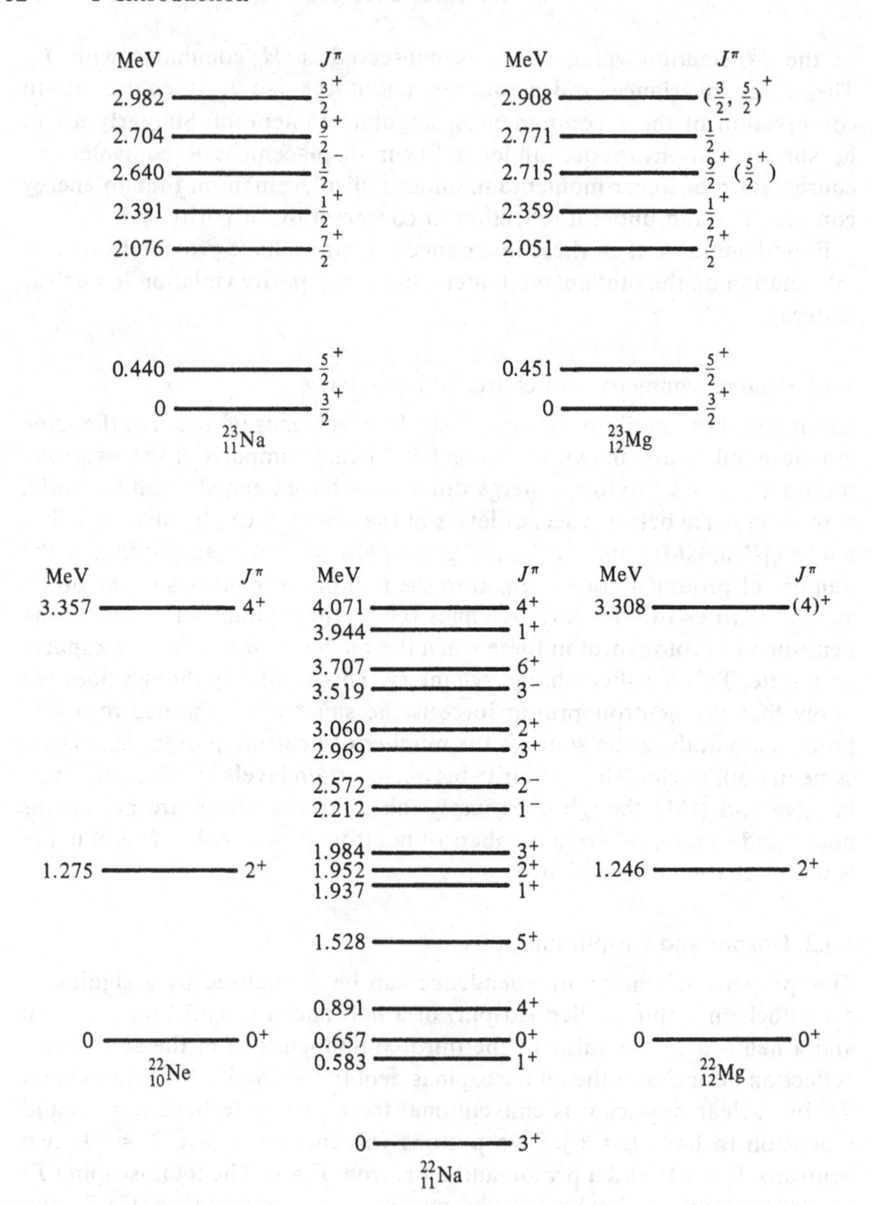

1.5. A comparison of the energy levels of $^{23}_{11}$Na and $^{23}_{12}$Mg, and of $^{22}_{10}$Ne, $^{22}_{11}$Na and $^{22}_{12}$Mg.

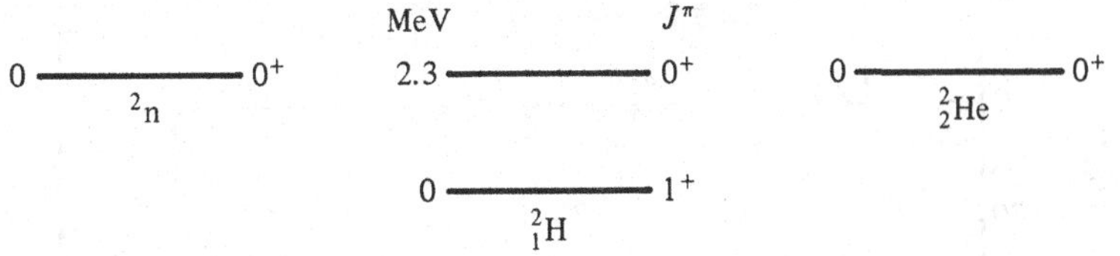

MeV J^{π}

0 —— 0^+ ^{2}n   2.3 —— 0^+ 0 —— 0^+ $^2_2$He

0 —— 1^+ $^2_1$H

1.6. The lowest energy levels of two nucleons.

deuteron has $J^{\pi}=0^+$ and only differs from the dineutron ^{2}n and ^{2}He 0^+ ground states by the change of a proton to a neutron or vice versa. These three 0^+ states form what is called an isospin triplet and have the same energy of interaction allowing for the neutron–proton mass and Coulomb energy differences. They are an example of analogue states.

Charge independence can now be expressed by saying that the energy of a two-nucleon state only depends on the total isospin and not on its third component. This is just like the invariance of the energy of an atom under a rotation in space – the energy does not depend on m, the z component of angular momentum – and, as discussed above, this invariance is equivalent to conservation of angular momentum. Likewise invariance under rotations in isospin space leads to conservation of isospin, which for nucleons is equivalent to charge independence of the nuclear force.

This invariance under rotation in isospin space also holds for collections of nucleons. For the three $A=22$ nuclei, $^{22}_{10}$Ne, $^{22}_{11}$Na and $^{22}_{12}$Mg shown in figure 1.5, the corresponding levels which only differ by changing a neutron to a proton or vice versa are members of an isospin triplet all of which have $T=1$ but different T_z values.

The masses $M(T_z)$ of members of an isospin multiplet only differ in their Coulomb energy and neutron–proton mass differences. The Coulomb energy E_c depends on the expectation value of $E_c=\sum_p (e^2/4\pi\varepsilon_0 r_{ij})$ summed over all protons. This sum can be extended over all nucleons using the isospin formalism:

$$E_c=\sum_{i>j}(\tfrac{1}{2}-t_{z_i})(\tfrac{1}{2}-t_{z_j})e^2/4\pi\varepsilon_0 r_{ij}$$

$$=\sum_{i>j}\{\tfrac{1}{4}-(t_{z_i}+t_{z_j})/2+t_{z_i}t_{z_j}\}e^2/4\pi\varepsilon_0 r_{ij}$$

since $(\frac{1}{2}-t_z)=0$ for neutrons and $=1$ for protons. The quantity $\sum_{i>j}(t_{z_i}+t_{z_j})$ transforms under rotations in isospin space in the same way as T_z. Therefore, the expectation value of the term involving $\sum_{i>j}(t_{z_i}+t_{z_j})$ is proportional to the expectation value of T_z, with the constant of proportionality independent of T_z. (This is just like the expectation value of the magnetic moment operator μ_z being proportional to the expectation value of J_z.) Likewise the term involving $\sum_{i>j} t_{z_i}t_{z_j}$ is proportional to the expectation value of T_z^2.

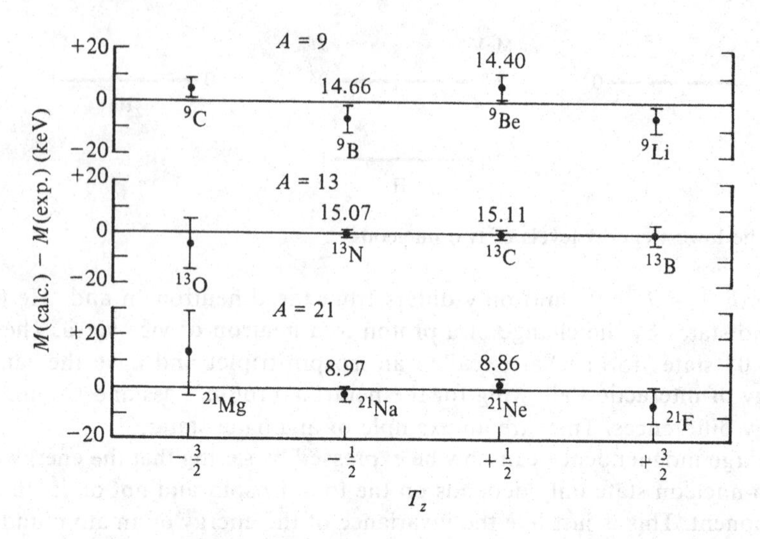

1.7. Isospin quartets illustrating the IMME. The masses are of the ground states of the $T_z = \pm\frac{3}{2}$ nuclei, and of the first excited $T = \frac{3}{2}$ states of the $T_z = \pm\frac{1}{2}$ nuclei. (After Wilkinson, D. H., *Nuclear Physics Lecture Notes.* Oxford Nuclear Physics Laboratory.)

The neutron-proton mass difference depends on T_z therefore:

$$M(T_z) = a + bT_z + cT_z^2$$

where a, b and c depend on the multiplet, as they are dependent on the wavefunctions of the levels. Three examples of this relation, called the isobaric multiplet mass equation (IMME), are shown in figure 1.7 and demonstrate how well it works.

The ground states of all the nuclei shown in the above figures have the minimum possible value of isospin i.e. $T_{gs} = |T_z|$. This is generally the case and in the main part is a consequence of the Pauli exclusion principle. Nucleons within a nucleus move, to a good approximation, independently in a potential well which is roughly square in shape. Figure 1.8 shows the levels occupied by the last few nucleons in a nucleus with $Z = N + 3$, e.g. $^{19}_{11}$Na. There are two per level corresponding to spin up and down. As protons are successively changed to neutrons the $T = \frac{3}{2}$ states in the $T_z = \pm\frac{1}{2}$ nuclei correspond to excited states and in this model it can be seen that the lowest energy states have the minimum possible isospin value. This is also found when there are several nucleons in the same energy level and is then a consequence of the nucleon-nucleon residual interaction favouring minimum T_{gs}. This is because this maximises the spatial symmetry of the wavefunction and hence also the expectation value of the short-range

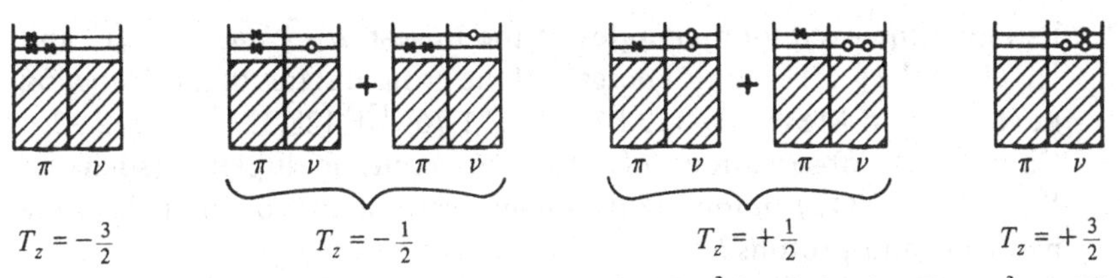

1.8. A simple picture of an isospin quartet. The $T=\frac{3}{2}$ levels in the $T_z = \pm\frac{3}{2}$ nuclei are ground states and in the $T_z = \pm\frac{1}{2}$ nuclei are excited states.

attractive residual interaction (an analogous argument gives rise to Hund's rules in atomic spectra).

The use of isospin allows a generalisation of the Pauli exclusion principle for wavefunctions describing nucleons. For two identical nucleons the wavefunction must be antisymmetric under exchange of the nucleons, i.e. under space and spin exchange. Introducing an isospin part to the wavefunction gives rise to the rule that the total wavefunction for two nucleons must be antisymmetric under space, spin and isospin exchange. Therefore, for two nucleons in an L state and with intrinsic spins coupled to $S(0$ or $1)$, the factor $(-1)^{L+S+T}$ must be odd.

1.5 Questions

1. Show that the expressions for $F(q^2)$ given by equations 1.4 and 1.5 follow from the definition of $F(q^2)$ (equation 1.3).
2.* In the elastic scattering of 200 MeV electrons through 11° by a gold foil, it is found that the scattered intensity is 70% of that expected for point nuclei. Calculate the r.m.s. radius of the gold nucleus.
3. Estimate the smallest structure that can be seen in the charge distribution of a heavy nucleus using 500 MeV incident electrons. Does your estimate agree with what is shown in figure 1.2?
4. The radius, r_n, and energy, E_n, of a particle of mass m in the nth Bohr orbit about an infinitely heavy point nucleus of charge Z are given by $r_n = a_0 n^2/Z$ and $E_n = -Z^2e^2/8\pi\varepsilon_0 a_0 n^2$ where $a_0 = 4\pi\varepsilon_0\hbar^2/me^2$ is the Bohr radius. Comment on the effect of the finite size of a lead nucleus ($Z=82$), radius $=7.1$ fm, on the energies of muonic X-rays corresponding to $n \to n-1$ transitions.
5. Account for the difference in the energies of protons elastically scattered off ^{12}C and ^{10}B seen in figure 1.3.
6. In high-energy physics a system of units in which $\hbar = c = 1$ is sometimes used. In this system show that length $\propto 1/m$, time $\propto 1/m$, energy $\propto m$ and momentum $\propto m$. If m is taken as the mass of a proton, what are the magnitudes of the units of length and of time?

7. Estimate the excitation energies of the lowest $T = \frac{3}{2}$ states in $^{17}_{8}O$ and $^{17}_{9}F$ given the total binding energies of the ground states of the following nuclei are, in MeV: $^{17}_{7}N$ 123.86; $^{17}_{8}O$ 131.76; $^{17}_{9}F$ 128.22; $^{17}_{10}Ne$ 112.92.
8. Could the observation of isotopic spin multiplets (such as $^{14}C_{gs}$, $^{14}N^*$, $^{14}O_{gs}$) imply that nuclei were made of particles like neutrons and protons?
9.* An electron with energy E scatters off a stationary target of mass M, transferring momentum p and energy $\nu = E - E'$, where E' is the electron's final energy. The 4-momentum transfer q is given by $q^2 = p^2 - \nu^2/c^2$. Find an expression for W, the mass of the recoiling hadronic system in an *inelastic* collision in terms of M, ν and q^2. Show that for an *elastic* $(W \equiv M)$ collision $M = q^2/2\nu$. If the electron scattering angle is θ show that, neglecting the electron mass, $q^2c^2 = 4EE' \sin^2(\theta/2)$. In electron scattering off carbon at $E = 194$ MeV and $\theta = 135°$ a peak at $\nu = 5.58$ MeV and a broad peak near $\nu = 51$ MeV are observed. Account for their origin and explain why the peak near 51 MeV is broad.

2 Nuclear models

Soon after the discovery of the neutron by Chadwick in 1932, Heisenberg proposed that nuclei consisted of neutrons and protons bound together by a strong nuclear force. This model avoided the difficulties of the proton plus electron model of the nucleus and was gradually accepted. By the early thirties a considerable number of nuclear masses had been measured by Aston and others using mass spectrometers and it was found that the binding energy per nucleon was approximately constant for all nuclei. The volumes of nuclei had also been determined from scattering experiments to be roughly proportional to the number of nucleons in the nucleus, which implied an approximately constant nuclear density and that the nuclear radius was proportional to $A^{1/3}$. Numerically the radius of a nucleus is given by $R = r_0 A^{1/3}$ fm where $r_0 \approx 1.2$.

In this chapter a number of models of nuclei are described which account for different aspects of nuclear behaviour. It is shown that for the bulk properties of nuclei a liquid drop model of a nucleus is very useful. However, to understand the spins of nuclei and the occurrence of magic numbers a description of the motion of individual nucleons is required and this is provided by the simple shell model. The existence of large nuclear electric quadrupole moments indicates that nuclei are generally deformed in shape and the generalisation of the simple shell model to account for this is described. Several examples of nuclear level schemes are then given and discussed in terms of the simple shell model. While many levels can be explained, it is seen that there are several states which are not easily described by the simple shell model. These are shown to correspond to the collective motion of many nucleons in rotational and vibrational states and the description of these in the collective model is then explained.

2.1 The liquid drop model

Both the behaviour of the binding energy and the size of nuclei are similar to those of a liquid drop where the interaction between molecules is a short-range attractive one with a repulsive core. This Van der Waals interaction is a saturating interaction in that a molecule only interacts with neighbouring, and not all of the other, molecules within the liquid drop. This saturating property of the interaction means that the binding energy of the drop is proportional to the number of molecules in the drop.

In the thirties the proton and neutron were thought to be elementary particles and the idea of a Van der Waals type of strong nuclear force was rejected in favour of a simpler type of saturating short-range interaction called an exchange force. Such a force is found in covalent chemical bonding, which arises through the exchange of electrons. An analogous strong nuclear exchange force was postulated to be a significant component of the interaction between nucleons in order to account for the observed saturation of the nuclear force (see chapter 7 for a discussion of exchange forces). Detailed calculations, however, failed to reproduce the actual values of the observed binding energies although they did account rather well for the variation of the binding energy per nucleon as a function of the number of neutrons, N, and protons Z.

In 1935 Von Weizsacker proposed a semi-empirical approach where the form of the dependence of the nuclear binding energy on N and Z was taken from theory but the coefficients were adjusted to give a best fit to observed binding energies. Bethe and Bacher in 1936 somewhat simplified Von Weizsacker's formula to give what is now known as the Bethe-Weizsacker semi-empirical mass formula.

2.1.1 The semi-empirical mass formula

The semi-empirical mass formula for the mass of a nucleus $M(A, Z)$ is

$$M(A, Z) = Zm_{\mathrm{p}} + (A - Z)m_{\mathrm{n}} - a_{\mathrm{v}}A + a_{\mathrm{s}}A^{2/3} + a_{\mathrm{c}}Z^2A^{-1/3} + a_{\mathrm{a}}(A-2Z)^2A^{-1} + \delta \tag{2.1}$$

where the first two terms are the mass of the protons and neutrons in the nucleus. The third term, called the volume term, $-a_{\mathrm{v}}A$, arises from the saturation of the nuclear force, which gives a binding energy proportional to the number of nucleons caused by nucleons interacting only through a short-range force with neighbouring nucleons. Nucleons near the surface of a nucleus, however, interact with fewer nucleons and are therefore less bound. The number of nucleons near the nuclear surface is proportional to the surface area so the total loss in binding is minimised if nuclei are spherical, as a sphere has the minimum surface area for a given volume. (This behaviour is just like that of a liquid drop.) The correction to the volume term from this loss in binding is therefore proportional to R^2 and hence to $A^{2/3}$, and is accounted for by the fourth term, called the surface term, $a_s A^{2/3}$.

The Coulomb repulsion of the protons in the nucleus is represented by the term $a_c Z^2 A^{-1/3}$. Classically this is the form expected for this term since a uniformly charged sphere has an electrostatic potential energy equal to $\frac{3}{5}(Z^2 e^2/4\pi\varepsilon_0 R)$. A quantum-mechanical calculation using antisymmetrised wavefunctions for the protons also gives a term proportional to $Z^2 A^{-1/3}$ but in addition a smaller term proportional to Z. This latter term is neglected in the mass formula as its effect is included to a large degree in the volume term since for stable nuclei Z is approximately proportional to A.

The term $a_a(A-2Z)^2 A^{-1}$ reflects the tendency for stable nuclei to have approximately equal numbers of protons and neutrons. This tendency can be understood as a consequence of the Pauli exclusion principle. In a nucleus each nucleon moves in an approximately spherical containing potential which represents the average interaction of each nucleon with all the other nucleons. As a first approximation the fluctuations about this average potential (the residual interaction) can be neglected, which is equivalent to treating each nucleon as moving independently in a common potential well. This is unlike the motion of molecules in a liquid drop where the mean free path is short. However, the success of the semi-empirical mass formula reflects the constant density of nuclei and the short range of the nuclear force. It does not require the nuclear mean free path to be short and in this respect the name 'liquid drop model' is misleading.

As the range of the nuclear force is only ~1.4 fm the shape of the well is very similar to that of the nuclear density. The levels of this potential well are filled up satisfying the Pauli exclusion principle and this is illustrated in figures 2.1*a* and 2.1*b*. One half of the well labelled π represents protons, the other half labelled ν represents neutrons. In both figures the Coulomb interaction has been neglected.

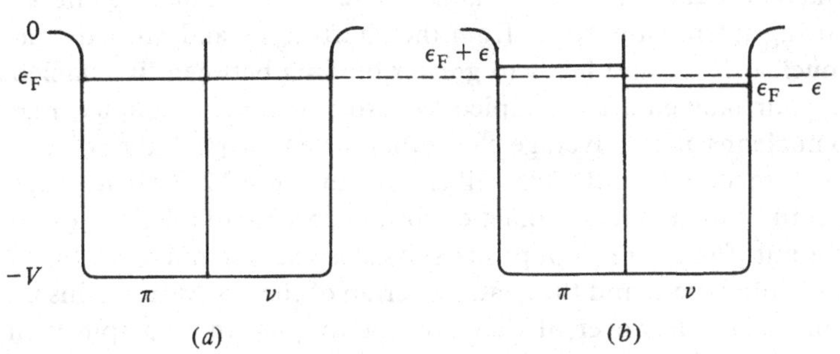

2.1. Diagram illustrating the effect of changing some neutrons to protons on the occupied levels in a nucleus.

Consideration of figure 2.1 shows that for a nucleus with equal numbers of protons and neutrons, changing neutrons to protons (or vice versa) causes a loss in binding as the exclusion principle results in there being an overall increase in the kinetic energy of the nucleons. The magnitude of this effect can be estimated from the following argument. In figure 2.1*a* let the kinetic energy of the protons and neutrons in the highest filled energy level, called the Fermi level, be ε_F. If the density of levels, $dn/d\varepsilon$, at the Fermi level is ρ then, noting that $dn = \rho\, d\varepsilon$ and that each level can contain two protons and two neutrons (spin up or down), the loss in binding energy B_a in changing a few neutrons to protons is given by:

$$B_a = 2\int_{\varepsilon_F}^{\varepsilon_F+\varepsilon} \varepsilon\rho\, d\varepsilon - 2\int_{\varepsilon_F-\varepsilon}^{\varepsilon_F} \varepsilon\rho\, d\varepsilon = 2\rho\varepsilon^2$$

where $\varepsilon_F + \varepsilon$ is the maximum proton kinetic energy and $\varepsilon_F - \varepsilon$ the maximum neutron kinetic energy. The connection between the difference in the number of protons and neutrons $(Z-N)$ and ε is given by:

$$\varepsilon = \frac{1}{4\rho}(Z-N)$$

so

$$B_a = \frac{1}{8\rho}(Z-N)^2 \equiv \frac{1}{8\rho}(A-2Z)^2 \qquad (2.2)$$

which has the quadratic dependence on $(A-2Z)$ shown by the asymmetry term, $a_a(A-2Z)^2A^{-1}$.

When the containing potential in the above model is taken to be a square well and both N and Z are assumed to be much greater than one, the model is called the Fermi gas model of the nucleus and this model (see below) predicts the density of levels at the Fermi level, ρ, to be proportional to A, thus accounting for the whole form of the asymmetry term.

A very important component of the interaction between nucleons which is not taken account of by the containing potential is the pairing interaction. This pairing interaction arises from the short-range and attractive nature of the nuclear force and leads to greater binding between like nucleons if their angular momenta are coupled to zero spin. When coupled like this the two nucleons are on average closer than when coupled to non-zero spin and hence are more bound. This is illustrated in figure 2.2. Two like nucleons cannot both be in the same magnetic substate because of the Pauli exclusion principle but if they are in opposite substates (i.e. m_j and $-m_j$) then their combined spin is zero and the spatial overlap of their wavefunctions is high (see figure 2.2*a*); however, if they are not coupled to zero spin then the overlap is less and the nucleons are less bound (see figure 2.2*b*). The overlap (figure 2.2*a*) increases with increasing orbital angular momentum l, as the orbitals become more localised, with the result that the pairing interaction increases.

This pairing interaction between like nucleons is responsible for the last term, the pairing term δ, in the semi-empirical mass formula. The function δ has been parametrised by:

$$\delta = \begin{cases} +a_{\mathrm{p}}A^{-3/4} & \text{for odd-odd nuclei} \\ 0 & \text{for even-odd nuclei} \\ -a_{\mathrm{p}}A^{-3/4} & \text{for even-even nuclei} \end{cases}$$

and the term accounts for the increased stability observed for nuclei which have an even number of like nucleons (see below). The reason the pairing interaction is not so effective between neutrons and protons in stable nuclei with $A > 40$ is because such nuclei have $N > Z$ because of the Coulomb repulsion of the protons which displaces the neutron and proton potential wells as shown in figure 2.3.

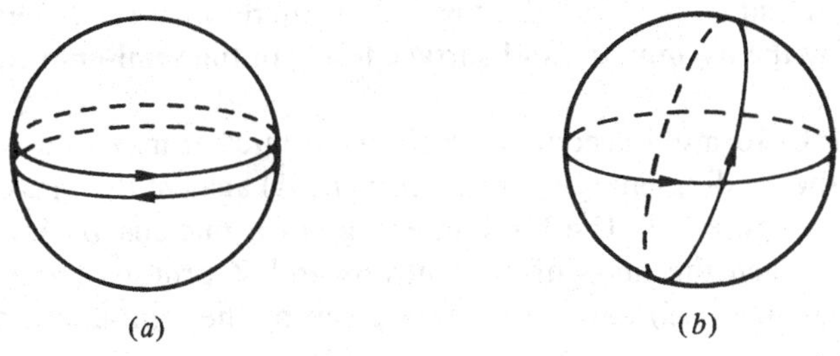

2.2. Illustration of the greater degree of overlap when two nucleons (a) are coupled up to spin zero, than (b) when they are not.

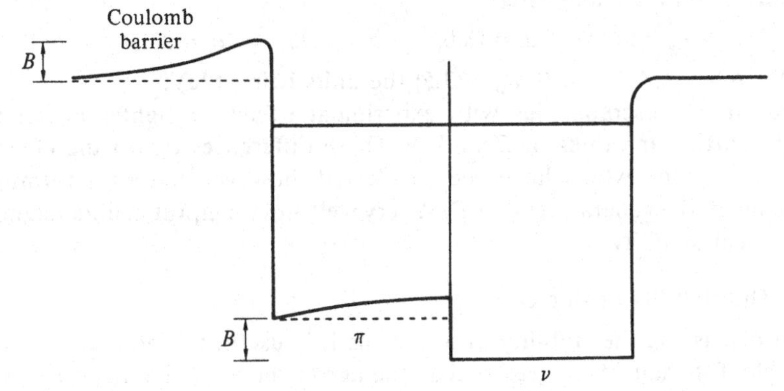

2.3. The effect of the Coulomb energy of the protons on the relative depth and shape of the proton and neutron potential wells.

The neutron and proton wells are both filled up to the same level (the Fermi level) otherwise it would be energetically favourable for a neutron to beta-decay to a proton or vice versa, and therefore there are more neutrons than protons. The residual interaction between nucleons after the containing interaction has been allowed for only acts between nucleons near the Fermi level; so interacting neutrons and protons will be in different states and hence the overlap of their wavefunctions will tend to be less than the overlap of two like nucleon wavefunctions coupled to zero spin. For $A<40$ the residual interactions combine to give an increased stability to even–even nuclei compared with even-odd or odd-odd nuclei.

The form of the semi-empirical mass formula therefore reflects several important features of the interaction between nucleons in nuclei. Not all of these, however, have been mentioned in the discussion above; in particular, the exchange character of the nuclear force (see chapter 7) and the surface diffuseness of nuclei have not been explicitly considered. The first of these gives rise to a term proportional to $(N-Z)^2$ and the second to a term proportional to $A^{2/3}$ and their effect has therefore already been taken account of by the asymmetry and surface terms in the semi-empirical mass formula.

How well the formula accounts for observed nuclear masses can be seen in a plot of the binding energy per nucleon (B/A) against the mass number (A) shown in figure 2.4. The binding energy of a nucleus $B(A, Z)$ is the difference between the mass of N neutrons and Z protons and the mass of the nucleus $M(A, Z)$ and is therefore given by the semi-empirical mass formula as:

$$B(A, Z)=a_vA-a_sA^{2/3}-a_cZ^2A^{-1/3}-a_a(A-2Z)^2A^{-1}-\delta \quad (2.3)$$

A typical set of values of the coefficients obtained by fitting $B(A, Z)$ to measured binding energies is

$$a_v=15.8 \qquad a_s=18.0, \qquad a_c=0.72, \qquad a_a=23.5$$

and for $\delta=\pm a_pA^{-3/4}$ or 0, $a_p=33.5$; the units being MeV.

The largest discrepancies with experiment occur for lighter nuclei at certain particular values of Z and N. These differences reflect the effects of shell structure which have been neglected; however, the mass formula accounts for the general trend of B/A very well and is helpful in understanding nuclear stability.

2.1.2 The stability of nuclei

When discussing the stability of a nucleus it is useful to define a nucleus as stable if its half-life is greater than the age of the earth, i.e. $t_{1/2}\gtrsim 10^9$ yrs, and figure 2.5 shows a plot, called a Segrè chart, of all such stable nuclei. As the mass number A of these nuclei increases so does the ratio of neutrons

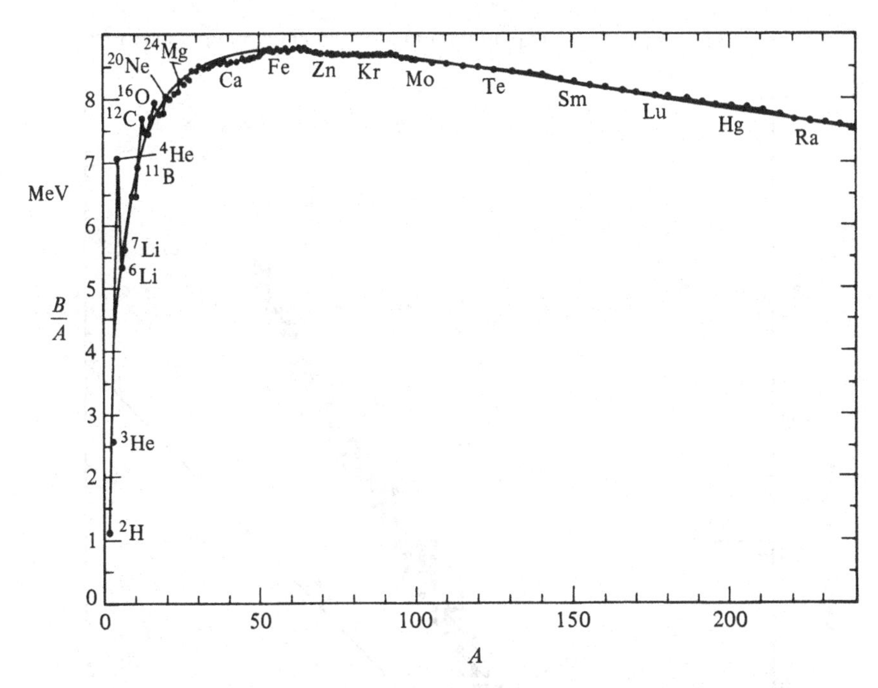

2.4. The binding energy per nucleon as a function of mass number A. The smooth curve is from a semi-empirical mass formula similar to equation 2.3. (From Leighton, R. B., *Principles of Modern Physics.* McGraw-Hill (1959).)

to protons and this reflects a balance between the Coulomb and asymmetry contributions to the nuclear mass. An analysis of the chart shows that there are 177 even–even, 121 even–odd and 6 odd–odd stable nuclei and for each A only one, two or three stable isobars.

Why this occurs is a result of the pairing interaction and can be understood by looking at the dependence of $M(A, Z)$ (equation 2.1) on Z for constant A. For odd-A isobars, for which the pairing term δ is zero, there is a single quadratic dependence of $M(A, Z)$ on Z while for even-A isobars there are two mass parabolas corresponding to $\pm\delta$ (see figure 2.6). Since a nucleus will β-decay if it is heavier than its adjacent isobar (either by β^-, β^+ or electron capture), only one isobar is expected to be stable for odd-A nuclei (figure 2.6a) unless the mass difference is so small that the β-decay half-life is $>10^9$ yrs, as is found for ^{113}In and ^{113}Cd, ^{115}S and ^{115}In and for ^{123}Sb and ^{123}Tb. For the even-A isobars shown in figure 2.6b both of the even–even isobars $^{106}_{48}$Cd and $^{106}_{46}$Pd are stable while the pairing-energy difference makes all of the odd–odd isobars unstable. Energetically $^{106}_{48}$Cd could double

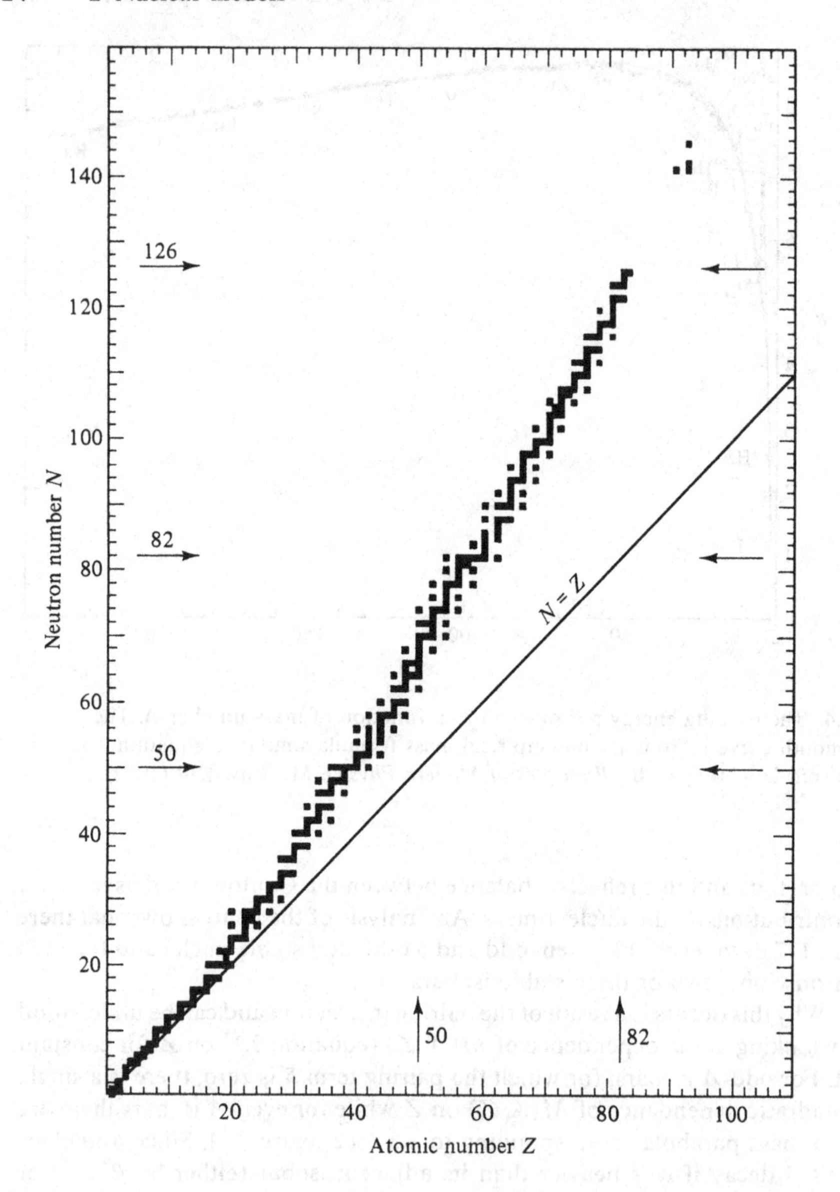

2.5. A plot of N versus Z for all stable nuclei (i.e. $t_{1/2} \gtrsim 10^9$ yrs). (From Cottingham, W. N. and Greenwood, D. A., *An Introduction to Nuclear Physics*. Cambridge University Press (1986).)

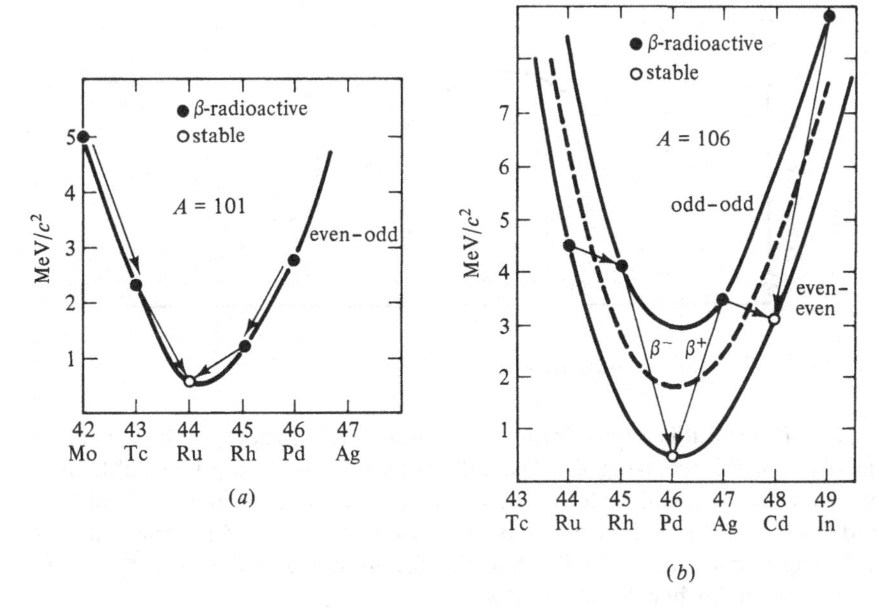

2.6. Variation of mass with Z for (a) odd-A isobars $(A=101)$; (b) even-A isobars $(A=106)$. (From Segrè, E., *Nuclei and Particles*. Benjamin (1977).)

β-decay to $^{106}_{46}\text{Pd}$ but the half-life for this process is much greater than the age of the earth, being $\sim 10^{20}$ yrs, so $^{106}_{48}\text{Cd}$ occurs naturally. There are some even-A isobars $(A=40, 96, 124, 130, 136, 176, 180)$ for which the mass parabolas are such that three isobars are stable. The semi-empirical mass formula would predict no stable odd-odd nuclei while there are a few (^{2}H, ^{6}Li, ^{10}B, ^{14}N, ^{40}K, ^{50}V, ^{138}La, ^{176}Lu, ^{180}Ta) found. However, four of these are very light and for small A the mass formula is unreliable as neglected shell and residual interaction effects are significant; the other five have very long β-decay lifetimes associated with large angular momentum changes (see chapter 3, p. 90).

Another significant feature of the Segrè chart is that there are no stable nuclei above $Z=92$ $(A=238)$, heavier nuclei being unstable $(t_{1/2}<10^9$ yrs) to α-decay. Such nuclei can also break up into two smaller nuclei. This form of decay, called spontaneous fission, is energetically possible for nuclei with $A \gtrsim 90$ (for $A/Z=2.3$), as can be seen from the semi-empirical mass formula, and the reason it is not observed for nuclei with $Z<92$ is that such nuclei are stable with respect to a small deformation from their equilibrium shape. This is because such a change causes an increase in surface area and consequent loss in binding energy which is not offset by the decrease in Coulomb energy arising from the increased separation of

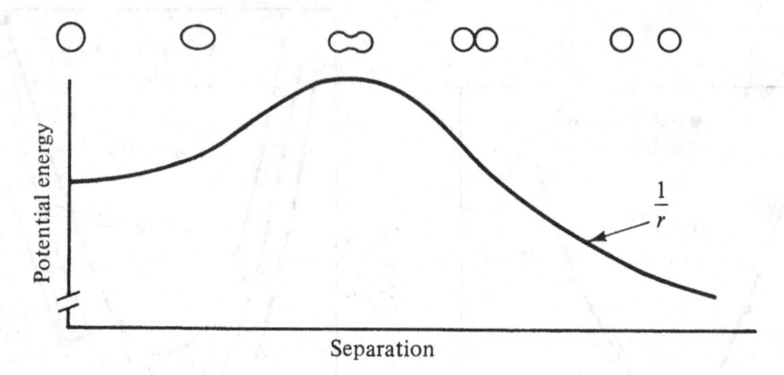

2.7. The form of the fission barrier.

charge. The result is a barrier of a form indicated in figure 2.7 whose height decreases with increasing Z. Classically such a nucleus would be stable but quantum mechanically decay by fission can take place through tunnelling and for $Z \geq 92$ the height of the barrier is sufficiently low for fission decay to be significant; for ^{238}U the half-life for fission decay is $\sim 10^{16}$ yr (see chapter 5 for further details on fission).

2.2 The shell model

2.2.1 Evidence for shell structure in nuclei

There is on the Segrè chart an absence of stable nuclei in the region $209 < A < 232$. An inspection of the binding energy per nucleon (B/A) curve (see figure 2.4) shows that above ^{208}Pb there is a significant decrease in B/A which causes the nuclei in the region above ($209 < A < 232$) to be unstable to α-decay. Similar changes in binding occur at N or $Z = 2, 8, 20, 28, 50, 82$ and $N = 126$ and these numbers are called 'the magic numbers'. There are also characteristic jumps in the single particle separation energies S_p and S_n at the magic numbers as illustrated for S_n in figure 2.8. The separation energies S_p and S_n are given by:

$$S_p = B(Z, N) - B(Z-1, N) \quad \text{and} \quad S_n = B(Z, N) - B(Z, N-1)$$

These changes are similar to those found to occur in atoms at closed electronic shells (the inert gases) and this strongly suggests that they are caused by shell structure in nuclei.

2.2.2 Independent-particle models

The existence of shell structure in atoms arises from the atomic electrons moving to a good approximation independently in a central potential. The short-range interaction between nucleons means that in a nucleus each

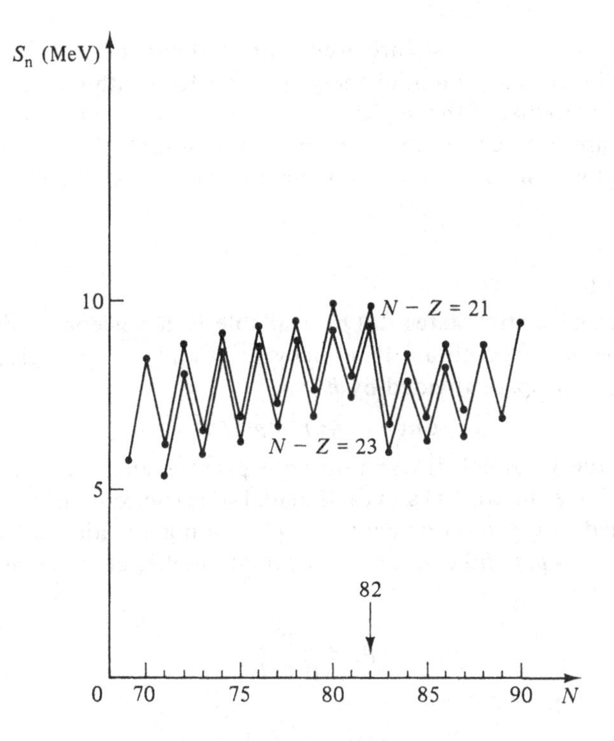

2.8. A plot of the neutron separation energies, S_n. (From Bohr, A. and Mottleson, B. R., *Nuclear Structure* Vol. I. Benjamin (1969).)

nucleon moves in an approximately spherical containing potential, which represents the average interaction of each nucleon with all the other nucleons, of a form illustrated in figure 2.1. Why the fluctuations about this average potential are small and have little effect is because the average separation between nucleons in a nucleus is sufficiently large (~2.4 fm) that their average potential energy is quite weak. This means that any interactions between nucleons can only involve levels quite close in energy and such interactions are inhibited by the Pauli exclusion principle. This is because only nucleons close to the Fermi level can interact as only these have levels which are sufficiently close and unoccupied into which they can scatter. The Pauli exclusion principle is also partly responsible for the large average nucleon separation of ~2.4 fm but as is discussed in chapter 7 the tensor and exchange character of the nuclear force are also important in explaining this separation.

It is therefore a reasonable first approximation to treat a nucleon in a nucleus as moving independently in a central potential. The potential has roughly the shape of a square well with a depth of ~50 MeV, which represents the average potential energy of a nucleon within a nucleus, with a radius equal to that of the nucleus. When the containing potential is taken to be a square well, and both the number of neutrons, N, and protons, Z, are assumed very much greater than one, the model is called the Fermi gas model.

2.2.3 Fermi gas model

The number of spatial states $dn(p)$ available to a nucleon confined to a volume Ω with a momentum lying between p and $p+dp$ is given by the volume in phase space divided by h^3:

$$dn(p) = 4\pi p^2 \, dp\Omega/h^3 \tag{2.4}$$

Neglecting the Coulomb interaction both protons and neutrons move in the same well (see figure 2.1) and each spatial state can contain four nucleons (spin up or down, proton or neutron). Integrating equation 2.4 up to the Fermi level ($p = p_F$) and equating to the total number of nucleons A yields the relation:

$$p_F^3 = \frac{3Ah^3}{16\pi\Omega}$$

and substituting

$$\Omega = 4\pi r_0^3 A/3$$

gives

$$p_F^3 = 9\hbar^3\pi/8r_0^3$$

and the Fermi energy ε_F as:

$$\varepsilon_F = p_F^2/2m$$

$$\varepsilon_F = \hbar^2(9\pi)^{2/3}/8mr_0^2$$

where m is the mass of a nucleon. Taking $r_0 = 1.2$ fm makes $\varepsilon_F = 33$ MeV. Adding the nucleon binding energy of ~16 MeV gives a potential well depth of ~49 MeV, in good agreement with experiment. (There is no allowance for surface, Coulomb or n-p asymmetry effects, so the binding energy is the coefficient of the volume term in the semi-empirical mass formula.)

As mentioned above, the level density at the Fermi level $\rho = (dn/d\varepsilon)_{\varepsilon_F}$ can be estimated with this model. Noting that $dp/p = d\varepsilon/2\varepsilon$ gives:

$$\rho = 3A/8\varepsilon_F = 0.011A \text{ MeV}^{-1}$$

Taking this value for ρ in the expression for the asymmetry energy B_a (equation 2.2 above) gives approximately half the observed value in the semi-empirical mass formula. The difference reflects a dependence of the potential energy on the number of nucleons in spatially symmetric states which in turn depends on $(N-Z)^2$ through the Pauli exclusion principle.

2.2.4 The single-particle shell model

Although the Fermi gas model predicts a level density and a Fermi energy ε_F, no attempt is made in this model to calculate the energy levels exactly. For well bound levels the energies are close to those of an infinite square well and in figure 2.9 the levels for an infinite square well and for a 3-D harmonic oscillator are given. As in the Fermi gas model the separation between levels decreases with increasing A and the order remains the same. The levels are filled up satisfying the Pauli exclusion principle to ~8 MeV, the nucleon separation energy, from the top of the well. A magic number of nucleons occurs when there is a large gap between the last filled level and the next unfilled level.

States of the same l are distinguished by a serial number so, for example, the second p state is labelled 2p. In the harmonic oscillator well (figure 2.9) the levels are equally spaced and for each oscillator quantum number N, $l \leq N$ and l is even or odd as N is even or odd. This means that there are degeneracies: for example, for $N=2$ the $l=0$ (2s) and $l=2$ (1d) states have the same energy. For the square well (figure 2.9) this degeneracy is lifted with the 1d state being lower than the 2s state corresponding to a nucleon having a larger expectation value of r and hence a lower potential energy in the 1d than in the 2s state. As the spin of the nucleon is a half, the maximum number of protons or neutrons each level can take is $2(2l+1)$ so the sequence of magic numbers predicted by the harmonic oscillator well is 2, 8, 20, 40, 70, 112 and by the square well is 2, 8, 20, 40, 92. Although the first three agree with experiment neither of these wells accounts for the higher magic numbers. The shape of the nuclear potential is intermediate between a square and a harmonic oscillator well, so using a more accurate well shape does not solve the problem.

2.2.5 The spin-orbit interaction and magic numbers

These predictions, however, of the magic numbers are based on the assumption of independent nucleon motion. In 1949 it was independently pointed out by Mayer and by Haxel, Jensen and Suess that if there were a significant spin-orbit component to the residual interaction between nucleons, i.e. the interaction which is not taken account of by the containing potential, then all the magic numbers could be predicted. Figure 2.9 shows how the energy levels for a realistically shaped potential well can be altered by the introduction of an attractive spin-orbit potential of the form $-V_{so}(r)\mathbf{l}\cdot\mathbf{s}$ to give the observed magic numbers. This potential gives rise to a jj coupling scheme in which j, m_j, l and s are good quantum numbers. The expectation value of $\mathbf{l}\cdot\mathbf{s}$ can be evaluated by using the relation $\mathbf{j}^2=(\mathbf{l}+\mathbf{s})^2=\mathbf{l}^2+\mathbf{s}^2+2\mathbf{l}\cdot\mathbf{s}$. With

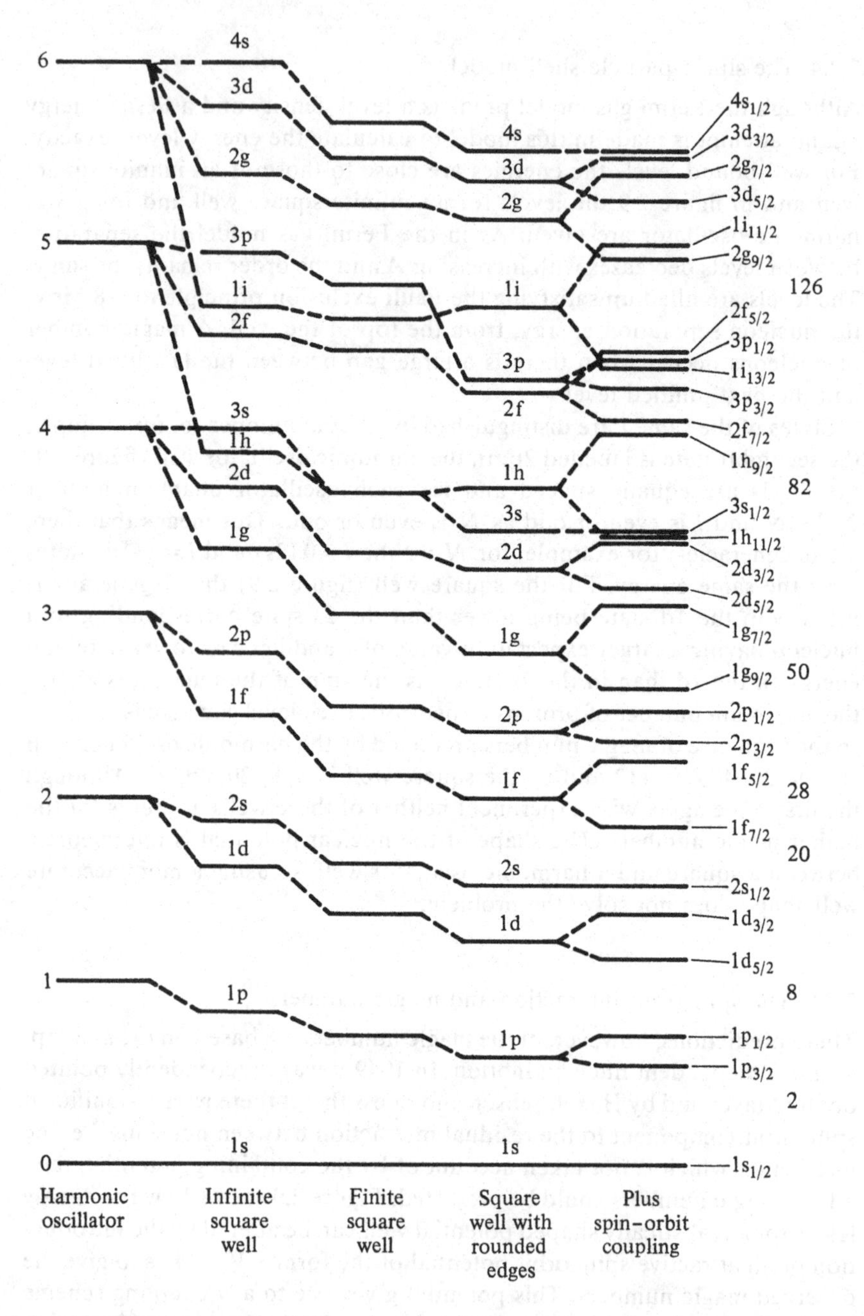

2.9. Ordering of states according to the shell model, using various potentials (schematic). (After Feld, B. T., *Ann. Rev. Nuclear Sci.* **2** (1953) 239.)

j, **l** and **s** in units of $\hbar$, then:

$$\begin{aligned}\langle \mathbf{l}\cdot\mathbf{s}\rangle &= \tfrac{1}{2}\{\langle \mathbf{j}^2\rangle - \langle \mathbf{l}^2\rangle - \langle \mathbf{s}^2\rangle\} \\ &= \tfrac{1}{2}j(j+1) - \tfrac{1}{2}l(l+1) - \tfrac{1}{2}s(s+1) \\ &= \tfrac{1}{2}l, \qquad j = l+\tfrac{1}{2} \\ &= -\tfrac{1}{2}(l+1), \qquad j = l-\tfrac{1}{2} \qquad (l \neq 0)\end{aligned}$$

The splitting of the levels with $j = l+\frac{1}{2}$ and $j = l-\frac{1}{2}$ increases with increasing l with the larger j being more strongly bound.

The origin of this potential is the spin-orbit component in the interaction between two nucleons. How this gives rise to the spin-orbit potential can be seen with a semi-classical argument. In figure 2.10 the nucleon labelled x, which is in a level with orbital angular momentum l, is moving within the nuclear volume with its spin s and l in the same direction (into the paper). The relative angular momenta of the nucleons x and z and of x and y are in opposite directions so there is only a net spin-orbit interaction experienced by the nucleon x if there are more nucleons on one side than on the other, and this occurs when x is near the surface of the nucleus. (The spin-orbit interaction observed in atoms, which is a relativistic electromagnetic effect, is of the opposite sign and is also much weaker.)

2.2.6 Ground state spin and parity

In the single-particle (sp) shell model the remaining residual interaction is assumed to be just the pairing interaction. With this assumption the spins and parities (J^π) of many ground states can be understood. For a nucleus with just one nucleon outside a closed shell then the spin of a nucleus is predicted to be equal to that nucleon's spin as the spins of the nucleons in the closed shells couple up to spin zero; e.g., ${}^{17}_{8}\mathrm{O}$ has one neutron outside the doubly closed shell nucleus ${}^{16}_{8}\mathrm{O}$ and from figure 2.9 its spin is 5/2, which is the observed spin of ${}^{17}_{8}\mathrm{O}$. When a nucleus has levels which are partly

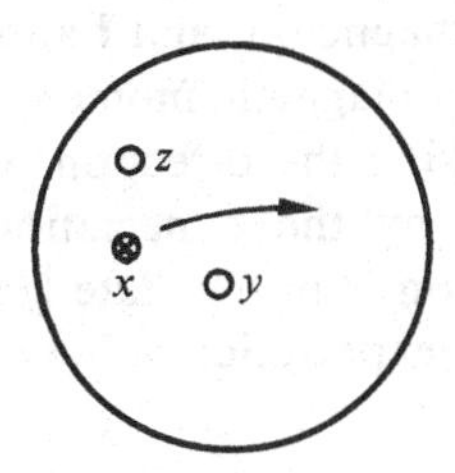

2.10. A diagram illustrating the motion of nucleon x within a nucleus. The relative orbital angular momenta of x with z and with y are opposite, so a spin-orbit interaction only occurs when x is near the surface of the nucleus.

filled, e.g. $^{49}_{22}$Ti with two protons and seven neutrons in the $f_{7/2}$ level, the *sp* shell model predicts that the pairing interaction couples the spin of pairs of like nucleons to spin zero so for an odd-A nucleus the ground state spin would equal the unpaired nucleon's spin, e.g. for $^{49}_{22}$Ti then $J=7/2$. For an odd–odd nucleus the *sp* shell model can only predict that the spin will lie between $|j_p - j_n|$ and $(j_p + j_n)$, where j_p and j_n are the odd proton's and odd neutron's spin, respectively.

The parity of the ground state of a nucleus is determined by the parity of the many-particle wavefunction describing the nucleon state. In the *sp* shell model the total wavefunction involves the product of single-particle wavefunctions, i.e. $\psi_t = \psi_1 \cdot \psi_2 \cdot \psi_3 \ldots$, where ψ_i is the wavefunction of the ith nucleon. The parity of ψ_t is given by the product of the parities of the *sp* wavefunctions so for even–even nuclei the parity is predicted to be always positive; for odd-A nuclei to be equal to the parity of the odd nucleon's wavefunction; and for odd–odd nuclei to be given by the product of the parities of the odd proton and odd neutron's wavefunction. This means, for example, that $^{17}_{8}$O is predicted to have positive parity as the odd neutron is in a $d_{5/2}$ state and $^{40}_{19}$K negative parity as the odd neutron is in a $f_{7/2}$ state and the odd proton in a $d_{3/2}$ state. Besides the spin and parity, the magnetic and electric moments of the nucleus can also be calculated with the *sp* shell model.

2.2.7 Magnetic dipole moment

For even–even nuclei the magnetic dipole moment $\boldsymbol{\mu}$ is zero as the nuclear spin is zero. When there are an odd number of nucleons then $\boldsymbol{\mu}$ is predicted to arise from the magnetic moment of the unpaired nucleon. If this is a neutron then there is only a spin contribution to the moment while if it is a proton then there is both a spin and an orbital contribution, i.e.:

$$\boldsymbol{\mu} = \boldsymbol{\mu}_l + \boldsymbol{\mu}_s = g_l \mu_N \mathbf{l} + g_s \mu_N \mathbf{s}$$

where $g_l = 1$ and $g_s = 5.586$ for a proton, $g_l = 0$ and $g_s = -3.826$ for a neutron, $\mu_N = e\hbar/2m$ is one nuclear magneton, and $\mathbf{l}$ and $\mathbf{s}$ are in units of $\hbar$.

The interaction energy of a magnetic moment $\boldsymbol{\mu}$ in a magnetic field $\mathbf{B}$ is $\boldsymbol{\mu} \cdot \mathbf{B}$, which equals $\mu_z B$, taking the direction of $\mathbf{B}$ as the z axis. So the observed moment is given by the expectation value of μ_z, which is $\langle jm|\mu_z|jm\rangle$, if the odd nucleon is in the state $|jm\rangle$. The magnetic moment $\boldsymbol{\mu}$ is a vector quantity so this expectation value can be calculated using the relation:

$$\langle jm|\boldsymbol{\mu}|jm'\rangle = k\langle jm|\mathbf{j}|jm'\rangle$$

where k is independent of m. The value of k may be evaluated by considering the expectation value of $\boldsymbol{\mu}\cdot\mathbf{j}$:

$$\begin{aligned}\langle jm|\boldsymbol{\mu}\cdot\mathbf{j}|jm\rangle &= \sum_{m'}\langle jm|\boldsymbol{\mu}|jm'\rangle\langle jm'|\mathbf{j}|jm\rangle \\ &= k\sum_{m'}\langle jm|\mathbf{j}|jm'\rangle\langle jm'|\mathbf{j}|jm\rangle \\ &= k\langle jm|\mathbf{j}^2|jm\rangle \\ &= kj(j+1)\end{aligned}$$

where $\mathbf{j}$ is in units of $\hbar$, so the expectation value of μ_z is given by:

$$\begin{aligned}\langle jm|\mu_z|jm\rangle &= \frac{\langle\boldsymbol{\mu}\cdot\mathbf{j}\rangle}{j(j+1)}\cdot\langle jm|j_z|jm\rangle \\ &= \frac{m\langle\boldsymbol{\mu}\cdot\mathbf{j}\rangle}{j(j+1)}\end{aligned}$$

where $\langle\boldsymbol{\mu}\cdot\mathbf{j}\rangle\equiv\langle jm|\boldsymbol{\mu}\cdot\mathbf{j}|jm\rangle$. This is an example of the projection theorem for vector operators, which is the basis of the vector model.

The magnetic moment μ is defined to be the maximum value of $\langle jm|\mu_z|jm\rangle$ and is therefore given by:

$$\begin{aligned}\mu &= \frac{1}{(j+1)}\langle\boldsymbol{\mu}\cdot\mathbf{j}\rangle \\ &= \frac{\mu_N}{(j+1)}\langle(g_l\mathbf{l}+g_s\mathbf{s})\cdot\mathbf{j}\rangle \\ &= \mu_N[g_l(j-\tfrac{1}{2})+\tfrac{1}{2}g_s] \qquad j=l+\tfrac{1}{2} \\ &= \mu_N[g_l(j+\tfrac{3}{2})-\tfrac{1}{2}g_s]\frac{j}{j+1} \qquad j=l-\tfrac{1}{2}\end{aligned}$$

These values, known as the Schmidt limits, are shown joined by lines in figure 2.11 together with the measured magnetic moments of odd-A nuclei.

While there are a few nuclei in good agreement, e.g. ^{17}O and ^{39}K, most nuclei fall between these limits. This shows that in general the wavefunction describing the nuclear ground state cannot be described simply by the *sp* wavefunction of the odd nucleon but must also contain other terms. The deviations, however, do not mean that the *sp* shell model wavefunction is not necessarily a good approximation as the following argument shows. Consider a nucleus whose wavefunction ψ (with c_i real) is:

$$\psi = c_0|0\rangle + \sum_{i=1,n} c_i|i\rangle$$

where c_0 is the amplitude of the *sp* wavefunction $|0\rangle$ and c_i are the amplitudes of the other terms. Then the expectation value of $\mu_z=\langle\mu_z\rangle$ becomes:

$$\langle\mu_z\rangle = c_0^2\mu_{00} + c_0\sum_{i=1,n} c_i(\mu_{0i}+\mu_{i0}) + \cdots$$

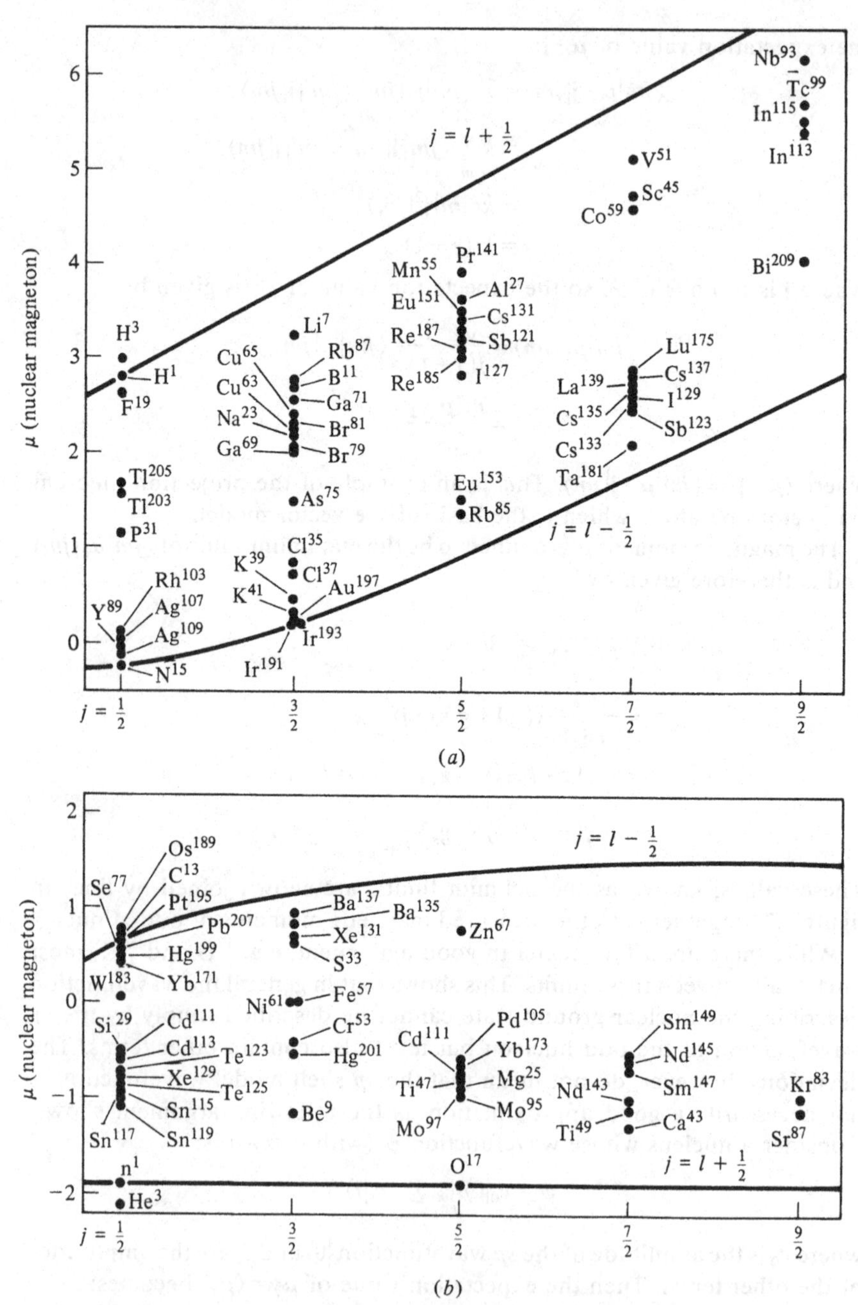

2.11. Magnetic moments for (*a*) odd-*Z* nuclei and (*b*) odd-*N* nuclei, as a function of angular momentum. (From Mayer, M. G. and Jensen, J. H. D., *Elementary Theory of Nuclear Shell Structure*. Wiley, N.Y. (1955).)

where $\mu_{ab} = \langle a|\mu_z|b\rangle$. Now suppose $c_0^2 = 0.8$ and there are 20 other terms each with $c_i = 0.1$, so $\sum_i c_i^2 = 1$, then only the first two terms in $\langle\mu_z\rangle$ are significant with the coefficient of the second, $c_0 \sum_i c_i = 1.78$, greater than that of the first, $c_0^2 = 0.8$. Each μ_{ab} will in general not be of the same sign nor of the same magnitude but it is clear from this expression that large deviations from the *sp* shell model predictions are possible even when the amplitude of the model wavefunction is close to one (0.89).

Another reason why there could be deviations from the Schmidt lines is that the magnetic moment of the nucleon has been assumed to be equal to that of the free nucleon even though the nucleon is within the nucleus and interacting with other nucleons. The mean separation (~2.4 fm), however, is quite a lot larger than the mean nucleon radius (~0.8 fm) so the perturbation from the free nucleon value might be expected to be small and the corrections are only of the order of $\frac{1}{10}$ of a nuclear magneton.

2.2.8 Electric dipole moment

The electric dipole moment of a many-particle system described by a wavefunction ψ is given by the expectation value of $\mathbf{d} = \sum_i q_i \mathbf{r}_i$ where q_i and $\mathbf{r}_i$ are the charge and position of the *i*th particle. If ψ is a non-degenerate state of good parity then $\langle \mathbf{d} \rangle$ will be zero. This is because $\langle \mathbf{d} \rangle$ involves the sum of terms like $\langle \psi_i|\mathbf{r}_i|\psi_i\rangle$, where ψ_i is the *sp* wavefunction of the *i*th particle, all of which are zero as $\mathbf{r}_i$ is odd under the parity operator, i.e. $P\mathbf{r}_i|\psi_i\rangle = -\mathbf{r}_i P|\psi_i\rangle$, so:

$$\begin{aligned} \langle \mathbf{r}_i \rangle &\equiv \langle \psi_i|\mathbf{r}_i|\psi_i\rangle \\ &= \langle \psi_i|\mathbf{r}_i P^2|\psi_i\rangle \qquad P^2 \equiv 1 \\ &= -\langle \psi_i|P\mathbf{r}_i P|\psi_i\rangle \\ &= -\langle \psi_i|\mathbf{r}_i|\psi_i\rangle \\ \therefore \quad \langle \mathbf{r}_i \rangle &= 0 \end{aligned}$$

The interaction between nucleons in a nucleus is not exactly parity conserving, because of the weak nucleon–nucleon interaction, so the ground states of nuclei are only described to a very good approximation by non-degenerate states of good parity; however, their electric dipole moments are still expected to be zero if the nucleon–nucleon interaction is time-reversal invariant. This is because the energy of interaction $\mathbf{d} \cdot \mathbf{E}$ of a nucleus with a dipole moment $\mathbf{d}$ in an electric field $\mathbf{E}$ is not invariant under time-reversal. The dipole moment $\mathbf{d}$ is a vector quantity so $\mathbf{d}$ is proportional to the angular momentum $\mathbf{J}$ and the interaction energy is therefore proportional to $\mathbf{J} \cdot \mathbf{E}$.

Under time-reversal (T):

$$\mathbf{J} \to -\mathbf{J} \text{ and } \mathbf{E} \to \mathbf{E}$$

so for time-reversal invariance $\mathbf{d}$ must be zero. As $\mathbf{J} \cdot \mathbf{E}$ also changes sign under parity $\mathbf{d}$ must be zero for invariance under parity, as shown above. For comparison, the interaction energy of a magnetic dipole in a magnetic field $\mathbf{B}$ is proportional to $\mathbf{J} \cdot \mathbf{B}$ and this is invariant under time-reversal, since $\mathbf{B} \to -\mathbf{B}$, and also under parity.

Although there is no evidence that the strong or electromagnetic interactions violate T there is evidence (K_L and K_S decay) that the weak interaction does to a very small degree, so the nucleon–nucleon interaction is not expected to be strictly T invariant. However, this is an exceedingly small effect so nuclei are not expected to have measurable electric dipole moments, unlike certain molecules such as ammonia.

The reason some molecules have an electric dipole moment is because there are two essentially degenerate ground states of opposite parity arising from the molecule having two ground state configurations with the same energy. For example, in ammonia, NH_3, the two configurations described by wavefunctions ϕ_1 and ϕ_2 are illustrated in figure 2.12. The nitrogen atom can tunnel through the electrostatic potential barrier formed by the three hydrogen atoms so there is a very small probability amplitude for an ammonia molecule to change from one configuration to the other.

As a result the two configurations describing the ground state of NH_3 are not exactly degenerate and are $\phi_+ = (\phi_1 + \phi_2)/\sqrt{2}$ with positive parity and $\phi_- = (\phi_1 - \phi_2)/\sqrt{2}$ with negative parity. The energy difference is very small (10^{-4} eV with ϕ_+ the lower) and in a small electric field the two states ϕ_+ and ϕ_- are essentially totally mixed, i.e. the ammonia molecule is described by $\phi_1 = (\phi_+ + \phi_-)/\sqrt{2}$ and $\phi_2 = (\phi_+ - \phi_-)/\sqrt{2}$, both of which have a non-zero expectation value for the electric dipole moment. Put another way, the ammonia molecule in an electric field experiences a linear Stark effect just like a hydrogen atom does when it is in the metastable 2s state which is almost degenerate with the 2p state. (Note that in a very small electric field NH_3 will exhibit a quadratic Stark effect.)

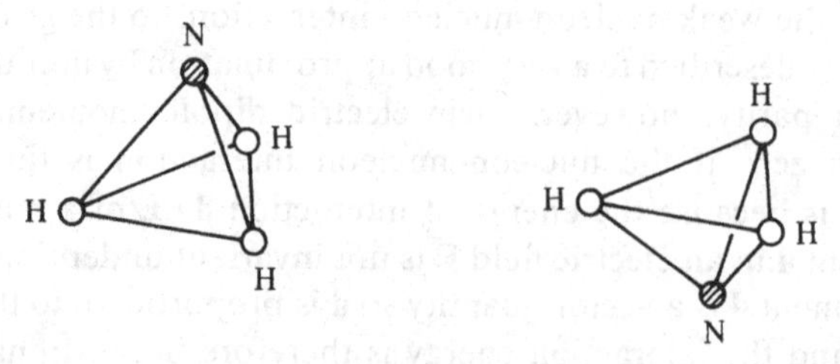

2.12. Illustration of the two degenerate configurations of the NH_3 molecule.

The ammonia molecular wavefunctions ϕ_1 and ϕ_2 are mixtures of an even and an odd function and both have non-zero electric dipole moments. The weak interaction between nucleons means that an essentially positive parity nuclear wavefunction will actually be of the form:

$$\psi = \psi_+ + \delta\psi_-$$

where ψ_+ is even, ψ_- is odd and δ is a very small constant (typically 10^{-6}). This appears to be like ϕ_1 or ϕ_2 in form and therefore it might also be expected to have a non-zero electric dipole moment $\langle\mathbf{d}\rangle$; however, if the weak interaction is time reversal invariant then if the constant δ is purely imaginary and the off-diagonal matrix elements of $\mathbf{d}$ are real then:

$$\begin{aligned}\langle\mathbf{d}\rangle &= \langle\psi_+|\mathbf{d}|\psi_+\rangle + |\delta|^2\langle\psi_-|\mathbf{d}|\psi_-\rangle \\ &\quad + \delta\langle\psi_+|\mathbf{d}|\psi_-\rangle + \delta^*\langle\psi_-|\mathbf{d}|\psi_+\rangle \\ &= 0\end{aligned}$$

since $\delta = -\delta^*$ and $\langle\psi_+|\mathbf{d}|\psi_-\rangle = \langle\psi_-|\mathbf{d}|\psi_+\rangle$.

2.3 Deformed nuclei and nuclear electric quadrupole moments

In the liquid drop model nuclei are assumed to be spherical as the surface tension effect of the short-range interaction between nucleons makes a spherical shape the most stable. However, this model neglects the effects of individual nucleons and in the single-particle shell model odd-A nuclei are generally not expected to be exactly spherical. While pairs of like nucleons coupling up to $J = 0$ give rise to a spherical mass distribution and hence spherical potential for the odd nucleon to move in, the distribution of the odd nucleon is not spherically symmetric unless its spin is a half. The resultant deviation from a sphere, though, is expected to be small as only one nucleon is involved.

A measure of the deviation of a charge distribution from a spherical shape is the electric quadrupole moment of the distribution. Consider a localised classical charge distribution whose charge density is given by $\rho(r, \theta, \psi)$ within a volume τ in a non-uniform electric field $\mathbf{E}$ which has an axis of symmetry along the z axis. Take the centre of the charge distribution as the origin; then the potential V which gives rise to the electric field can be expanded as:

$$V = V_0 + \left(\frac{\partial V}{\partial z}\right)_0 z + \frac{1}{2}\left(\frac{\partial^2 V}{\partial x^2}\right)_0 x^2 + \frac{1}{2}\left(\frac{\partial^2 V}{\partial y^2}\right)_0 y^2 + \frac{1}{2}\left(\frac{\partial^2 V}{\partial z^2}\right)_0 z^2 + \cdots$$

The symmetry of $\mathbf{E}$ means that $(\partial V/\partial x)_0 = (\partial V/\partial y)_0 = 0$ and $\partial^2 V/\partial x^2 = \partial^2 V/\partial y^2$. The source of $\mathbf{E}$ is external to the volume τ so div $\mathbf{E} = 0$. Therefore $\partial^2 V/\partial x^2 = -\frac{1}{2}\partial^2 V/\partial z^2$ and V becomes:

$$V = V_0 - (E_z)_0 z - \frac{1}{4}\left(\frac{\partial E}{\partial z}\right)_0 (3z^2 - r^2) + \cdots$$

The energy U of the charge distribution in the field $\mathbf{E}$ is:

$$U = \int \rho V \, d\tau$$

$$= V_0 q - (E_z)_0 d_z - \frac{1}{4}\left(\frac{\partial E_z}{\partial z}\right)_0 eQ + \cdots$$

where $q = \int \rho \, d\tau$ is the total charge, $d_z = \int \rho z \, d\tau$ is the z component of the electric dipole moment, e is the electronic charge and $Q = (1/e)\int \rho(3z^2 - r^2) \, d\tau$ is called the electric quadrupole moment of the charge distribution. As defined Q has the dimensions of area and is measured in m^2 or barns (10^{-28} m^2). The value of Q depends on the shape of the charge distribution, which is given by ρ and is zero if it is spherically symmetric for then $\int \rho x^2 \, d\tau = \int \rho y^2 \, d\tau = \int \rho z^2 \, d\tau$. For a positively charged distribution with an axis of symmetry along the z axis then $Q > 0$ if it is cigar-shaped and $Q < 0$ if it is discus-shaped (see figure 2.13).

For a nucleus described by a many-body wavefunction ψ then $q = Ze$ is the charge of the nucleus; $d_z = \sum_i \int \psi^* q_i z_i \psi \, d\tau = 0$, as nuclei have zero electric dipole moments (see above); and $Q(\psi)$, the electric quadrupole moment, is given by:

$$Q(\psi) = \frac{1}{e}\sum_i \int \psi^* q_i (3z_i^2 - r_i^2) \psi \, d\tau$$

The quadrupole moment depends on ψ as it depends on the shape of the charge distribution and, in particular, if $\psi^*\psi$ is spherically symmetric then $Q(\psi) = 0$ as for a classical charge distribution. The electrostatic energy U of a nucleus in an electric field $\mathbf{E}$ is therefore:

$$U = V_0 Ze - \frac{1}{4}\left(\frac{\partial E}{\partial z}\right)_0 eQ(\psi)$$

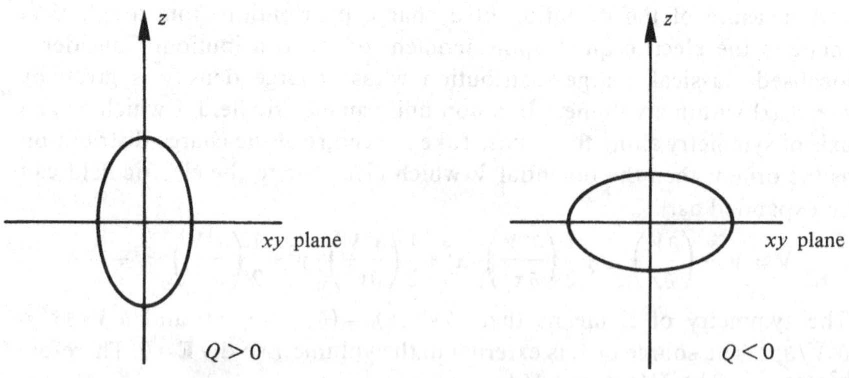

2.13. Illustrating the shape of nuclei with positive ($Q > 0$) and negative ($Q < 0$) quadrupole moments. The axis of symmetry is along the z axis.

If the nucleus has a spin J then the quadrupole moment $Q(JM)$ will depend on M as it depends on the shape and hence the orientation of the charge distribution. The values of $Q(JM)$ for different M are related by:

$$Q(JM_2) = \{[3M_2^2 - J(J+1)]/[3M_1^2 - J(J+1)]\}Q(JM_1) \qquad (2.5)$$

and the quadrupole moment of the nucleus Q is defined as the value of $Q(JJ)$.

In the single-particle shell model an odd-A odd-Z nucleus is described by a single proton of spin j moving in a spherical potential and Q is given by:

$$Q = \int \psi_{jj}^*(3z^2 - r^2)\psi_{jj}\, d\tau$$

If j is large then the wavefunction $\psi(jj)$ is quite localised in the xy plane and $Q \approx -\langle r^2 \rangle \approx -R^2$ where R is the nuclear radius. So, for example, for nuclei with $A \sim 125$ then $R \sim 6$ fm and $Q \sim -0.4\ 10^{-28}\ \mathrm{m}^2$. The exact value of Q is:

$$Q = -\langle r^2 \rangle \frac{(2j-1)}{(2j+2)}$$

so for $J = \frac{1}{2}$, $Q = 0$. For odd-A odd-N nuclei then Q is predicted to be zero as the unpaired neutron has zero charge.

In the 1930s observations on the hyperfine structure of some atomic spectra indicated that some nuclei had an electric quadrupole $Q(JM)$ as well as a magnetic dipole moment $\mu(JM)$. The effect of an electric quadrupole moment is to alter the electron energy levels by an amount proportional to $Q(JM)$ and as $Q(JM)$ has a quadratic dependence on M (equation 2.5) while $\mu(JM)$ has a linear dependence on M, deviations from the Landé interval rule will occur if Q is not zero. By 1949 the data shown in figure 2.14 on nuclear quadrupole moments were available. The arrows indicate where closed shells are predicted by the single-particle shell model. The change of sign of Q at shell closures is as expected in the *sp* shell model as a nucleus with one proton less than a closed shell behaves like a closed shell nucleus with a negatively charged proton, called a proton hole.

While odd-A odd-Z nuclei with only a few nucleons outside a closed shell do have quadrupole moments of the order of $-R^2$, the quadrupole moments are generally much larger and there is no significant difference between odd-Z and odd-N odd-A nuclei. This suggests that for these nuclei many protons are in non-spherically symmetric states (giving rise to a large Q) and as a result the whole nucleus is not spherically symmetric. A non-spherical nucleus, however, is in conflict with the liquid drop model of a nucleus where a spherical shape is the most stable. This difficulty was resolved by Rainwater in 1950 who showed that combining the single-

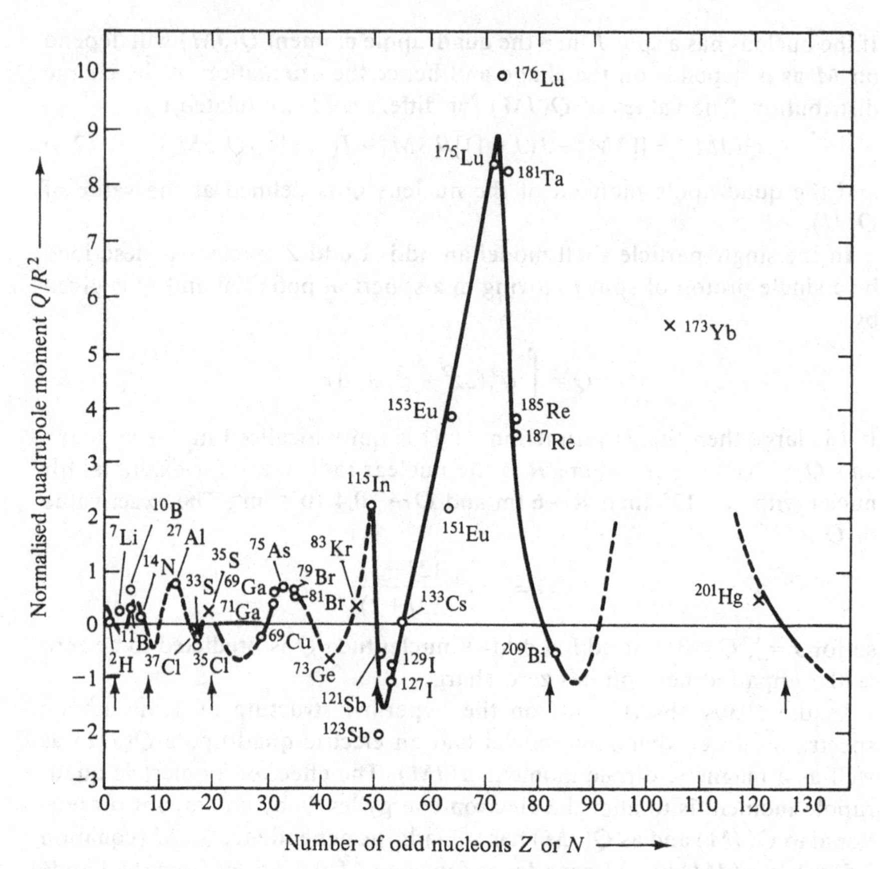

2.14. Measured quadrupole moments Q (expressed in units of the square of the nuclear radius, $R^2 \equiv (1.5 \times 10^{-13} A^{1/3})^2$ cm^2) of odd-proton nuclei (excepting ^{6}Li and ^{36}Cl) plotted as circles against the number of protons Z, and odd-neutron nuclei, plotted as crosses against the number of neutrons N. Arrows indicate the closure of major shells. The solid curve represents the regions where the quadrupole data appear to be well established, the dashed part indicates more doubtful regions. (From Townes, C. H., Foley, H. M. and Low, W., *Phys. Rev.* **76** (1949) 1415.)

particle shell model and the liquid drop model would predict that nuclei were in general deformed, in which case nucleons moved in a deformed potential.

2.3.1 The deformed shell model

Consider first the change in binding energy predicted by the liquid drop model when a spherical nucleus is deformed to an ellipsoid, as illustrated

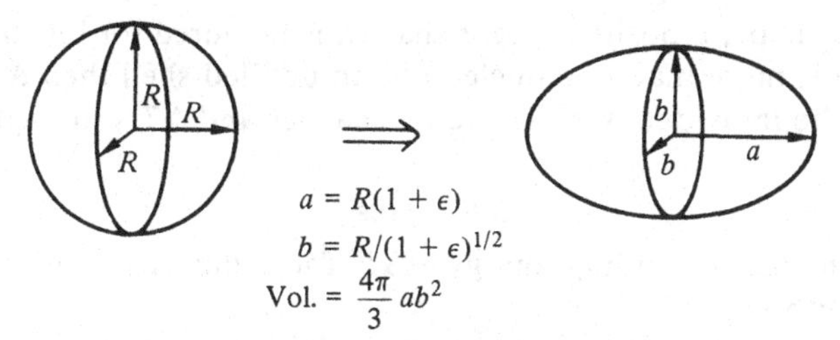

2.15. Diagram illustrating the deformation of a sphere to a prolate (cigar-like) ellipsoid of the same volume. (The parameter ε is negative for an oblate deformation, positive for prolate.)

in figure 2.15. The nucleus is assumed incompressible so the ellipsoid has the same volume. The major and minor axes are given by:

$$a = R(1+\varepsilon) \quad \text{and} \quad b = R/(1+\varepsilon)^{1/2} \tag{2.6}$$

To second order in ε, the surface area of the nucleus increases by $(1+\frac{2}{5}\varepsilon^2)$ while the Coulomb energy, assuming a uniform charge distribution, decreases by $(1-\frac{1}{5}\varepsilon^2)$. Substituting these factors into the semi-empirical mass formula (equation 2.3) gives the change in binding energy $-\Delta E_B$ as:

$$\Delta E_B = a_r \varepsilon^2 \text{ MeV}$$

where

$$a_r = 0.14 A^{2/3}\left(49 - \frac{Z^2}{A}\right) \tag{2.7}$$

predicting that nuclei with $Z^2/A \lesssim 49$ will be stable.

Rainwater assumed that the expression for ΔE_B held for closed shell nuclei but not for nuclei with an unfilled shell. For such nuclei he showed that the change in kinetic energy of the nucleons in the unfilled shell upon distortion gives rise to an additional term linear in ε. To see how this comes about consider first one nucleon outside a closed shell in a spherical potential well of radius R with orbital angular momentum l in the $m = l$ substate (a Bohr orbit). If the momentum of the nucleon is p_0 then, for $l \gg 0$, $p_0 R \approx l\hbar$, so its kinetic energy is proportional to $1/R^2$. The nucleon is at the Fermi level in the nucleus so the Fermi gas model (see above) can be used to estimate its kinetic energy, T_0. The change ΔT in kinetic energy caused by the deformation $R \to R/(1+\varepsilon)^{1/2}$ is therefore:

$$\Delta T = \varepsilon T_0 \quad \text{for } m = l \gg 0$$

i.e. an oblate deformation, ε negative (discus-like), gives a lowering in the kinetic energy for the $m = l$ state. For $m \neq l$ ΔT is given in the limit of large l by:

$$\Delta T = -\frac{3}{2l^2}[\tfrac{1}{3}l(l+1) - m^2]\varepsilon T_0 \tag{2.8}$$

so a prolate shape, ε positive (cigar shape), is favoured for low m^2 values. When there is more than one nucleon in an unfilled shell then ΔT will be the sum of the individual kinetic energy changes and ΔT will in general be given by:

$$\Delta T = +b_r\varepsilon$$

The total change in binding energy $-\Delta E$ for a nucleus with an unfilled shell is therefore:

$$\Delta E = a_r\varepsilon^2 + b_r\varepsilon$$

which has a minimum value of $-b_r^2/4a_r$ when $\varepsilon = -b_r/2a_r$, corresponding to an increase of binding energy of $b_r^2/4a_r$, so the nucleus will be deformed in its ground state and not spherical.

In general these considerations predict both an oblate and a prolate minimum energy shape for a nucleus, the one lower in energy corresponding to the ground state. As an example consider the nucleus ${}^{173}_{70}\text{Yb}$, which has a nearly full $l = 3$ neutron shell with a closed 3s proton subshell, in the approximation of neglecting the spin–orbit interaction (see figure 2.9). There are eleven neutrons in the $l = 3$ state (the 2f levels) and if ten are paired off in the $m = 0$, ± 1 and ± 2 substates with the eleventh neutron in the $m = \pm 3$ substate, then substituting in equation 2.8 for the individual kinetic energy changes and summing gives a value for b_r of $-\frac{5}{2}T_0$, which corresponds to a prolate (cigar) equilibrium shape. The alternative pairing, which gives the most oblate rather than prolate shape, is with the neutrons in the $m = \pm 1$, ± 2 and ± 3 substates with the unpaired neutron in the $m = \pm 1$ substate. This yields a value of b_r equal to $\frac{11}{6}T_0$. But this gives less of an increase in binding $(= b_r^2/4a_r)$ than the other pairing so ${}^{173}_{70}\text{Yb}$ is predicted to have a prolate ground state. The value of a_r for ${}^{173}_{70}\text{Yb}$ $J = \frac{5}{2}$ is 90 MeV, which with $b_r = -\frac{5}{2}T_0$ and $T_0 = 33$ MeV gives an $\varepsilon = +0.46$. The deformation parameter ε is related to the intrinsic or classical quadrupole moment for small ε by

$$Q_{\text{int}} = +\tfrac{6}{5}ZR^2\varepsilon \tag{2.9}$$

This is the quadrupole moment of a deformed nucleus whose orientation is fixed in space. Quantum-mechanically a deformed nucleus and its orientation in space are described by a wavefunction and the observed quadrupole moment is related to Q_{int} by:

$$Q = \frac{J(2J-1)}{(J+1)(2J+3)}\,Q_{\text{int}} \tag{2.10}$$

which for $\varepsilon = +0.46$, $r_0 = 1.2$ and $J = \frac{5}{2}$ gives $Q = 6.1 \times 10^{-28}$ m^2, which is more than twice as large as the measured value of 2.8×10^{-28} m^2.

This overestimation of Q using the Rainwater model is typical and comes about from underestimating the coefficient of ε^2, but shows very clearly that quadrupole moments many times the simple shell model estimate can be understood by simply generalising the simple shell model to include a deformed potential well. Moreover, the greatest contribution to the quadrupole moment comes from the distortion of the core and explains why odd-A nuclei with an unpaired neutron are found with large quadrupole moments.

The distortion of the core produces a potential well for the nucleons in an unfilled shell which is closer in shape to the shape of the nucleon orbitals, i.e., it is more self-consistent and hence lower in energy (see chapter 8, p. 245 for discussion of self-consistent solutions). Near magic numbers, the pairing interaction which pairs off like nucleons in $J=0$ (spherically symmetric) states tends to keep nuclei nearly spherical. The resultant shape of any nucleus can therefore be viewed as a balance between the overlap of non-spherical orbitals which encourages deformation and the pairing of like nucleons which favours a spherical shape.

2.3.2 The Nilsson model

The model proposed by Rainwater treated the core by the liquid drop model and only the outer nucleons quantum-mechanically. A significant advance was made by Nilsson, who considered the single-particle energy levels of all the nucleons moving in an axially deformed harmonic oscillator potential $V(x, y, z)$ with spin–orbit and orbital angular momentum dependent terms giving a Hamiltonian of the form:

$$H=\frac{-\hbar^2}{2m}\nabla^2+V(x,y,z)-C\mathbf{l}\cdot\mathbf{s}-D\mathbf{l}^2 \tag{2.11}$$

where $V(x, y, z)=\frac{1}{2}m(\omega_x^2x^2+\omega_y^2y^2+\omega_z^2z^2)$ and $\omega_x=\omega_y$.

The frequencies ω_x and ω_z are inversely proportional to the half-axes b and a, respectively, of the ellipsoid (see figure 2.15): $\omega_x=\omega_0R/b$ and $\omega_z=\omega_0R/a$, so $\omega_x=\omega_0(1+\varepsilon)^{1/2}$ and $\omega_z=\omega_0(1+\varepsilon)^{-1}$, for small ε. A deformation parameter δ is usually used which is equal to $3\varepsilon/2$ for small ε (n.b. many different parametrisations of nuclear deformation are used). To second order in δ, $\omega_x=\omega_0(\delta)(1+\frac{1}{3}\delta)$ and $\omega_z=\omega_0(\delta)(1-\frac{2}{3}\delta)$, where $\omega_0(\delta)=\omega_0(1+\frac{1}{9}\delta^2)$ to ensure $\omega_x\omega_y\omega_z=\omega_0^3$. The oscillator constant ω_0 is determined by the r.m.s. radius of the nucleus and is given by:

$$\hbar\omega_0\approx 41A^{-1/3}\ \text{MeV}$$

Consider first the energy eigenvalues ε_0 of the deformed harmonic oscillator potential well alone. These are given by:

$$\begin{aligned}\varepsilon_0(n_xn_yn_z)&=\hbar\omega_x\left(n_x+\frac{1}{2}\right)+\hbar\omega_y\left(n_y+\frac{1}{2}\right)+\hbar\omega_z\left(n_z+\frac{1}{2}\right)\\&=\left[\left(N+\frac{3}{2}\right)-\delta\left(n_z-\frac{N}{3}\right)+\tfrac{1}{9}\delta^2\left(N+\frac{3}{2}\right)\right]\hbar\omega_0\end{aligned} \tag{2.12}$$

where $N=n_x+n_y+n_z$. For $\delta=0$ these are just the eigenvalues of a 3-D harmonic oscillator well. For δ positive (prolate) then if $n_z>N/3$ (i.e. more probability amplitude for large z than large x or y) the energy is lower than for $\delta = 0$, and vice versa for δ negative. The axial symmetry means that m_l, m_s and $m_j = \Omega$ are good quantum numbers and it is usual to characterise the eigenstates by the quantum numbers $\Omega\pi[Nn_z m_l]$ where π is the parity of the state ($\pi = (-1)^l = (-1)^N$).

In the Nilsson Hamiltonian (equation 2.11) there is in addition an $\mathbf{l} \cdot \mathbf{s}$ and an $\mathbf{l}^2$ term. The $\mathbf{l}^2$ term has the effect of making the potential well more like a deformed square well, as it makes higher l states have a lower energy, which is a better approximation to the nucleon potential well. The parameter C gives the strength of the spin–orbit force and C and D are chosen to fit the observed sequence of single-particle levels. Because of the $\mathbf{l} \cdot \mathbf{s}$ and $\mathbf{l}^2$ terms the only good quantum numbers are Ω and π but $[Nn_z m_z]$ can also be used to characterise the levels as in the 3-D deformed oscillator potential. Figure 2.16 shows the predicted energy levels as a function of δ for light nuclei. At $\delta = 0$ there is the usual sequence of levels: $1s_{1/2}$, $1p_{3/2}$, $1p_{1/2}$, $1d_{5/2}$, $2s_{1/2}$, $1d_{3/2}$, $1f_{7/2}$, $2p_{3/2}$.... For non-zero δ each level splits up into $(2j+1)/2$ levels each of which is two-fold degenerate with eigenvalues $\pm\Omega$. As in the Rainwater model, levels with low Ω (equivalent to low m) are shifted lower in energy for prolate deformation and higher for oblate deformation.

The spin and parity of the ground state of a deformed odd-A nucleus can be found with the Nilsson model in the following way. From the observed quadrupole moment δ is determined using equations 2.9 and 2.10. Each level is two-fold degenerate so can accommodate two protons and two neutrons and these are assumed to be coupled to spin zero through the residual pairing interaction. The spin and parity of the odd-A nucleus is therefore determined as in the simple shell model by that of the unpaired nucleon. The wavefunction of the unpaired nucleon is for a nucleon in a deformed potential well fixed in space with the result that $\mathbf{J}^2$ does not commute with the Hamiltonian. This is unphysical, however, as the nucleus is free to orientate itself in any direction in space, with the result that the true wavefunction for the unpaired nucleon has a spin $J = \Omega$. As an example consider the nucleus ^{23}Na: its measured quadrupole moment is 0.10 barns and its $J^\pi = \frac{3}{2}^+$ corresponding to a δ of ~0.33. Filling the levels up pairwise leaves the unpaired proton in a $\frac{3}{2}+[211]$ state, i.e. a $J^\pi = \frac{3}{2}^+$ state, in agreement with what is observed. In a similar way the observed spins and parities of many odd-A deformed nuclei are able to be accounted for.

For even–even nuclei the predicted spin and parity is 0^+ in the Nilsson model and hence the wavefunction for the nucleus is spherically symmetric although the nucleus is deformed. This can be thought of as an intrinsically deformed state having an equal probability of pointing in any direction giving rise to an isotropic probability distribution. The diffuseness of the

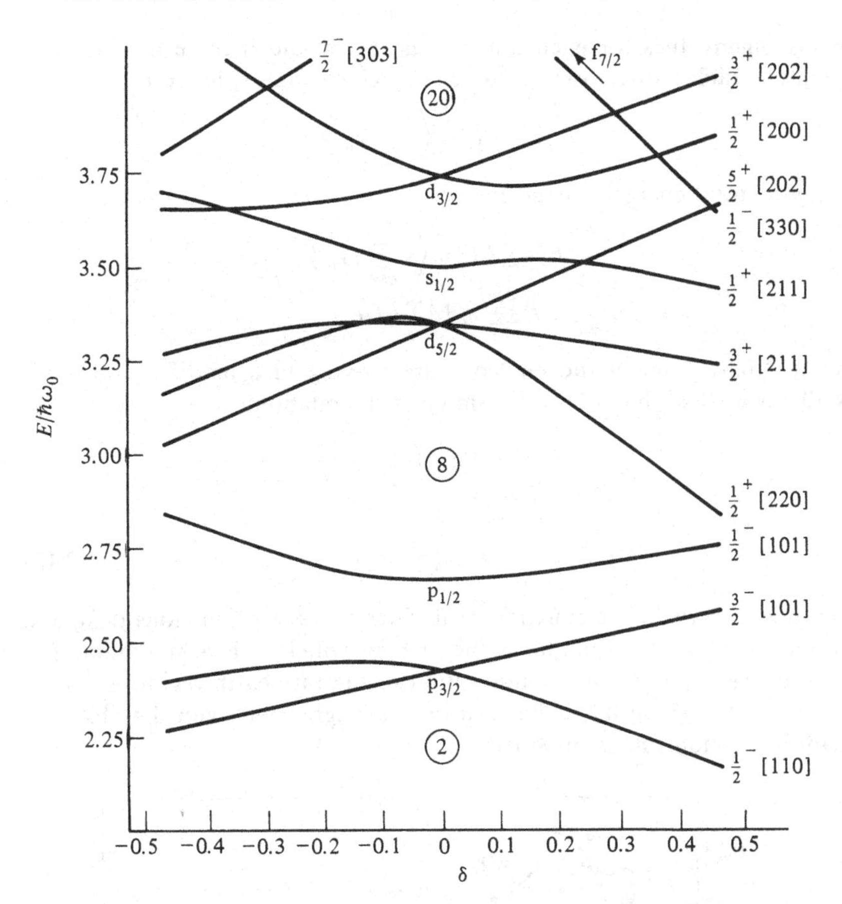

2.16. Predicted energy levels in the Nilsson model as a function of the deformation parameter δ for light nuclei. (From Nathan, O. and Nilsson, S. G., in *α-, β- and γ-ray Spectroscopy* (ed. Siegbahn, K.). North-Holland (1965), p. 601.)

surface of a deformed even–even nucleus would however be expected to be larger than for a spherical nucleus and this has indeed been seen in electron–nucleus scattering.

2.3.3 Ground state quadrupole moment

The ground state deformation and hence quadrupole moment can be calculated by finding the deformation for which the total energy of the nucleus is a minimum. This cannot be done by simply summing the individual energy eigenvalues for each nucleon as this would include the potential energy of each particle twice. The energy eigenvalue ε_i is given by:

$$\varepsilon_i = \langle T_i \rangle + \left\langle \sum_{i,j \neq i} V_{ij} \right\rangle$$

while the total energy E is given by:

$$E = \sum_i \langle T_i \rangle + \left\langle \tfrac{1}{2} \sum_{i,j \neq i} V_{ij} \right\rangle$$
$$= \tfrac{1}{2} \sum_i \varepsilon_i + \tfrac{1}{2} \sum_i \langle T_i \rangle$$

In the Nilsson model the nucleons are moving in a modified harmonic oscillator well so the virial theorem gives the relation:

$$\langle T_i \rangle = \langle V_i \rangle = \left\langle \sum_{i,j \neq i} V_{ij} \right\rangle$$

so

$$E = \tfrac{3}{4} \sum_i \varepsilon_i \qquad (2.13)$$

The total energy E is a function of δ, since the ε_i are functions of δ, and the ground state deformation is the one for which E is a minimum. The result of carrying out this minimisation in the rare earth region is shown in figure 2.17, where it can be seen that fair agreement with the observed quadrupole moments is obtained.

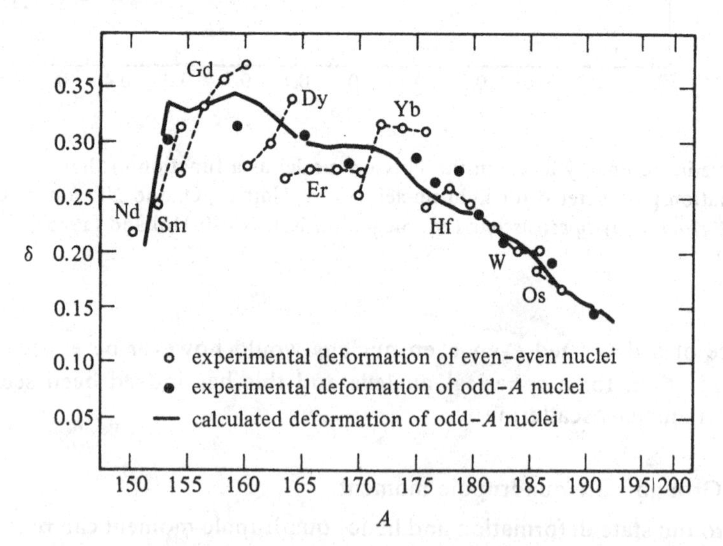

2.17. Comparison of theoretical and experimental deformation. (From Mottelson, B. R. and Nilsson, S. G., *Mat. Fys. Skr. Dan. Vid. Selsk.* **1** (1959) no. 8.)

The change in Coulomb energy with deformation was not allowed for in the original calculation. However, later work which included both the Coulomb interaction and the pairing interaction found that these two residual interactions tended to cancel one another (the Coulomb tending to increase distortion while the pairing interaction tending to make a nucleus more spherical) with the result that the original agreement was maintained.

The total binding energy given by this procedure, however, is in poor agreement with the observed value. This is because no residual interactions have been taken into account and these cannot be neglected for such a bulk property as even small shifts in ε_i can be significant. The shape, however, is just determined primarily by the nucleons in the unfilled shells (see chapter 8 for a further discussion).

2.4 Excited states of nuclei

The shell model, both deformed and spherical, is able to account rather well for the spins, parities and quadrupole moments of nuclear ground states. Excited states in both the single-particle and Nilsson models, however, are not generally described so well. For odd-A and odd-odd nuclei excited states are predicted corresponding to the promotion of an unpaired nucleon to a higher level; while for even-even nuclei levels corresponding to exciting or breaking one or more of the pairs of nucleons are expected. In general this is far too simple a description and the residual interaction between the nucleons gives rise to a much more complex level spectrum. In some cases the complexity can be understood qualitatively quite simply. Several examples of nuclear level schemes are shown below and the extent to which simple shell model ideas can explain the features of the level schemes is discussed.

2.4.1 Closed-shell nuclei

Figure 2.18 shows the level schemes of the even-even doubly closed shell nuclei ${}^{16}_{8}$O and ${}^{208}_{82}$Pb. Their first excited states are at 6.05 MeV and 2.61 MeV, respectively. While this is roughly what would be expected for $A = 16$ using the pairing term $\delta(A)$ in the semi-empirical mass formula to estimate the energy required to break a pair of nucleons as $34A^{-3/4}$ MeV ($\approx$4 MeV), for $A = 208$ the formula gives a considerable underestimate ($\approx\frac{1}{2}$ MeV) and there is a much larger gap. To say much more in the shell model about these level schemes requires detailed calculation of the residual interaction.

The nucleus ${}^{207}_{82}$Pb, whose level scheme is shown in figure 2.19, has one neutron less than the magic number of 126 and can be described as one neutron hole in the doubly closed shell nucleus ${}^{208}_{82}$Pb. As ${}^{208}_{82}$Pb has $J^{\pi} = 0^{+}$ the spin and parity of the neutron hole nucleus ^{207}Pb is given by the J^{π} of the neutron hole state. The single-particle levels of figure 2.9 would predict

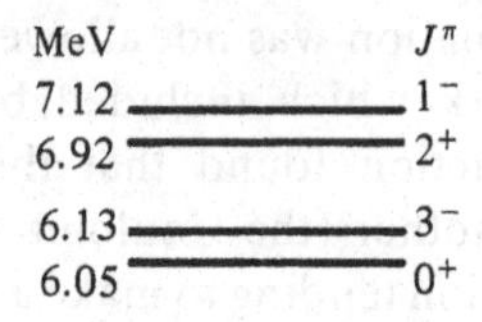

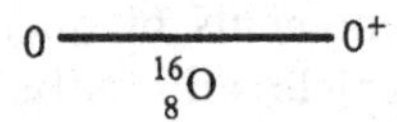

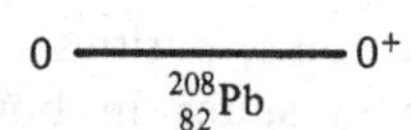

2.18. The low-lying levels in the doubly closed shell nuclei $^{16}_{8}O$ and $^{208}_{82}Pb$.

a sequence of levels: $\frac{5}{2}^-$, $\frac{1}{2}^-$, $\frac{13}{2}^+$, $\frac{3}{2}^-$, $\frac{7}{2}^-$ and the observed order is explained by the l-dependence of the pairing energy. This increases with increasing l (see p. 20) so it is energetically favourable for the ground state spin to be $\frac{1}{2}^-$ rather than $\frac{5}{2}^-$ and for the $\frac{3}{2}^-$ to be below the $\frac{13}{2}^+$.

In $^{17}_{8}O$ (see figure 2.19) the single-particle shell model would predict the J^π of the ground state and first two excited states as $\frac{5}{2}^+$, $\frac{1}{2}^+$ and $\frac{3}{2}^+$, corresponding to the unpaired proton in the $d_{3/2}$, $s_{1/2}$ and $d_{3/2}$ levels (see figure 2.16); these states are found at 0, 0.871 and 5.086 MeV excitation. Between the $\frac{1}{2}^+$ and $\frac{3}{2}^+$, however, lie three negative parity levels which are not explained by the single-particle shell model; these correspond to a significant extent to nucleons being excited from the $p_{1/2}$ and $p_{3/2}$ levels of the core (see below).

In the closed proton shell nucleus $^{117}_{50}Sn$ (see figure 2.19) the number of neutrons is not close to a magic number but the single-particle shell model explains the observed sequence of low-lying levels quite well. From figure 2.9 the predicted levels are $\frac{3}{2}^+$, $\frac{11}{2}^-$, $\frac{1}{2}^+$ and then a gap until a $\frac{9}{2}^-$ level. Apart from the change in order this sequence is correct for the first three excited states; however, the next five states are not accounted for and arise, as in ^{17}O, to a significant degree from nucleons excited out of the core.

2.4.2 Core-excited states

In both $^{17}_{8}O$ and $^{117}_{50}Sn$, excited states of the opposite parity to that expected in the *sp* shell model are seen at quite low excitation. A possible component

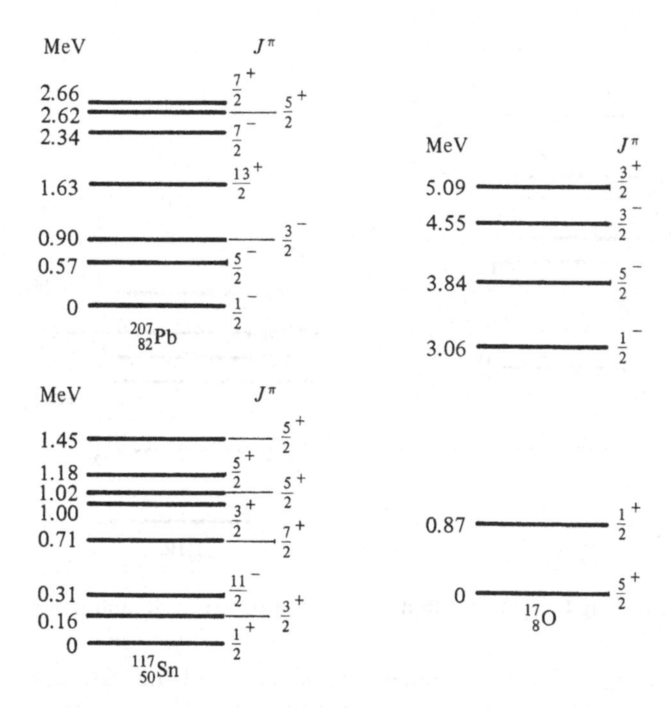

2.19. The low-lying levels in the closed proton shell odd-A nuclei $^{207}_{82}$Pb, $^{17}_{8}$O and $^{117}_{50}$Sn.

to the wavefunction of these states is a single nucleon promoted to the next shell, but often these levels correspond to a significant extent to nucleons excited from levels of the core: in $^{17}_{8}$O from the $p_{1/2}$ and $p_{3/2}$ levels, and in $^{117}_{50}$Sn from the $d_{5/2}$ and $g_{7/2}$ levels. A possible $\frac{3}{2}^-$ configuration in $^{17}_{8}$O is:

$$(\pi p_{3/2})^{-1}_{3/2}(\nu d_{5/2})^{2}_{0}\,{}_{\frac{3}{2}}^{-}$$

where $\pi \equiv$ proton, $\nu \equiv$ neutron, $(\ )^{-1} \equiv$ a hole state and $(\ )^2_0 \equiv$ coupled to spin 0. Such a state would be called a two-particle one-hole (2p–1h) state. Other examples of $\frac{3}{2}^-$ configurations are: $(\pi p_{1/2})^{-1}_{1/2}(\nu d_{5/2})^2_2\,{}^{-}_{\frac{3}{2}}$ and $(\pi p_{3/2})^{-1}_{3/2}(\nu d_{3/2}\nu d_{5/2})_1\,{}^{-}_{\frac{3}{2}}$. The relative amplitudes of these depend on diagonalising the residual interaction and this is discussed in chapter 8.

2.4.3 Mid-shell nuclei

The nucleus $^{165}_{67}$Ho ($J^\pi = \frac{7}{2}^-$) lies between magic numbers of protons and neutrons and is deformed ($Q = 3.5 \times 10^{-28}$ m^2) so the Nilsson model with the appropriate δ ($=0.32$) would be expected to describe its excited states. As is indicated in figure 2.20, three of the lowest ten states are predicted by the Nilsson diagram shown in figure 2.21 (the order is altered by the

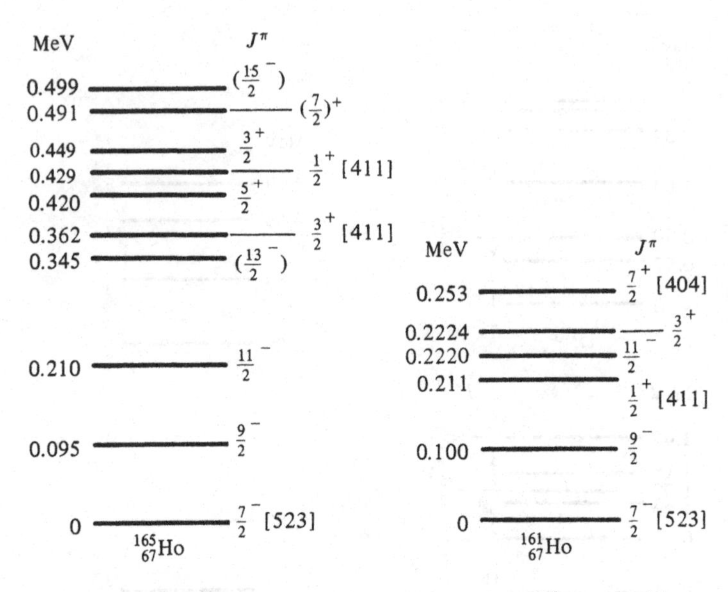

2.20. The low-lying levels in the mid-shell odd-A nuclei $^{165}_{67}$Ho and $^{161}_{67}$Ho.

effect of pairing), but the remainder are not explained and these correspond to collective motion (see below). The nearby nucleus $^{161}_{67}$Ho shows a similar spectrum (see figure 2.20).

While even–even doubly closed shell nuclei such as $^{16}_{8}$O and $^{208}_{82}$Pb have first excited states at a high excitation, as expected on the single-particle shell model, the mid-shell even–even nuclei $^{176}_{70}$Yb and $^{106}_{46}$Pd have rather different spectra of excited states (figure 2.22). In $^{176}_{70}$Yb there are several low-lying levels, forming a sequence $0^+, 2^+, 4^+, 6^+, 8^+$ below 1 MeV excitation whilst in $^{106}_{46}$Pd the first excited state, also a 2^+, is much higher at 0.512 MeV excitation, with a triplet of levels, $0^+, 2^+, 4^+$ at approximately twice the excitation. The levels in $^{106}_{60}$Pd could correspond to the excitation of pairs of nucleons; however, the $^{176}_{70}$Yb low-lying levels are not so simply explained and are reminiscent of molecular rotational spectra as their excitation energies are proportional to $J(J+1)$.

2.5 The collective model

From the above examples of nuclear spectra there are clearly excited states in nuclei which are not simply described in terms of single-particle motion in a spherical or a deformed potential well. There are, however, as will be seen, relatively simple motions of all the nucleons which do provide an explanation for many of these excited states; these are collective rotational

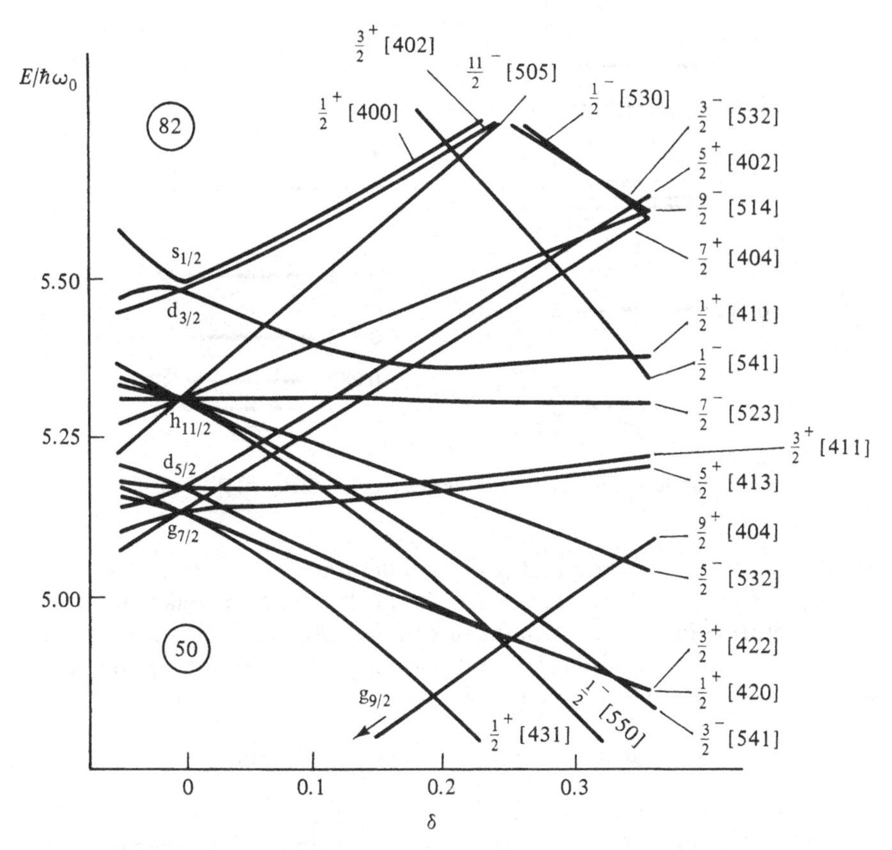

2.21. Nilsson proton levels for $50 \leq Z \leq 82$. (From Nathan, O. and Nilsson, S. G., in *α-, β- and γ-ray Spectroscopy* (ed. Siegbahn, K.). North-Holland (1965), p. 601.)

and vibrational motions. The analysis of these collective motions was developed in particular by Bohr and Mottelson.

2.5.1 Vibrational levels

If a nucleus were quite easy to deform about its equilibrium shape then, like in a liquid drop, vibrations about this shape could be important in describing the low-lying excited states of the nucleus. The shape of a nucleus can be parametrised by the length of the radius vector $R(\theta, \phi)$ from the origin to the surface, which for small deformation at constant volume is given by:

$$R(\theta, \phi) = R_0 \left[1 + \sum_{\lambda=2}^{\infty} \sum_{\mu=-\lambda}^{\lambda} \alpha^*_{\lambda\mu} Y_{\lambda\mu}(\theta, \phi) \right]$$

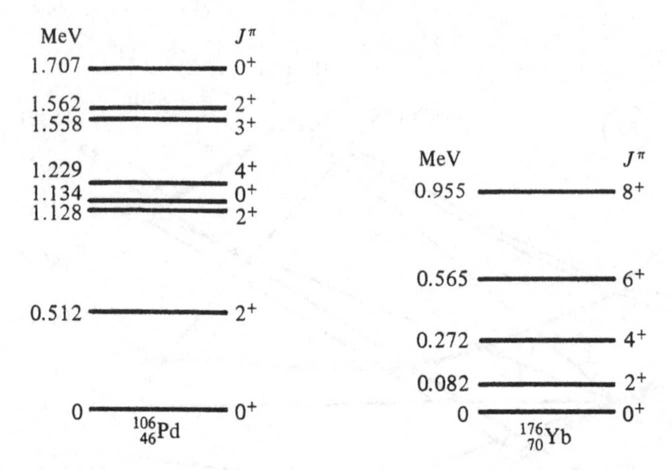

2.22. The low-lying levels in the mid-shell even–even nuclei $^{106}_{46}$Pd and $^{176}_{70}$Yb.

The $\alpha_{1\mu}$ parameters correspond (for small deformation) to a translation of the nucleus and are therefore zero. If the nucleus has a permanent quadrupole deformation ($\lambda = 2$) then the orientation of the coordinate axes may be chosen so that $\alpha_{21} = \alpha_{2-1} = 0$ and $\alpha_{22} = \alpha_{2-2}$. The parameters $\alpha_{2\mu}$ are sometimes related by the expressions:

$$\alpha_{20} = \beta \cos \gamma \quad \text{and} \quad \alpha_{22} = \alpha_{2-2} = \frac{1}{\sqrt{2}} \beta \sin \gamma$$

An axially-symmetric nucleus has $\gamma = 0$.

Oscillations in α_{20} and α_{22} are called β and γ vibrations, respectively. A permanently deformed nucleus with axial symmetry has three independent vibrational coordinates: α_{20}, which fluctuates about some finite value, and $\alpha_{2\pm2}$, which fluctuate about zero. These oscillations are drawn in figure 2.23.

For small axially symmetric quadrupole ($\lambda = 2$) oscillations about a spherical equilibrium shape, $\alpha_{20} = \beta = (4\pi/5)^{1/2}\varepsilon$, where ε is defined as above through $b = R/(1+\varepsilon)^{1/2}$ and $a = R(1+\varepsilon)$ (equation 2.6) and the other α parameters are zero. In the liquid drop model of the nucleus the change in binding energy ΔE of a spherical nucleus when it is deformed is given by the expression: $\Delta E = a_r\varepsilon^2$ (equation 2.7) $\equiv 0.4a_r\alpha_{20}^2$. This change in potential energy is like in a harmonic oscillator and the Hamiltonian H describing α_{20} as a function of time, corresponding to quadrupole surface oscillations, is given by:

$$H = \tfrac{1}{2}D_2(\dot{\alpha}_{20})^2 + \tfrac{1}{2}C_2(\alpha_{20})^2$$

where $C_2 \equiv 0.80a_r$ is the restoring force parameter and D_2 is the mass parameter. The frequency ω is equal to $\sqrt{C_2/D_2}$ and the excitation energy

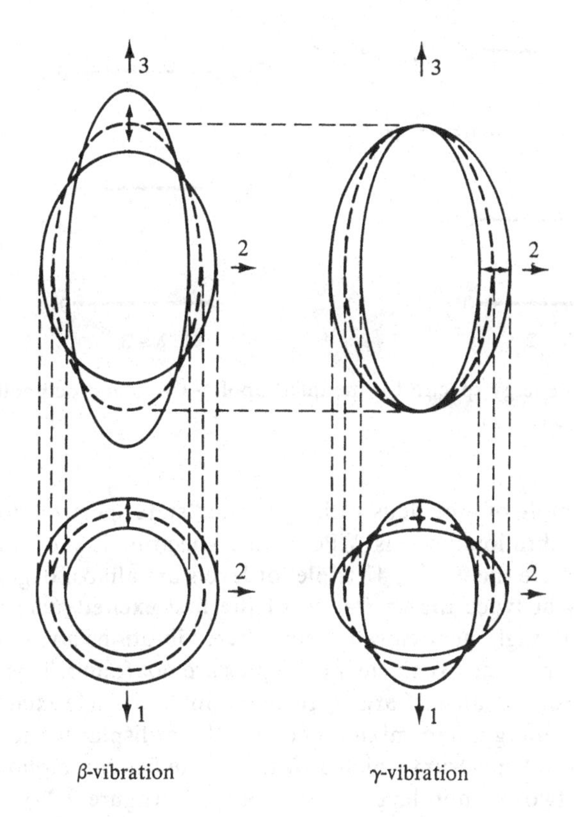

2.23. Schematic representation of β- and γ-vibrations. The axis of symmetry of the deformed nucleus is along the 3-axis. (From Ring, P. and Shuck, P., *The Nuclear Many-Body Problem.* Springer-Verlag (1980).)

of the first excited vibrational state is $\hbar\omega$. To determine ω the parameter D_2 must be estimated. The velocity of the nucleons on the surface of the nucleus is of the order $\dot{\alpha}_{20}R_0$ so their kinetic energy is $\sim\frac{1}{2}mR_0^2(\dot{\alpha}_{20})^2$. On the assumption of irrotational flow of the nucleons, which means motion in the radial direction only, the average kinetic energy of the nucleons is approximately one eighth of this value so:

$$\tfrac{1}{2}D_2(\dot{\alpha}_{20})^2 \approx \tfrac{1}{2}\frac{A}{8}mR_0^2(\dot{\alpha}_{20})^2$$

Therefore $D_2 \approx \frac{1}{8}AmR_0^2$ and $\hbar\omega \approx \hbar c(8C_2/Amc^2R_0^2)^{1/2}$. In general, oscillations described by the other $\alpha_{\lambda\mu}$ parameters are possible and each vibrational quantum (a phonon) associated with a variation in $\alpha_{\lambda\mu}$ has an angular momentum of λ. The energy spectra for $\lambda = 2$ and $\lambda = 3$ vibrations are shown in figure 2.24.

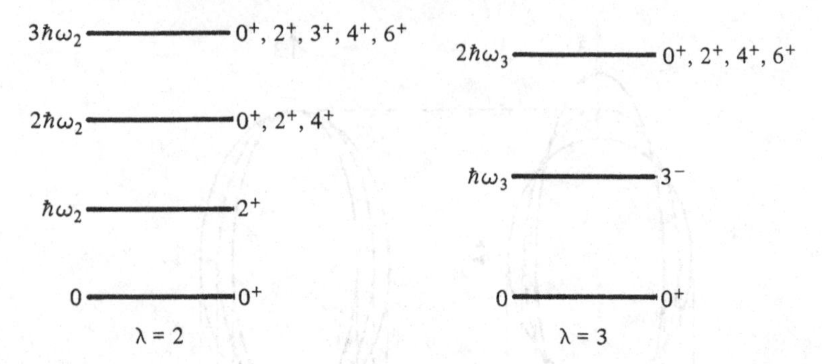

2.24. Harmonic energy spectra for the quadrupole ($\lambda = 2$) and octupole ($\lambda = 3$) surface oscillations.

A good example of a nucleus with an excited state spectrum looking like a quadrupole vibrational one is $^{106}_{46}Pd$, whose spectrum is shown above (see figure 2.22), where the 0^+, 2^+, 4^+ triplet of levels are almost degenerate and approximately at twice the excitation of the first excited state which is a 2^+. Generally though, near closed shells where vibrations about a spherical shape could be expected such clear examples are not found. This is because configurations of the same J^π arising through single-particle excitation often have a similar energy, and mixing occurs, which displaces the levels. An example of a level spectrum which also shows the five three-phonon levels, as well as the two-phonon levels, is that of ^{118}Cd (figure 2.25).

This mixing with other configurations of the same J^π results in poor agreement being found with the predictions of the liquid drop model given by $\hbar\sqrt{C_\lambda/D_\lambda}$. For example, for the nucleus discussed above, $^{106}_{46}Pd$, the coefficient $a_r \equiv 1.25C_2$ equals 91 MeV and with $R_0 = r_0A^{1/3}$ fm and $r_0 = 1.2$ this makes $\hbar\omega = 2.67$ MeV. This poor agreement can be seen in figure 2.26 for both the lowest 2^+ and 3^- states and the disagreement is most noted between closed shells where the 2^+ is much lower in excitation than expected.

2.5.2 Giant resonances

Although the low-lying excited states of a nucleus are not well described by surface oscillations there is a high-lying excited state in all nuclei which is well described as a vibrational mode. This is the giant dipole resonance state and arises from an oscillation of the protons with respect to the neutrons in the nucleus illustrated in figure 2.27. The vectors $\mathbf{R}_N$ and $\boldsymbol{R}_Z$ are the centre-of-mass of the neutrons and protons, respectively. The restoring force in the liquid drop model interpretation is calculated from the asymmetry term in the semi-empirical mass formula, $a_a(Z-N)^2/A$, and

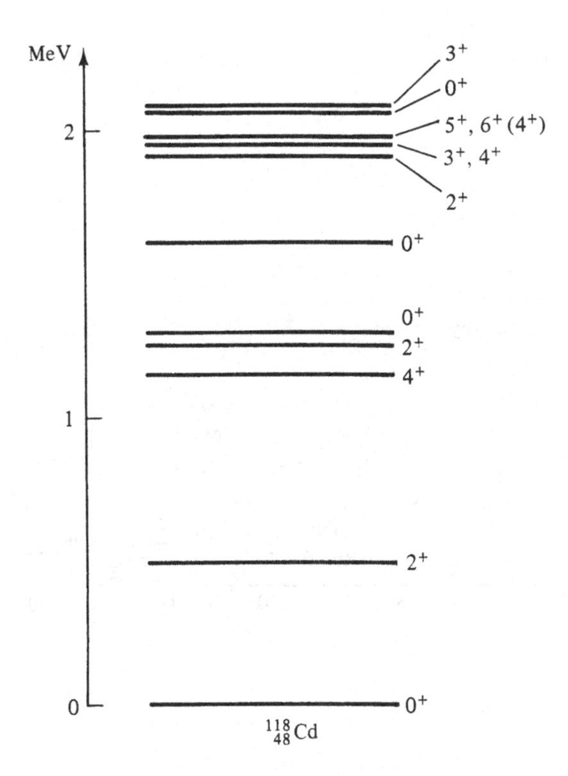

2.25. The low-lying level spectrum of ^{118}Cd showing five levels identified with the three-phonon levels ($0^+, 2^+, 3^+, 4^+, 6^+$), together with the one- and two-phonon levels and an additional 0^+ single-particle excitation.

the predicted energy for such oscillations is:

$$E_{\text{dipole}} = 76A^{-1/3}\ \text{MeV}$$

which is in very good agreement with experimental values. It is called a dipole state as the displacement of the protons with respect to the neutrons gives rise to a large electric dipole transition moment. In self-conjugate ($N = Z$) even–even nuclei these states have $J^\pi = 1^-$ and $T = 1$ and there is a very strong electric dipole gamma transition to the ground state. As is discussed later in chapter 8, these states can also be described, as pointed out by Wilkinson, by an independent-particle model and in this model their wavefunctions are a superposition of many particle–hole configurations.

These giant dipole states are seen in reactions in which a photon is absorbed by the nucleus, examples of which are inelastic electron scattering and from photons produced in the annihilation of positrons in flight. The cross-sections for two such reactions are shown in figure 2.28. In the $\gamma + {}^{208}$Pb

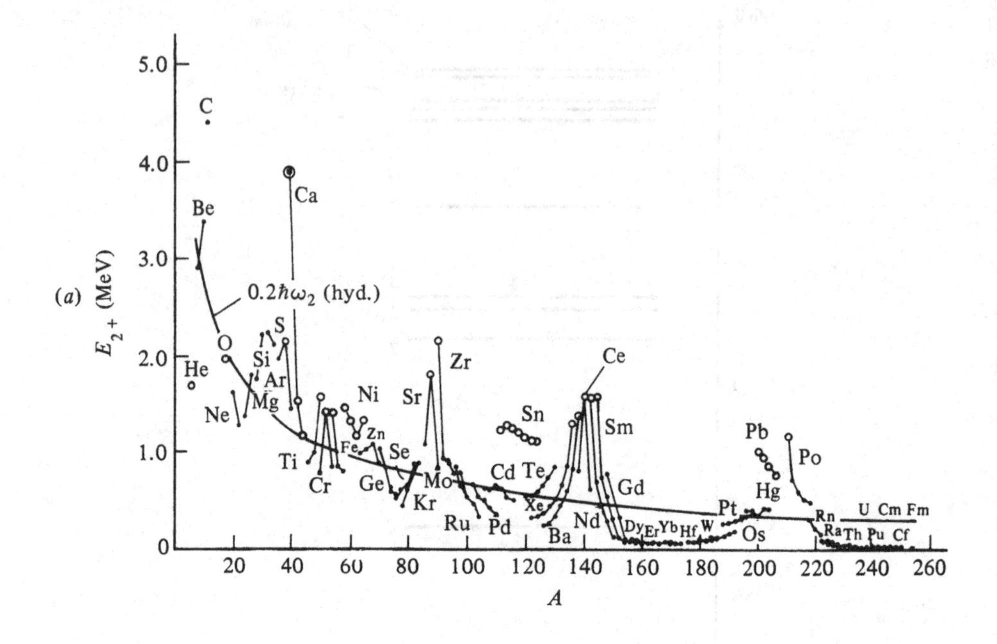

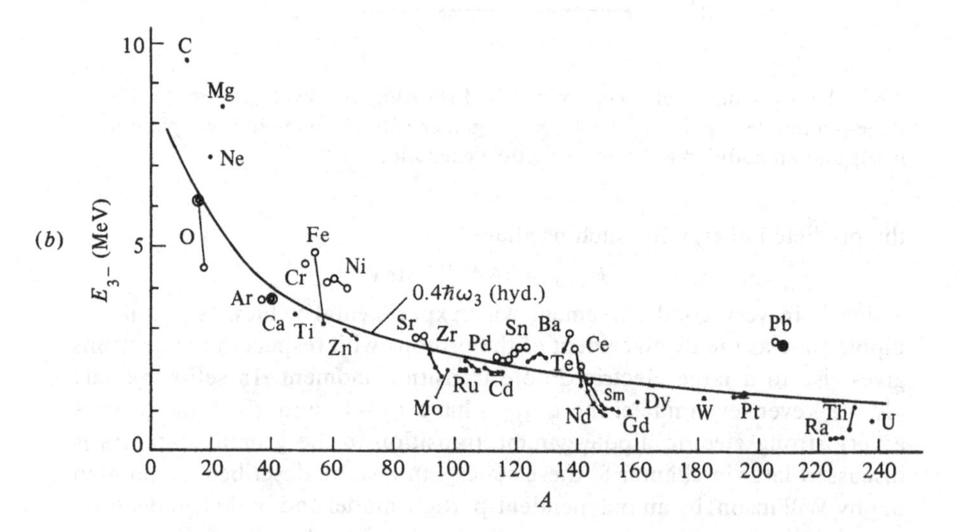

2.26. Excitation energies in even–even nuclei of (*a*) the first excited 2^+ state; (*b*) the collective vibrational 3^- state. Nuclei with closed proton or neutron shells are denoted by open circles. (From Nathan, O. and Nilsson, S. G., in *α-, β- and γ-ray Spectroscopy* (ed. Siegbahn, K.). North-Holland, (1965), p. 601.)

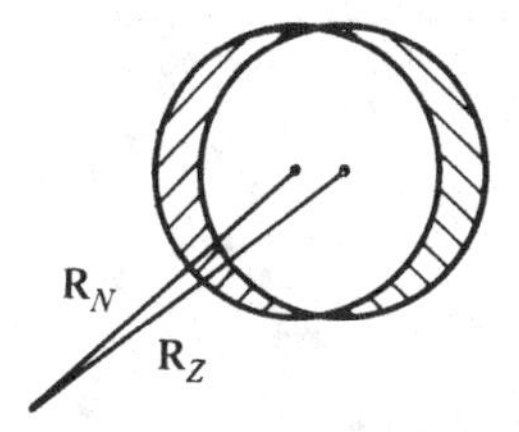

2.27. Representation of the oscillation of protons with respect to neutrons in a nucleus. The vectors $\mathbf{R}_N$ and $\mathbf{R}_Z$ are the centre of mass of the neutrons and protons, respectively.

reaction there is a broad peak centred on 14 MeV excitation. The width of the peak does not arise from the existence of a single level with a very short lifetime but rather from the dipole mode of excitation being shared amongst many levels, or equivalently from the dipole strength being split by residual interactions.

In the $\gamma + {}^{165}$Ho reaction there are two bumps in the cross-section curve corresponding to two different excitation energies. These arise because the holmium nucleus is quite strongly axially deformed and has different stiffnesses with respect to deformation along and at right angles to its symmetry axis. Its prolate (cigar) shape makes oscillations along its symmetry axis have a lower frequency and hence lower energy than oscillations at right angles. The areas under the two bumps differ by a factor of two because there are two oscillation modes (x and y) at right angles while only one (z) along the symmetry axis. There are also other giant resonances that have been identified recently such as the electric quadrupole E2 giant resonance associated with quadrupole surface oscillations and resonances built upon excited states.

2.5.3 Rotational levels of even–even nuclei

The plot in figure 2.26 of the excitation energies of the first excited 2^+ states versus A shows significantly lower-lying 2^+ states for nuclei between closed shells. For even–even nuclei in these regions the spectra of excited states show a characteristic 0^+ ground state, 2^+, 4^+, 6^+... sequence of levels, analogous to a molecular rotational band of levels, and lend themselves to a simple interpretation as the eigenstates of a rotating deformed nucleus.

The shape of mid-shell nuclei is expected to be deformed, as discussed above. If the nucleus has an ellipsoidal shape then classically two different principal moments of inertia $\mathcal{I}_1(=\mathcal{I}_2)$ and $\mathcal{I}_3$ would be expected; these are the moments about the x' axis (or y' axis) and z' axis, respectively, where

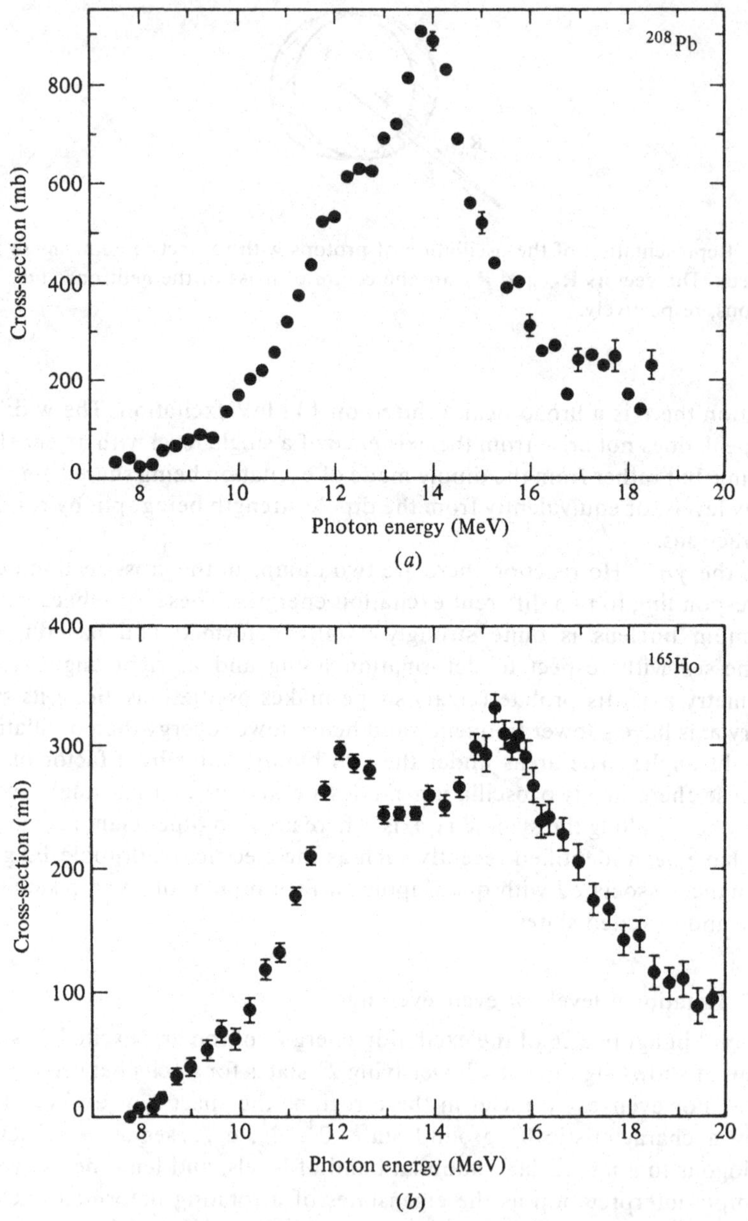

2.28. Gamma-absorption cross-sections for (*a*) the spherical lead nucleus; (*b*) the deformed holmium nucleus. (^{165}Ho data from Bramblett, R. L. *et al.*, *Phys. Rev.* **129** (1963) 2723; ^{208}Pb data from Fuller, E. G. and Hayward, E., *Nucl. Phys.* **33** (1962) 431.)

the z' axis is along the axis of symmetry of the ellipsoid. Classically, rotation about all three axes is possible but quantum-mechanically the energy of rotation about the z' axis must be zero for an axially symmetric nucleus. Axial symmetry means that the energy of the nucleus's wavefunction rotated by an arbitrary angle about the symmetry axis is the same as its unrotated value, as the effect of rotation is only to introduce a phase change, so there is no energy associated with such a rotation; put another way there is no way of measuring a rotation of the whole nucleus about the z' axis so there is no associated collective (rotational) energy. This means that any rotational angular momentum **R** is perpendicular to the symmetry axis.

The energy associated with such a rotation is just like that associated with a spinning symmetric rotor, which is $R^2/2\mathcal{I}$, where $\mathcal{I}$ is the moment of inertia of the rotor. The Hamiltonian, H_{rot}, corresponding to the rotation of an even–even deformed nucleus is therefore given by:

$$H_{\text{rot}} = \frac{1}{2\mathcal{I}} \mathbf{R}^2$$

where **R** is the rotational angular momentum. Since the intrinsic angular momentum **J** of an even–even nucleus is zero the total angular momentum **I** equals **R** hence:

$$H_{\text{rot}} = \frac{1}{2\mathcal{I}} \mathbf{I}^2$$

where $\mathcal{I}$ is the moment of inertia of the nucleus. The total Hamiltonian H for the nucleus is $H_{\text{rot}} + H_{\text{int}} + H_{\text{coup}}$, where H_{int} describes the internal motion of the nucleons and H_{coup} represents the coupling between rotational and intrinsic motion.

2.5.4 The adiabatic approximation

If the characteristic frequency of rotation of the whole nucleus is much less than that of the internal nucleon motion then to a first approximation the rotational motion will not affect the intrinsic wavefunction, i.e. H_{coup} can be neglected. This condition is equivalent to the excitation energy of the first rotational excited state being much less than that of a level of the same J^π arising through single-particle excitation. Making the adiabatic approximation gives:

$$H = H_{\text{rot}} + H_{\text{int}}$$

so the total wavefunction ψ and energy E will be given by:

$$\psi = \phi_{\text{int}} D(\boldsymbol{\theta}) \quad \text{and} \quad E = E_{\text{int}} + E_{\text{rot}}$$

where $H_{\text{int}}\phi_{\text{int}} = E_{\text{int}}\phi_{\text{int}}$, $H_{\text{rot}}D(\boldsymbol{\theta}) = E_{\text{rot}}D(\boldsymbol{\theta})$ and $\boldsymbol{\theta}$ describes the orientation of the deformed nucleus's symmetry axis. The wavefunction $D(\boldsymbol{\theta})$ is an eigenfunction of $\mathbf{I}^2$. The rotational energy of the levels E_{rot} is given by:

$$E_{rot} = \langle H_{rot} \rangle = \frac{\hbar^2}{2\mathscr{I}} I(I+1) \tag{2.14}$$

This gives a sequence of levels just like in a molecular rotational band.

2.5.5 The moment of inertia

An example of a rotational spectrum in an even-even nucleus is that of ^{238}U shown in figure 2.29. Using equation 2.14 for the excitation energies gives good agreement, and a value of $2\mathscr{I}/\hbar^2 \approx 137$ MeV^{-1}. This value of $\mathscr{I}$ can be compared with the moment of inertia of a rigid sphere:

$$\mathscr{I}_{rig} = \tfrac{2}{5} MR^2$$

This gives for ^{238}U, using $r_0 = 1.2$, $2\mathscr{I}_{rig}/\hbar^2 = 253$ MeV^{-1}. (The nucleus ^{238}U is axially deformed with a deformation of $\beta = 0.28$ but classically this would only make its moment of inertia slightly larger than $\mathscr{I}_{rig}$.) $\mathscr{I}$ is therefore only $\sim\frac{1}{2}\mathscr{I}_{rig}$. A fraction of between a quarter and three quarters is typical of nuclei with a rotational spectrum of excited states. A much smaller moment of inertia $\mathscr{I}_{irrot}$ is predicted using the hypothesis of irrotational flow of the nucleons in a deformed nucleus. If the deformation is given by

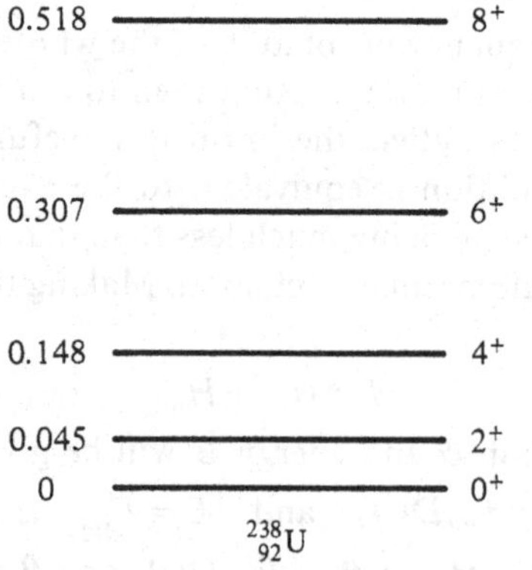

MeV	J^π
0.827	5^-
0.776	10^+
0.732	3^-
0.680	1^-
0.518	8^+
0.307	6^+
0.148	4^+
0.045	2^+
0	0^+

$^{238}_{92}$U

2.29. The low-lying levels of $^{238}_{92}$U.

β then:

$$\mathscr{I}_{\text{irrot}} \approx \tfrac{3}{8} MR^2\beta^2$$

For the nucleus ^{238}U $\beta = 0.28$ which gives $2\mathscr{I}_{\text{irrot}}/\hbar^2 = 19$ MeV^{-1} which is $\sim\frac{1}{7}(2\mathscr{I}/\hbar^2)$, so the nucleon motion is not irrotational. In general the observed moments of inertia of nuclei lie between $\mathscr{I}_{\text{irrot}}$ and $\mathscr{I}_{\text{rig}}$. A qualitative interpretation is to say that only a fraction of the nucleons are involved in the collective rotational motion and the rest are effectively inoperative. This reduction from the rigid body value, which is the value expected for nucleons moving independently in a deformed potential well, is caused by the pairing residual interaction and this is discussed in chapter 8.

2.5.6 The rotational wavefunction

A distinctive feature of rotational spectra in deformed even–even nuclei is the spin sequence 0^+, 2^+, 4^+, 6^+ .... The absence of odd spins cannot be explained by H_{rot} alone but requires understanding of the symmetries of the rotational wavefunction $D(\boldsymbol{\theta})$. The wavefunction $D(\boldsymbol{\theta})$ is an eigenfunction of $\mathbf{I}^2$, $\mathbf{I}_z$ and $\mathbf{I}_{z'}$ with eigenvalues $I(I+1)$, M and K, respectively, where I_z and $I_{z'}$ are the components of the total angular momentum along an arbitrary z-axis and along the symmetry axis z' which is orientated in a direction $\boldsymbol{\theta}$ to the x, y, z axes (see figure 2.30). The orientation $\boldsymbol{\theta}$ can be defined by two angles, but three (the Euler angles α, β and γ) are generally used. The expression for $\psi = \phi D$ involves the internal coordinates of ϕ plus the coordinates of D so there are redundant variables in this wavefunction. For well-deformed nuclei this complication can be ignored though the justification of this is difficult. Collective rotational motion can be formulated avoiding this problem, and in a more satisfactory quantum-mechanical way, using a variational method. This method also explains the absence of odd spins in the spectra of deformed even–even nuclei; it arises from the symmetry of these deformed nuclei. This feature can also be understood from the rotational transformation properties of $D(\boldsymbol{\theta})$ and ϕ. The variational method and the transformation properties of $D(\boldsymbol{\theta})$ and ϕ are discussed in chapter 8.

2.5.7 Odd-A nuclei

Deformed nuclei with an odd number of nucleons also exhibit rotational spectra. If the intrinsic angular momentum of the deformed nucleus is $\mathbf{J}$, and the rotational angular momentum $\mathbf{R}$, then the total angular momentum $\mathbf{I} = \mathbf{R} + \mathbf{J}$ can be denoted by the diagram shown in figure 2.30. The values of $I_{z'}$ and $J_{z'}$ are denoted K and Ω respectively. If the nucleus is axially symmetric then the rotational angular momentum $\mathbf{R}$ is perpendicular to the symmetry axis z', as for even–even axially symmetric nuclei, so $K = \Omega$. The excitation energies of the rotational levels are given by the eigenvalues of

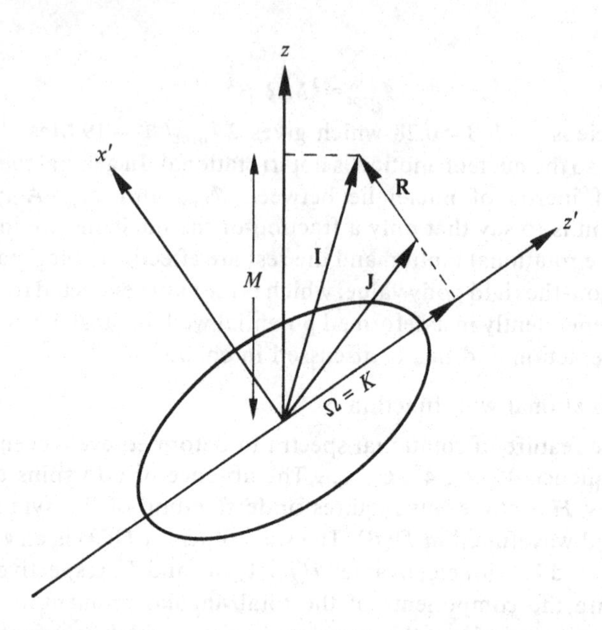

2.30. Diagram showing the body-fixed axes x' and z' for an axially-symmetric deformed nucleus and space-fixed axis z. **I** is the total, **R** the rotational and **J** the intrinsic angular momentum. The z' components of **I** and **J** are K and Ω, respectively, and the z component of **I** is M (in units of $\hbar$).

$\hbar^2\mathbf{R}^2/2\mathscr{I}$, where:

$$\begin{aligned}\mathbf{R}^2 &= (\mathbf{I}-\mathbf{J})^2\\ &= \mathbf{I}^2 - 2\mathbf{I}\cdot\mathbf{J} + \mathbf{J}^2\\ &= \mathbf{I}^2 + \mathbf{J}^2 - 2K^2 - (I'_+J'_- + I'_-J'_+)\end{aligned}$$

where $I'_\pm = I_{x'} \pm \mathrm{i}I_{y'}$, $J'_\pm = J_{x'} \pm \mathrm{i}J_{y'}$ and $J_{z'} = I_{z'} = K$. By symmetry the energies of states with $\pm K$ are degenerate and the wavefunction for a rotational state of given K involves a linear combination of the $\pm K$ configurations (see chapter 8 for details).

The total angular momentum I and its projection on the symmetry axis K are good quantum numbers but $H_{\text{coup}} = -(\hbar^2/2\mathscr{I})(I'_+J'_- + I'_-J'_+)$, which corresponds to the Coriolis force in a rotating frame of reference, can couple rotational and intrinsic motion. The operators $I'_\pm$, however, only link states with K differing by ± 1 and these correspond to different intrinsic states (except for $K = \frac{1}{2}$) and so can be neglected in a first approximation for low I (the Coriolis force increases with increasing I). For $K = \frac{1}{2}$ states, H_{coup} couples the $K = \pm\frac{1}{2}$ components of the wavefunction. The result for this case is derived in chapter 8.

Neglecting H_{coup} gives the following expression for the excitation energies:

$$E_{\text{rot}} = \frac{\hbar^2}{2\mathscr{I}}[I(I+1)+J(J+1)-2K^2]$$

Since K and J are constant for a rotational band of levels, E_{rot} depends only on $I(I+1)$, just as for even–even nuclei. An example of a rotational spectrum for an odd-A nucleus is shown in figure 2.31. In ^{239}Np two rotational bands can be seen built upon two different intrinsic states: a $J^\pi = \frac{5}{2}^+$ state (the ground state) and a $J^\pi = \frac{5}{2}^-$ excited state, each of which is well described by the Nilsson model with designations [642] and [523], respectively. The moments of inertia for each band are different (reflecting the difference in intrinsic states) and are $\sim\frac{7}{8}\mathscr{I}_{\text{rig}}$ and $\sim\frac{5}{8}\mathscr{I}_{\text{rig}}$.

2.5.8 β and γ bands

In molecular spectra rotational bands are found built upon vibrational levels. Similar levels are found in nuclei. For a permanently deformed even–even nucleus with axial symmetry the first excited α_{20} intrinsic vibrational state (a β-vibration, see above) has $K=0$ and the rotational band

MeV	J^π
0.241	$\frac{11}{2}^-$
0.220	$(\frac{1}{2}^+)$
0.180	$\frac{13}{2}^+$
0.173	$\frac{9}{2}^-$
0.123	$(\frac{11}{2}^+)$
0.118	$\frac{7}{2}^-$
0.075	$\frac{5}{2}^-$ [523]
0.071	$\frac{9}{2}^+$
0.031	$\frac{7}{2}^+$
0	$\frac{5}{2}^+$ [642]

$^{239}_{93}$Np

2.31. The low-lying states of $^{239}_{93}$Np.

built upon this state is called a β band. The first $a_{2\pm2}$ intrinsic vibrational states (γ-vibrations, see above) are degenerate and have $K = \pm 2$ and give rise to a $K = 2$ rotational band which is called a γ band. An example of these bands is found in ^{238}U (see figure 2.32). For an odd-A deformed nucleus, β and γ bands are also found built upon the ground state, which is generally well described by the Nilsson model.

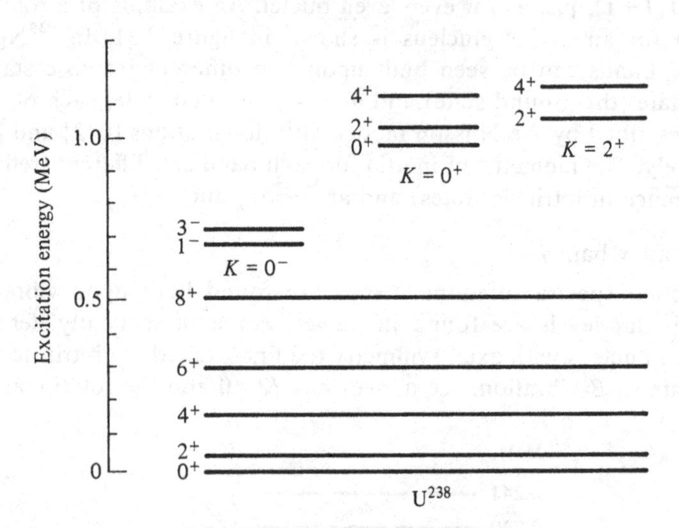

2.32. Low-lying rotational bands in $^{238}_{92}U$. The excited bands are a $K = 0^-$, $r = -1$ octupole vibration, a $K = 0^+$, $r = +1$ β-vibration and a $K = 2^+$, γ-vibration. (From Rogers, J. D., *Ann. Rev. of Nucl. Sci.* **15** (1965) 241.)

2.6 High-spin states

2.6.1 Backbending

The slight increase in $\mathscr{I}$ for high spin observed in ^{238}U is found in other deformed nuclei, with both odd and even A. Classically this might be thought to arise from a centrifugal stretching causing an increase in distortion and hence $\mathscr{I}$. However, this is a rather small effect and the principal cause is a decrease in the number of nucleons paired off, which results in an increase in $\mathscr{I}$. For higher spin states, typically $I \gtrsim 16$, there are several examples of a significant change in $\mathscr{I}$ with spin. This phenomenon is called backbending, after the shape of the plot of the moment of inertia versus the square of the angular frequency.

The ground state rotational band of levels in ^{158}Er and ^{174}Hf are shown in figure 2.33, where their excitation energies are plotted against their angular momenta. These levels are called Yrast levels as each level is the lowest in

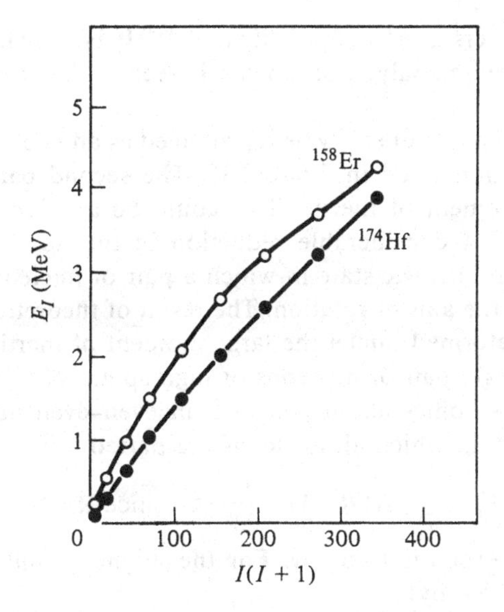

2.33. Plot of excitation energy versus $I(I+1)$ for the ground state rotational band in ^{158}Er and ^{174}Hf. (From Lieder, R. M. and Ryde, H., *Adv. Nucl. Phys.* **10** (1978) 1.)

excitation for its angular momentum. While the curve is almost straight for ^{174}Hf there is a slight kink in the curve for ^{158}Er near $I = 14$. This anomaly is seen much more clearly if the effective moment of inertia is plotted against the square of the angular frequency of the levels. Classically an object of moment of inertia $\mathcal{I}$ rotating with an angular frequency ω has an angular momentum $I = \mathcal{I}\omega$ and a rotational energy E given by:

$$E = \tfrac{1}{2}\mathcal{I}\omega^2 = I^2/2\mathcal{I}$$

$$\therefore \quad \omega = \frac{\mathrm{d}E}{\mathrm{d}I}$$

For a rotational band in an even–even nucleus ω may be defined by analogy by:

$$(\hbar\omega)_{I-1} = \tfrac{1}{2}(E_I - E_{I-2}) \tag{2.15}$$

and since

$$E_I = \frac{\hbar^2}{2\mathcal{I}} I(I+1)$$

then the effective moment of inertia $\mathcal{I}_{I-1}$ can be defined by:

$$\left(\frac{2\mathcal{I}}{\hbar^2}\right)_{I-1} = \left(\frac{4I-2}{E_I - E_{I-2}}\right)$$

A plot of $2\mathscr{I}/\hbar^2$ versus $\hbar^2\omega^2$ for ^{158}Er and ^{174}Hf is shown in figure 2.34, where a distinctive anomaly, a backbend, is seen in the shape of the curve for ^{158}Er.

Such a phenomenon can easily be reproduced as an effect of the crossing of two bands as illustrated in figure 2.35. The second band has a much larger effective moment of inertia. This could be a reflection of a larger deformation or of a considerable reduction in the number of nucleons paired off or of an intrinsic state in which a pair of nucleons of high spin are aligned along the axis of rotation. The result of theoretical calculations is that for well-deformed nuclei the large moment of inertia is due to the alignment of a single pair of nucleons of high spin.

To see how this comes about consider an even–even nucleus. For the ground state band, in which all nucleons are paired off:

$$E_{gs} = \frac{\hbar^2}{2\mathscr{I}} R(R+1) \approx \frac{\hbar^2}{2\mathscr{I}} I^2 \quad \text{since } \mathbf{R} = \mathbf{I} \tag{2.16}$$

where E_{gs} is the excitation energy. For the aligned band the excitation energies E_{al} are given by:

$$E_{al} = \frac{\hbar^2}{2\mathscr{I}} R(R+1) + E_j \tag{2.17}$$

$$\approx \frac{\hbar^2}{2\mathscr{I}} (I - j_{x'})^2 + E_j$$

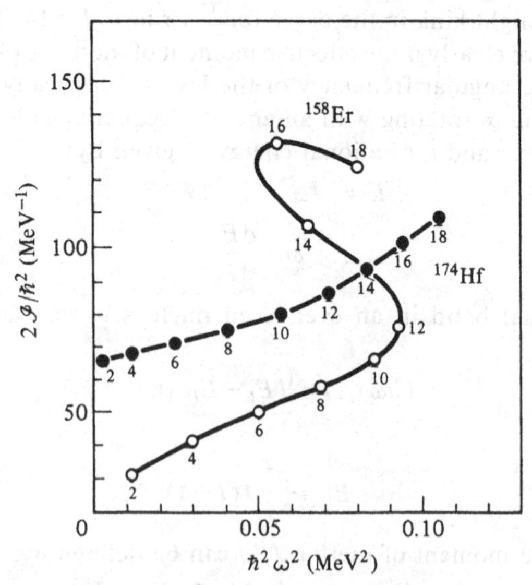

2.34. Plot of $2\mathscr{I}/\hbar^2$ versus $\hbar^2\omega^2$ for the ground state rotational band in ^{158}Er and ^{174}Hf. (From Lieder, R. M. and Ryde, H., *Adv. Nucl. Phys.* **10** (1978) 1.)

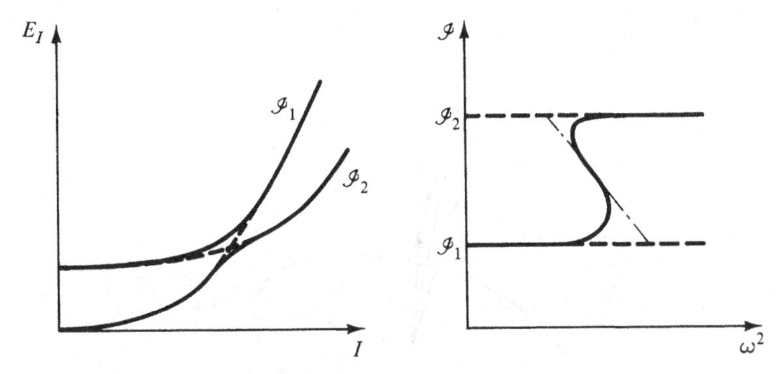

2.35. Schematic picture of two intersecting bands with different moments of inertia $\mathscr{I}_1$ and $\mathscr{I}_2$, and the corresponding backbending plot. (From Ring, P. and Shuck, P., *The Nuclear Many-Body Problem.* Springer-Verlag (1980).)

where E_j is the energy required to break a pair of nucleons and $j_{x'}$ is the total spin from the aligned nucleons. The breaking of the pair is assumed not to appreciably alter $\mathscr{I}$. The coupling of the angular momenta $\mathbf{I}=\mathbf{R}+\mathbf{j}_{x'}$ is shown in figure 2.36.

Therefore compared to the ground state band (no particle alignment) the rotational angular momentum $\mathbf{R}$ is much reduced if the total spin from the aligned nucleons is large. This happens in the nucleus ^{158}Er where the aligned nucleons are two $i_{13/2}$ neutrons (the $i_{13/2}$ neutron level lies in the vicinity of the Fermi level) coupled to $j_x=12$ ($j_x=13$ is not allowed by the Pauli principle). Now the effective moment of inertia is, from above, given by:

$$\frac{\hbar^2}{2\mathscr{I}_{\text{eff}}}(4I-2)=(E_I-E_{I-2})$$

To calculate this effective moment of inertia in terms of the actual moment of inertia for the nucleus ^{158}Er, consider two levels which are on the aligned band, the $I=16$ and $I=18$ levels.

$$\text{For } I=16,\ E_{\text{al}}\approx\frac{\hbar^2}{2\mathscr{I}}(16-12)^2+E_j=\frac{\hbar^2}{2\mathscr{I}}16+E_j$$

$$\text{For } I=18,\ E_{\text{al}}\approx\frac{\hbar^2}{2\mathscr{I}}36+E_j$$

so

$$(E_{18}-E_{16})_{\text{al}}=\frac{\hbar^2}{2\mathscr{I}}20=\frac{\hbar^2}{2\mathscr{I}_{\text{eff}}}70$$

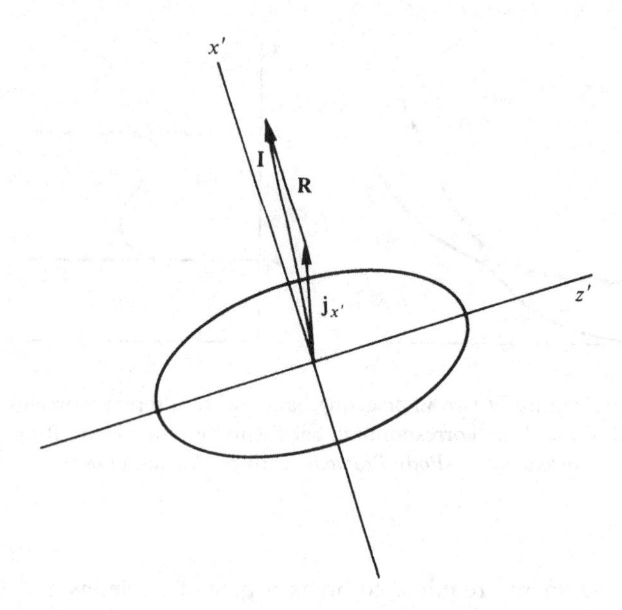

2.36. Diagram illustrating the rotational alignment of the spin $j_{x'}$ of a pair of nucleons along the axis (x') of rotation.

so

$$\mathscr{I}_{\text{eff}} = 3.5\mathscr{I}$$

This is in qualitative agreement with what is found (see figure 2.34). Moreover this approximate formula predicts a drop in $\mathscr{I}_{\text{eff}}$ for higher I as the spin of the rotating core $\mathbf{R}$ increases:

$$(E_{20} - E_{18})_{\text{al}} = \frac{\hbar^2}{2\mathscr{I}} 28 = \frac{\hbar^2}{2\mathscr{I}_{\text{eff}}} 78 \quad \text{so } \mathscr{I}_{\text{eff}} = 2.8\mathscr{I}$$

which is also in qualitative agreement.

The effective moment of inertia is approximately equal to the rigid body moment of inertia given by:

$$\mathscr{I}_{\text{rig}} = \tfrac{2}{5} MR^2$$

This is accidental and is not due to a total breakdown in pairing, which would give a moment of inertia equal to the rigid body value. The high effective moment of inertia is caused by the reduction in the spin of the rotating core at the backbend. The internal structure of the core in the aligned band is the same as in the ground state band except for the $i_{13/2}$ pair of neutrons. To obtain a large change in the effective moment of inertia as seen in ^{158}Er any coupling between the bands must be small, i.e. any mixing must extend only over a few I.

The phenomenon of backbending is seen in many nuclei and for some in the rare earth region a secondary irregularity, called the second backbend, has been seen, which has been interpreted as the alignment of an $h_{11/2}$ proton pair, see figure 2.37. From the position of the backbend the energy required to break a pair of nucleons can be estimated. For the nucleus ^{158}Er the energy E_b of the first backbend is at approximately 3.2 MeV and the spin $I = 14$ so:

$$\frac{\hbar^2}{2\mathcal{I}}(14-12)^2 + E_j \approx \frac{\hbar^2}{2\mathcal{I}} 14^2 = E_b$$

so

$$E_j \approx E_b = 3.2 \text{ MeV}$$

Detailed calculations of the shape of the curve of $\mathcal{I}$ versus spin have given a lot of information about the effect of high angular momentum upon the structure and deformation of nuclei.

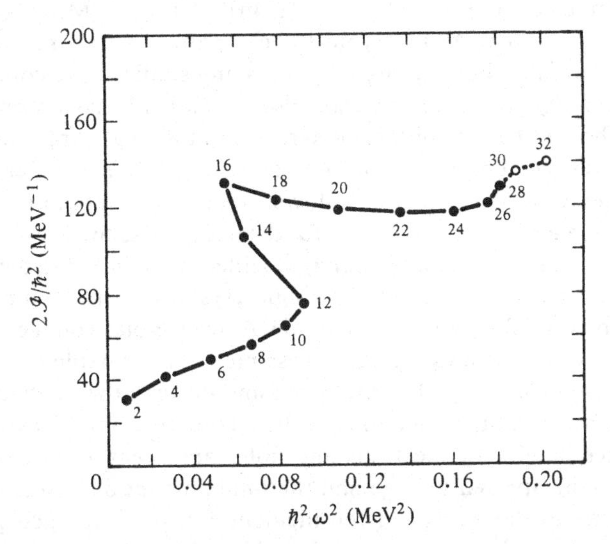

2.37. Plot of $2\mathcal{I}/\hbar^2$ versus $\hbar^2\omega^2$ for the ground state rotational band of ^{160}Yb showing evidence of a second backbend. (From Lee, I. Y. *et al.*, *Phys. Rev. Lett.* **38** (1977) 1454.

2.6.2 Very high spin states

As the spin of the nucleus increases the effect of pairing decreases because the Coriolis force generated by the rotation of the nucleus becomes sufficient to unpair individual pairs of nucleons. Nucleons with their spin aligned along the axis of rotation have an oblate deformation. This will tend to make a prolate nucleus triaxial in shape (i.e. $\mathscr{I}_1 \neq \mathscr{I}_2 \neq \mathscr{I}_3$) and eventually at very high spin oblate. If the nucleons are not paired, then the motion becomes like that of a nucleus of independent nucleons with a moment of inertia expected to be close to the rigid body value. This behaviour was first seen by Chapman *et al.* in 1983 in ^{168}Hf where the relation between energy and angular momentum was exactly that of a classical rotor in the spin range $I = 22$–34.

Rotational bands are also expected to be built on very deformed, called super-deformed, intrinsic states. The existence of such states with axis ratios as high as 2:1, for example, is predicted by the liquid drop model when the effects of shell structure are allowed for. This model is described in chapter 5, where it is used to explain the existence of fission isomers. The effect of including high spin in this model is discussed in chapter 8.

In order to study such high spin states, heavy-ion (HI) beams are used to bring in sufficient angular momentum. Typical examples are the fusion (compound nucleus) reactions ^{124}Sn(^{48}Ca, 4n)^{168}Hf and ^{100}Mo(^{36}S, 4n)^{132}Ce at energies of ~4 MeV/A. The typical decay pattern following compound nucleus formation is shown in figure 2.38. Many studies have concentrated on the gamma rays produced by transitions between levels on or near the yrast line. The number of coincident γ-rays and the total γ-ray energy (the multiplicity and sum energy) can be used to help select a particular reaction, e.g. the higher multiplicity favours a lower number of evaporated neutrons in a (HI, xn) reaction. A very powerful detector for such investigations is an array of bismuth germanate (BGO) shielded germanium detectors.

A germanium detector contains a single crystal of germanium suitably fabricated and doped so as to form a diode. A reverse-bias voltage is applied to the diode, which creates a region of semiconductor depleted in charge carriers. A γ-ray (or charged particle) passing through the depleted region loses energy by promoting electrons to the conduction band leaving a hole in the valance band. The electrons and holes are swept away to opposite electrodes by the applied voltage and the total amount of charge collected is proportional to the energy of the incident γ-ray (or charged particle). Both germanium and silicon are used in these detectors. Germanium has a slightly smaller band gap and must be cooled to liquid nitrogen temperature to reduce the leakage current across the diode. Because of its higher photoelectric cross-section germanium is used for γ-ray detectors.

The amount of energy ε lost per electron–hole pair is ~3 eV so the number N of pairs is $N \approx \frac{1}{3}E_\gamma$ (MeV)10^6. If there were no variation in ε there would be none in N; however there is some and the variation in N is given by $\sqrt{FN}$ where F is called the Fano factor. It is ~0.1 for germanium and silicon. The energy spread is therefore $\pm E_\gamma\sqrt{F/N}$, which for a 1 MeV γ-ray is ~±0.5 keV. At γ-ray energies below 1 MeV a γ-ray loses energy

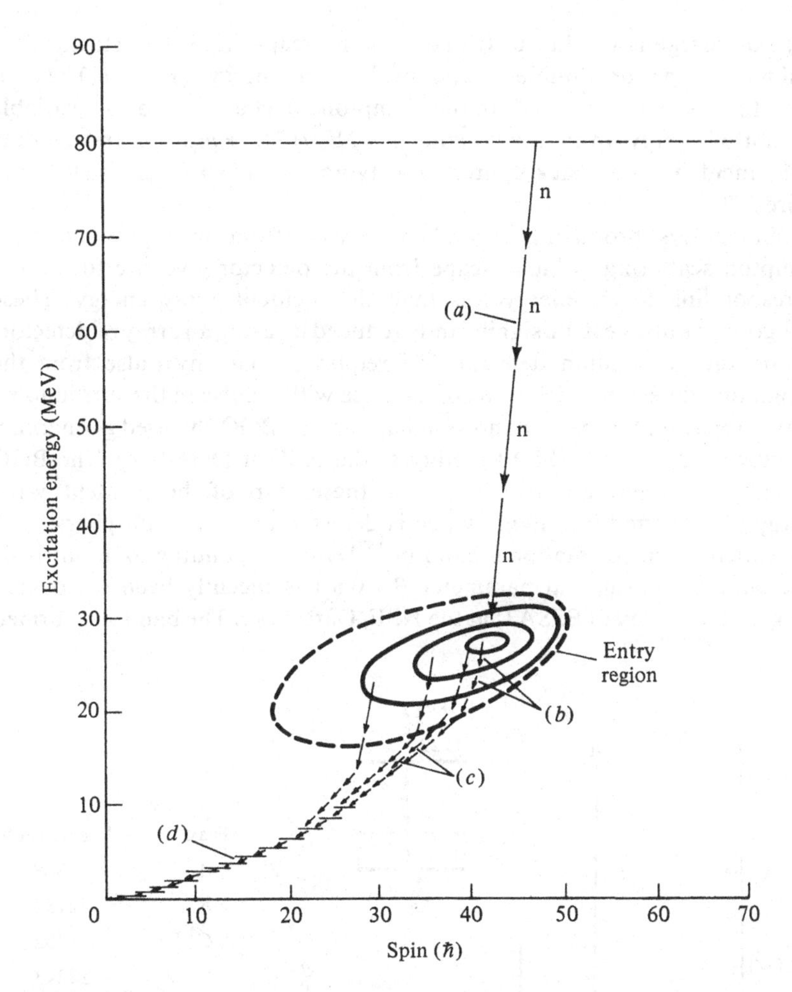

2.38 Typical heavy-ion reaction showing (*a*) particle emission (4n in this case); (*b*) statistical γ-rays; (*c*) continuum γ-rays; (*d*) yrast (normally discrete) γ-rays. (From *Proceedings of the International Nuclear Physics Conference*, Harrogate (1986)(ed. Nolan, P. J.) Institute of Physics Publishing Ltd, p. 155.)

through the photoelectric and Compton effects. Above 2 mc^2 (1.022 MeV) pair production is possible and this process dominates at high energies ($\geq$5 MeV). In a detector an e^-e^+ pair would initially have a kinetic energy ($E_\gamma - 2mc^2$). The e^- and e^+ would lose their kinetic energy through the production of e − h pairs. When the positron comes to rest it is annihilated with another electron producing two 511 keV γ-rays which may interact within the detector. If both are stopped within the detector a full energy pulse of charge is produced. If one or both escape from the detector then a single-escape or double-escape peak with energy $(E_\gamma - 511)$ keV or $(E_\gamma - 1022)$ keV is produced. In the Compton effect a γ-ray loses a variable amount of energy up to a maximum $E = 2E_\gamma^2/(2E_\gamma + m_e c^2)$ corresponding to the incident γ-ray back-scattering. A typical γ-ray spectrum is shown in figure 2.39.

Gamma-rays producing secondary γ-rays (from pair production or Compton scattering) which escape from the detector give rise to a pulse corresponding to an energy less than the incident γ-ray energy. These background pulses can be significantly reduced by using an array of detectors around the germanium detector and requiring that any pulse from the germanium detector which is in coincidence with a pulse in the surrounding array is rejected. Figure 2.40 shows a diagram of a BGO shielded germanium detector used in the TESSA3 facility at the NSF at Daresbury. The BGO detectors are scintillation detectors. In these, part of the incident γ-ray energy is converted into light, which is detected by photomultipliers.

A superdeformed rotational band in ^{152}Dy corresponding to an intrinsic state with a deformation parameter $\beta \approx 0.6$ has recently been discovered using such an array (TESSA3) at the NSF, Daresbury. The band was formed

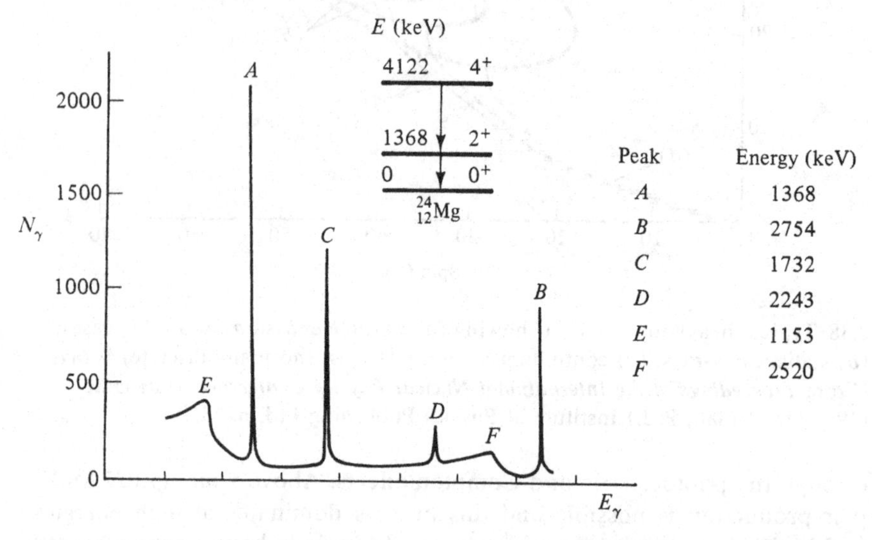

Peak	Energy (keV)
A	1368
B	2754
C	1732
D	2243
E	1153
F	2520

2.39. Typical γ-ray energy spectrum obtained when a small germanium detector is exposed to γ-radiation from $^{24}_{12}$Mg. The peaks *A* and *B* are the full-energy peaks; *D* is the single-escape and *C* the double-escape peak of the 2754 keV γ-ray, and *E* and *F* are the Compton edges corresponding to the 1368 keV and 2754 keV γ-rays, respectively, back-scattering out of the crystal.

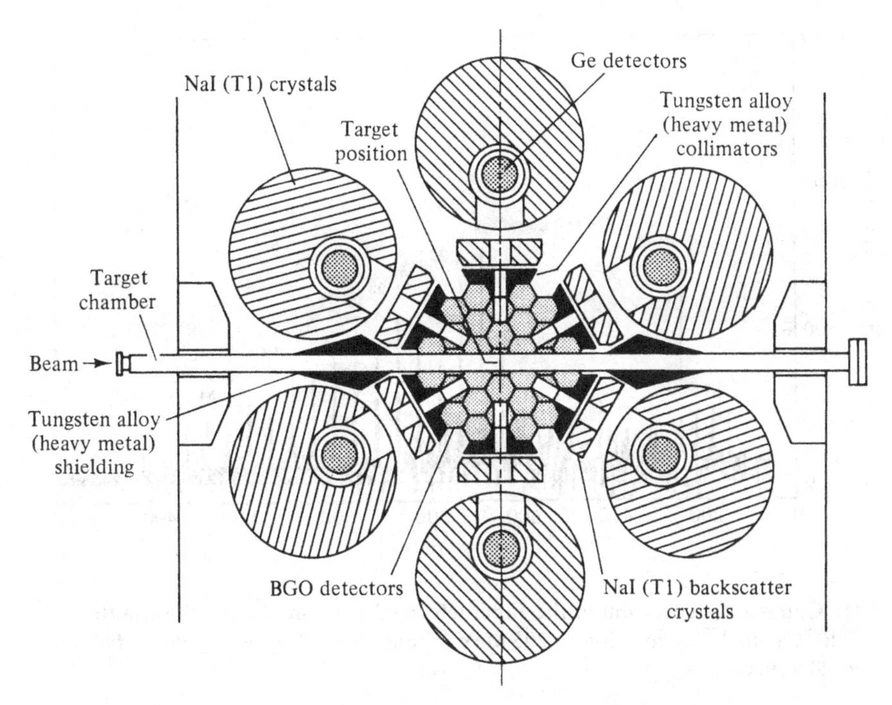

2.40. A schematic diagram of the Total Energy Suppression Shield Array (TESSA) at the NSF at Daresbury. It is a vertical centre line section and shows that the detector systems are cylindrically symmetric about a horizontal axis through the target position. (From Turk, P. J. *et al.*, *Nucl. Phys.* **A409** (1983) 343.)

in the reaction ^{108}Pd(^{48}Ca, 4n)^{152}Dy at a beam energy of 205 MeV. Over 200 million events were recorded corresponding to two or more germanium detectors registering a γ-ray. A coincident background subtracted γ-ray spectrum is shown in figure 2.41 in which a series of E2 transitions extending up to $I = 60$ can be seen. These transitions are down a superdeformed band. An angular momentum versus rotational frequency $\hbar\omega$ (equation 2.15 above) plot shows a constant slope at high spin both for the superdeformed and for the ground state band in ^{152}Dy. As mentioned above this classical rotational behaviour suggests that the pairing correlations have been significantly reduced.

2.7 Summary

Several models have been described which account for different aspects of nuclear behaviour. The masses and stability of nuclei are understood using

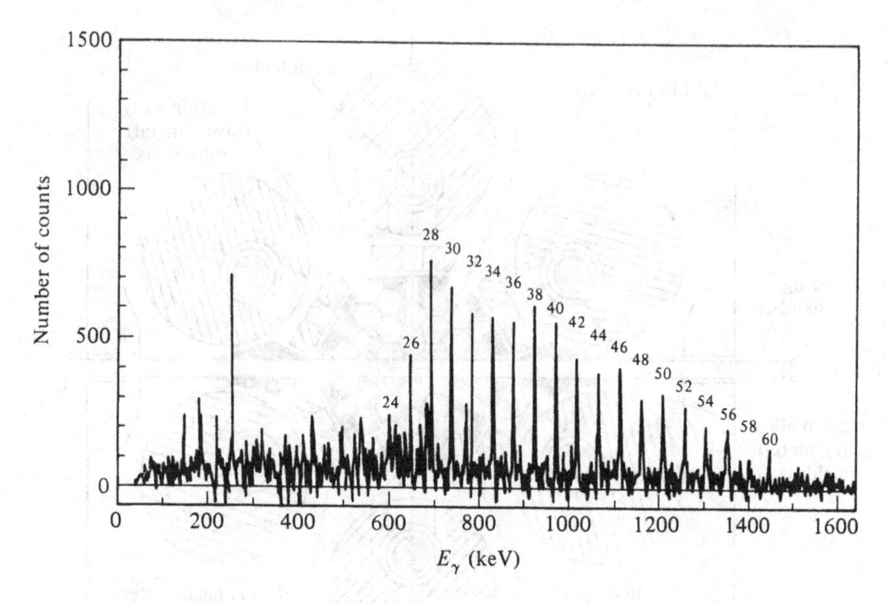

2.41. Gamma-ray spectrum in the superdeformed band in ^{152}Dy following the ^{108}Pd(^{48}Ca, 4n)^{152}Dy reaction at 205 MeV (From Twin, P. J., *et al., Phys. Rev. Lett.* **57** (1986) 811.)

the liquid drop model which accounts for the gross features of the interactions between nucleons within nuclei. The approximate independent-particle motion of nucleons leads to the single-particle shell model in which nucleons move in a spherical potential well. This explains the magic numbers and the spins and parities of many nuclear ground states.

Consideration of the effect of deformation on the kinetic and potential energies of nucleons explains why most nuclei are deformed. The Nilsson model provides a generalisation of the single-particle shell model to one in which nucleons move in a deformed potential well. Although many levels of nuclei can be understood using the simple shell model (spherical or deformed), several states are seen which are not easily described by this model and which are reminiscent of molecular rotational levels. They correspond to collective rotational motion of the nucleus.

Many of the low-lying levels in a rotational band are described in terms of a nucleus with a constant moment of inertia which is typically about $\frac{1}{4}$–$\frac{3}{4}$ of the classical value. This reduction reflects the effect of the pairing residual interaction between the nucleons. At high spin the effect of pairing is reduced and larger moments of inertia and the phenomenon of backbending are observed. Besides rotational levels, vibrational levels are also seen in nuclei. Prominent vibrational states in all nuclei are the 1^- giant dipole state and the 3^- octupole state.

These collective states correspond in the shell model to a superposition of many simple shell model configurations. This superposition leads to enhanced electromagnetic multipole transition rates and this is discussed in chapter 4. The connection between the collective and shell model descriptions of nuclear states is further developed in chapter 8.

2.8 Questions

1.* Using the semi-empirical mass formula show that the energy S_n required to separate a neutron from the nucleus (A, Z) is given approximately by:

$$S_n \approx a_v - \tfrac{2}{3}a_s A^{-1/3} - a_a[1 - 4Z^2/A(A-1)]$$

Estimate the mass number of the Na nucleus which is just stable against neutron emission. Use the values of the coefficients given on page 22.

2.* The total binding energies of the nuclei $^{15}_{8}$O, $^{16}_{8}$O and $^{17}_{8}$O are 111.95 MeV, 127.62 MeV and 131.76 MeV, respectively. Deduce the energies of the last occupied state and of the first unoccupied state of neutrons in $^{16}_{8}$O.
The total binding energy of $^{17}_{9}$F is 128.22 MeV. Deduce the energy of the first unoccupied proton state in $^{16}_{8}$O. Why is it different from the energy of the corresponding neutron state? Account for the order of magnitude of the difference.
The first excited state $(J^\pi = \frac{1}{2}^+)$ of $^{17}_{8}$O has an excitation energy of 0.87 MeV. The corresponding state in $^{17}_{9}$F has an excitation energy of 0.5 MeV. Suggest a reason for the difference in excitation energy.

3. For odd-odd nuclei define $N = j_p - l_p + j_n - l_n$ where p and n refer to the odd proton and odd neutron, respectively. Nordheim's rules predict that if $N = 0$ then the spin of the nucleus $J = |j_n - j_p|$ and if $N = \pm 1$ $J = |j_n - j_p|$ or $j_n + j_p$. Use these rules to predict the spins of $^{70}_{31}$Ga, $^{58}_{27}$Co and $^{208}_{81}$Tl.

4. Using the single-particle shell model predict the J^π of: $^{15}_{7}$N, $^{27}_{12}$Mg, $^{87}_{38}$Sr, $^{167}_{68}$Er and $^{195}_{80}$Hg. The measured J^π are $\frac{1}{2}^-$, $\frac{1}{2}^+$, $\frac{9}{2}^+$, $\frac{7}{2}^+$ and $\frac{1}{2}^-$, respectively. Comment on any discrepancies.

5. Use the expressions for the Schmidt limits to estimate the magnetic moments of $^{41}_{20}$Ca, $^{15}_{7}$N, $^{11}_{5}$B, $^{25}_{12}$Mg.
Compare with the measured values of (in units of μ_N): $^{41}_{20}$Ca -1.6; $^{15}_{7}$N -0.28; $^{11}_{5}$B 2.69; $^{25}_{12}$Mg -0.86.

6. The nucleus $^{235}_{92}$U $(J^\pi = \frac{7}{2}^-)$ would have a closed proton subshell and partially filled $l = 4$ neutron subshell if the spin-orbit interaction were

neglected. In this approximation use the Rainwater model to predict the observed quadrupole moment of $^{235}_{92}$U and compare with the measured value of $+4.1\times10^{-28}$ m^2. What value would the single-particle shell model predict?

7. Confirm that $\omega_0(\delta)=\omega_0(1+\frac{1}{9}\delta^2)$ ensures $\omega_x\omega_y\omega_z=\omega_0^3$ to second order in δ and derive equation 2.12.
8. The observed quadrupole moment of $^{177}_{71}$Lu is 3.33 barns and its ground state has $J^\pi=\frac{7}{2}^+$. Calculate the deformation parameter δ and show using figure 2.21 that the Nilsson model would account for this J^π.
9.* A shell model configuration that is a component of the 1.98 MeV 2^+ level's wavefunction in ^{18}O is $(\nu d_{5/2})\ (\nu d_{3/2})2^+$. What shell model configurations might be important in describing: (*a*) the 3.56 MeV 4^+ level, (*b*) the 3.63 MeV 0^+ level and (*c*) the 4.46 MeV 1^- and 5.10 MeV 3^- levels in ^{18}O?
10.* Some of the low-lying states of $^{177}_{71}$Lu, labelled by excitation energy in keV and J^π, are: 0 $\frac{7}{2}^+$; 122 $\frac{9}{2}^+$; 150 $\frac{9}{2}^-$; 269 $\frac{11}{2}^+$; 289 $\frac{11}{2}^-$; 441 $\frac{13}{2}^+$; 451 $\frac{13}{2}^-$; 636 $\frac{15}{2}^+$; 854 $\frac{17}{2}^+$.
    Interpret these states in terms of an appropriate nuclear model.
11. Moments of inertia $\mathcal{I}^{(1)}$ and $\mathcal{I}^{(2)}$ defined by $\mathcal{I}^{(1)}=[(2/\hbar^2)/(\mathrm{d}E/\mathrm{d}I^2)]^{-1}$ and $\mathcal{I}^{(2)}=[(1/\hbar^2)/(d^2E/dI^2)]^{-1}$ are sometimes used. Show that $\mathcal{I}^{(2)}$ equals the moment of inertia $\mathcal{I}$ given in the expression (equation 2.17) for the energies of the aligned band. Show also that $\mathcal{I}^{(2)}=\mathcal{I}^{(1)}$ for the ground state band (equation 2.16).
12. Deduce the moment of inertia of the superdeformed band in ^{152}Dy from the data shown in figure 2.41 and compare with the rigid body value.

3 Beta decay

3.1 Introduction

The early experiments showed that in β^--decay a nucleus decays to another nucleus of the same mass number with the emission of an electron with a varying amount of energy. The simplest example of such a decay is the β^--decay of a neutron to a proton and the β^--spectrum for this decay is shown in figure 3.1.

If just an electron were emitted then energy and momentum conservation would require all the electrons to have the same energy determined by E_0, the Q-value for the decay which is equal to the difference in rest mass energies of the initial and final state. The shape of the spectrum therefore implies that another particle is also emitted which shares the available decay energy and this particle is called an anti-neutrino. The β^--decay of a neutron therefore corresponds to:

$$\mathrm{n} \rightarrow \mathrm{p} + \mathrm{e}^- + \tilde{\nu}$$

There are two closely related processes to β^--decay, which are β^+-decay where a proton bound in a nucleus decays to a bound neutron with the emission of a positron and a neutrino, ν, and electron capture (EC) where a proton bound in a nucleus changes to a bound neutron through interacting with one of the atomic electrons and a neutrino is emitted. Examples of these processes are:

$$^{22}\mathrm{Na} \rightarrow {}^{22}\mathrm{Ne} + \mathrm{e}^+ + \nu \qquad \beta^+\text{-decay}$$

$$\mathrm{e}^- + {}^{7}\mathrm{Be} \rightarrow {}^{7}\mathrm{Li} + \nu \qquad \mathrm{EC}$$

and when energetically possible β^+-decay and EC compete.

3.1.1 The energetics of beta decay

Conservation of energy requires the initial state to be heavier than the final state before a decay can occur. The mass of a nucleus is given in terms of the atomic mass, ${}^{A}_{Z}M$, which equals the mass of the nucleus $M(A, Z)$ plus the mass of the atomic electrons Zm_e less the mass equivalence of their

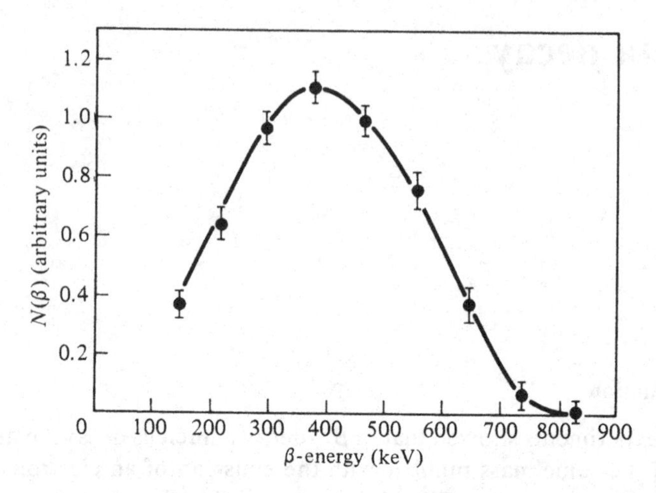

3.1. Electron energy spectrum from neutron β^--decay. (Below 300 keV the points are unreliable for experimental reasons.) (From Robson, J. M., *Phys. Rev.* **83** (1951) 349.)

binding energy B_e. Typically the differences in binding energies are small compared with the Q-value for a decay and can be neglected. Also the mass of the neutrino m_ν (which equals that of its anti-particle $m_{\bar{\nu}}$) can be set equal to zero to a very good approximation as its value is known to be very small (if not zero) ($\lesssim 30\ \mathrm{eV}/c^2$). With these approximations the conditions for β^--, β^+-decay and EC to be energetically possible are:

$$
\begin{aligned}
{}^{A}_{Z}M &> {}^{A}_{Z+1}M && \beta^-\text{-decay}\\
{}^{A}_{Z}M &> {}^{A}_{Z-1}M + 2m_e && \beta^+\text{-decay}\\
{}^{A}_{Z}M &> {}^{A}_{Z-1}M && \text{EC}
\end{aligned}
$$

In EC it is energetically possible for K-capture to be forbidden but L-capture to be allowed as the binding energy of a K-shell electron is greater than of an L-shell electron. Double β-decay is energetically possible for several nuclei but is very improbable, as it is a second-order weak process, with a half-life of $\sim 10^{22}$ yr.

3.2 The theory of beta decay

Beta decay is an example of a weak interaction and this interaction, like the strong, electromagnetic and gravitational interactions, can be understood as arising through the exchange of particles. In beta decay the exchanged particles are the charged intermediate vector bosons W^- and W^+, whose masses are $\sim 80\,\text{GeV}/c^2$ (cf. the mass of a proton $\sim 1\,\text{GeV}/c^2$), while in electromagnetic interactions the exchanged particles are photons, which have zero mass. There is a close similarity in form between the weak and electromagnetic interactions but their ranges are very different; this is related to the differences in masses of the exchanged particles.

3.2.1 The range of an interaction

The range of an interaction is related to the mass of the exchanged particle. A simple way to see how this comes about is to use the energy–time uncertainty relation: $\Delta E \Delta t \approx \hbar$. A particle can only create another particle of mass m for a time $t \approx \hbar/mc^2$, during which interval the created particle can travel at most a distance ct. Taking ct as an estimate of the range of the interaction, R, gives:

$$R \approx \hbar/mc$$

As an example the force between two neutrons arising from the exchange of a pion, which has a mass of $\sim 140\,\text{MeV}/c^2$, would be predicted to have a range $R \approx 1.4\,\text{fm}$, which is in good agreement with the observed range of the nuclear force between nucleons.

A more quantitative way of obtaining the range of an interaction is to look at the equation for the corresponding potential. In electromagnetism the equation for the electrostatic potential $\phi(x, t)$ in free space is: '

$$\nabla^2 \phi - \frac{1}{c^2}\frac{\partial^2 \phi}{\partial t^2} = 0$$

When there is a source of charge q at the origin this equation becomes

$$\nabla^2 \phi - \frac{1}{c^2}\frac{\partial^2 \phi}{\partial t^2} = -\frac{q}{\varepsilon_0}\,\delta(r)$$

where $\delta(r)$ is the Dirac delta function which has the property $\int f(r)\delta(r)\,\mathrm{d}^3 r = f(0)$. The time-independent solution ϕ satisfies the Poisson equation:

$$\nabla^2 \phi = -\frac{q}{\varepsilon_0}\,\delta(r)$$

which has the solution $\phi = q/4\pi\varepsilon_0 r$. The range of this potential is infinite and corresponds to the photon, which is the exchanged particle giving rise to the electromagnetic interaction, having zero mass.

For a massive spin zero particle the wave equation describing the motion of the particle in free space can be deduced from the relativistic energy–momentum relation:

$$E^2 = p^2c^2 + m^2c^4$$

by making the substitution $E \to +\mathrm{i}\hbar\partial/\partial t$ and $p \to -\mathrm{i}\hbar\nabla$, which yields the

Klein–Gordon equation:

$$\nabla^2\phi - \frac{m^2c^2}{\hbar^2}\phi - \frac{1}{c^2}\frac{\partial^2\phi}{\partial t^2} = 0$$

By analogy with the equation for the electrostatic potential, the equation for the potential associated with the exchange of a massive spin zero particle when there is a source of strength g at the origin is:

$$\nabla^2\phi - \mu^2\phi - \frac{1}{c^2}\frac{\partial^2\phi}{\partial t^2} = -g\delta(r)$$

where $\mu = mc/\hbar$. The time-independent solution ϕ satisfies the equation:

$$\nabla^2\phi - \mu^2\phi = -g\delta(r)$$

which has the solution $\phi = g[\exp(-\mu r)]/4\pi r$. The range of this potential is $\mu^{-1} = \hbar/mc$, which agrees with the previous estimate for the range of an interaction caused by the exchange of a massive particle.

The weak interaction arises through the exchange of intermediate vector bosons ($W^{\pm}$, Z^0) whose masses are ~80 GeV/c^2 so its range is very short indeed: ~10^{-3} fermis. Although the range of the electromagnetic interaction is infinite the forms of the weak and electromagnetic interaction are similar and it is helpful in understanding the weak interaction to consider first the form of the electromagnetic interaction.

3.2.2 The form of the electromagnetic interaction

A typical electromagnetic interaction is the scattering of a proton and an electron and shown in figure 3.2 is a diagrammatic representation (Feynman diagram) of this process. The proton, which initially has a momentum p_1, experiences a perturbation due to its interaction with the electromagnetic field arising from the electron. The result of this perturbation is that the proton's momentum is changed to p_3. Likewise the electron's momentum is altered from p_2 to p_4. The form of the perturbation can be inferred from considering the classical interaction of a current with an electromagnetic field.

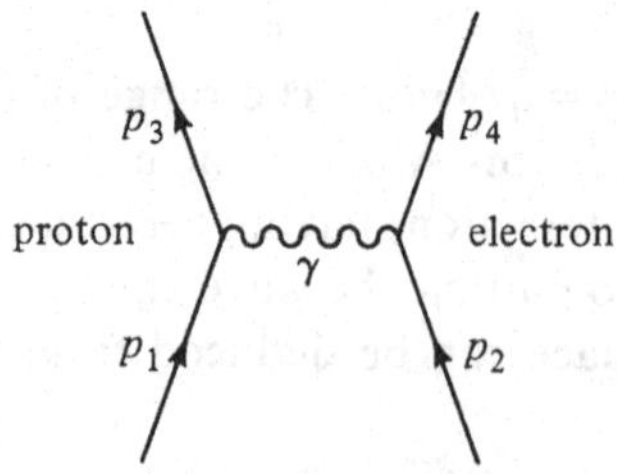

3.2. Electromagnetic interaction between a proton and an electron through the exchange of a photon.

Classically the energy of interaction or perturbation per unit volume, U, experienced by a current $\mathbf{J}$ in an electromagnetic vector potential field $\mathbf{A}$ is $-\mathbf{J}\cdot\mathbf{A}$. If the current is produced by a moving charged particle and an electrostatic potential ϕ is present there is an additional perturbation $\rho\phi$ where ρ is the charge density. The total perturbation per unit volume $U=\rho\phi-\mathbf{J}\cdot\mathbf{A}$ can be written more concisely and in a Lorentz covariant form as:

$$U=J_\mu A_\mu \equiv J_0A_0-J_1A_1-J_2A_2-J_3A_3$$

where J_μ is the four-vector current (ρ, J) and A_μ is the four-vector potential (ϕ, A).

The perturbation per unit volume U experienced by the proton current $J_\mu^{(\mathrm{p})}$ due to the field $A_\mu^{(\mathrm{e})}$ of the electron current $J_\mu^{(\mathrm{e})}$ is therefore $J_\mu^{(\mathrm{p})}A_\mu^{(\mathrm{e})}$. The vector potential $A_\mu^{(\mathrm{e})}$ is proportional to $J_\mu^{(\mathrm{e})}$ and so U is given by:

$$U \propto J_\mu^{(\mathrm{p})}J_\mu^{(\mathrm{e})}$$

which is of the form (current) · (current). An important approximation occurs if either of the currents arises from a very slow ($v^2 \ll c^2$) charged particle; for then the magnetic interaction $-\mathbf{J}\cdot\mathbf{A}$ is much weaker than the electrostatic interaction $\rho\phi$ and U is given to a good approximation by $U=\rho\phi$.

As an example consider an electron scattering off a stationary proton. For a stationary electron described by a wavefunction $\psi(r)$ its charge density $\rho(r)=-e\psi^*(r)\psi(r)$. For a scattered electron $\rho(r)$ becomes the transition charge density $\rho^{(\mathrm{e})}(r)$ given by (neglecting the electron spin):

$$\rho^{(\mathrm{e})}(r)=-e\psi_f^*(r)\psi_i(r)$$

where ψ_i and ψ_f are the initial and final electron wavefunctions. The matrix element M_{if} describing the scattering is

$$M_{if}=\int -e\psi_f^*(r)\psi_i(r)\phi^{(\mathrm{p})}(r)\,\mathrm{d}^3r,$$

where $\phi^{(\mathrm{p})}(r)$ is the electrostatic potential of the proton, and is the transition matrix element in the Born approximation.

If the scattering is from a distribution of charge described by a charge density $\rho(r')=e\psi^*(r')\psi(r')$ then the electrostatic potential $\phi^{(\mathrm{p})}(r)$ is given by:

$$\phi^{(\mathrm{p})}(r)=\int \rho(r')\phi(|\mathbf{r}-\mathbf{r}'|)\,\mathrm{d}^3r'$$

$$=\int e\psi^*(r')\psi(r')\phi(|\mathbf{r}-\mathbf{r}'|)\,\mathrm{d}^3r' \tag{3.1}$$

where ϕ is the electrostatic potential of unit charge.

3.2.3 Fermi's theory of the weak interaction

Returning now to the weak interaction, a typical weak process is neutron decay, which can be represented by the diagram shown in figure 3.3*a* where W is a charged intermediate vector boson. Stückelberg and Feynman showed that the emission of an anti-particle with four-momentum $-p_\mu$ is equivalent to the absorption of a particle with four-momentum p_μ. Therefore the neutron decay matrix element M_{if} is the same as that for the following process: $\mathrm{n}+\nu \rightarrow \mathrm{p}+\mathrm{e}^-$, shown in figure 3.3*b*. This process is very similar to the proton–electron scattering example discussed above (see figure 3.2).

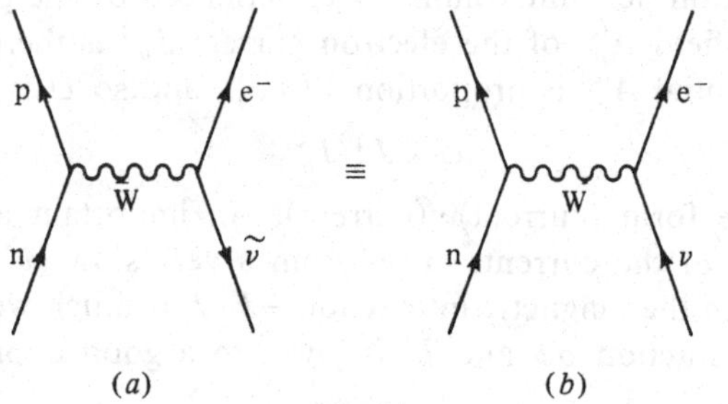

3.3. The equivalence of (*a*) neutron β^--decay: $\mathrm{n}\rightarrow \mathrm{p}+\mathrm{e}^-+\tilde{\nu}$ to (*b*) the process: $\mathrm{n}+\nu\rightarrow\mathrm{p}+\mathrm{e}^-$.

The neutron experiences a perturbation due to its interaction with the weak field of the neutrino. Assigning a weak charge g to the neutron and proton, then the motion of the neutron and proton gives rise to a weak nucleonic current $J_\mu^{(\mathrm{nucleon})}$. Similarly the motion of the neutrino and electron gives rise to a weak leptonic current $J_\mu^{(\mathrm{leptonic})}$. By comparison with the proton-electron electromagnetic interaction, Fermi postulated in 1934 that the perturbation per unit volume U_F due to the weak interaction would have the same (current) · (current) form and hence would be given by:

$$U_\mathrm{F} \propto J_\mu^{(\mathrm{n})} J_\mu^{(\mathrm{l})}$$

As U_F is formed from the scalar product of two four-vectors the Fermi interaction is also called the vector interaction.

In nuclear β-decay a further simplification is possible to U_F since $(v/c)^2$ for nucleons within a nucleus is small (typically $\lesssim 0.07$) so, just like in the electromagnetic example discussed above, it is a good approximation to take:

$$U_\mathrm{F} = \rho_\mathrm{w}^{(\mathrm{l})} \phi_\mathrm{w}^{(\mathrm{n})}$$

where $\phi_\mathrm{w}^{(\mathrm{n})}$ is the weak potential of a nucleon and the lepton weak transition charge density is given by (neglecting the leptons' spins):

$$\rho_{\mathrm{w}}^{(\mathrm{l})} = g\psi_{\mathrm{e}}^{*}\psi_{\nu}$$
$$= g\psi_{\mathrm{e}}^{*}\psi_{\tilde{\nu}}^{*}$$

using the equivalence of the absorption of a ν to the emission of a $\tilde{\nu}$ ($\psi_{\nu} = \exp(i\mathbf{k}_{\nu} \cdot \mathbf{r}) = \exp(-i\mathbf{k}_{\tilde{\nu}} \cdot \mathbf{r}) = \psi_{\tilde{\nu}}^{*}$). The matrix element M_{if} for the β^{-}-decay of a neutron is therefore given by:

$$M_{if} = \int g\psi_{\mathrm{e}}^{*}(r)\psi_{\tilde{\nu}}^{*}(r)\phi_{\mathrm{w}}^{(\mathrm{n})}(r)\,\mathrm{d}^{3}r$$

In a nuclear β^{-}-decay the neutron is not localised at $r = 0$ but is described by a wavefunction ψ_{n}. For example in tritium there are two neutrons and one proton in the $1\mathrm{s}_{1/2}$ state and only one of the neutrons can β^{-}-decay because of the Pauli exclusion principle. The distribution of weak charge is described by the weak transition charge density $\rho_{\mathrm{w}}^{(\mathrm{n})}(r') = g\psi_{\mathrm{p}}^{*}(r')\psi_{\mathrm{n}}(r')$ where in tritium ψ_{n} and ψ_{p} are $1\mathrm{s}_{1/2}$ wavefunctions. The weak potential $\phi_{\mathrm{w}}(r)$ is therefore given by (cf. equation 3.1 above):

$$\phi_{\mathrm{w}}^{(\mathrm{n})}(r) = \int g\psi_{\mathrm{p}}^{*}(r')\psi_{\mathrm{n}}(r')\phi_{\mathrm{w}}(|\mathbf{r}-\mathbf{r}'|)\,\mathrm{d}^{3}r'$$

where ϕ_{w} is the weak potential of unit weak charge. In the approximation that the weak potential ϕ_{w} arises from massive scalar particle exchange then $\phi_{\mathrm{w}}(r) = \exp(-\mu r)/4\pi r$. The range of the weak interaction is very short indeed ($\mu^{-1} \sim 10^{-3}$ fm) and over this range $\psi_{\mathrm{p}}(r')$ and $\psi_{\mathrm{n}}(r')$ are essentially equal to their value at r, so it is a very good approximation to take:

$$\phi_{\mathrm{w}}^{(\mathrm{n})}(r) = g\psi_{\mathrm{p}}^{*}(r)\psi_{\mathrm{n}}(r)\int \frac{\exp(-\mu|\mathbf{r}-\mathbf{r}'|)}{4\pi|\mathbf{r}-\mathbf{r}'|}\,\mathrm{d}^{3}r'$$
$$= \frac{g}{\mu^{2}}\psi_{\mathrm{p}}^{*}(r)\psi_{\mathrm{n}}(r)$$

This is equivalent to taking the weak force to have zero range (a contact interaction), which is what Fermi assumed.

The matrix element for a nuclear β^{-}-decay in which only one neutron, described by ψ_{n}, contributes is therefore:

$$M_{if} = \frac{g^{2}}{\mu^{2}}\int \psi_{\mathrm{e}}^{*}(r)\psi_{\tilde{\nu}}^{*}(r)\psi_{\mathrm{p}}^{*}(r)\psi_{\mathrm{n}}(r)\,\mathrm{d}^{3}r \qquad (3.2)$$
$$\equiv G_{\beta}M_{\mathrm{F}}$$

which defines the matrix element M_{F}. The constant G_{β}, the nuclear β-decay coupling constant, has the dimensions of (energy) · (volume). In this expression for M_{if} the lepton spins have been neglected. This has no effect on the overall rate and only affects the lepton angular correlation and the coupling of the leptons' spins. In a Fermi transition the lepton spins couple up to spin zero while in a Gamow–Teller transition (see below) they couple up to spin one.

In the Weinberg–Salam theory of the electroweak interaction the weak charge $g \approx e/\sqrt{\varepsilon_0}$, where e is the electric charge and ε_0 is the dielectric constant of free space, so in the approximation described above of massive scalar particle exchange G_β is given by:

$$G_\beta = \frac{g^2}{\mu^2} \approx \frac{e^2\hbar^2}{\varepsilon_0 M_W^2 c^2} = \frac{e^2}{4\pi\varepsilon_0 \hbar c} \frac{(\hbar c)^3 \cdot 4\pi}{(M_W c^2)^2}$$

where M_W is the mass of the $W^\pm$ particle (80 GeV/c^2). Using $\hbar c =$ 197 MeV fm, $e^2/4\pi\varepsilon_0\hbar c = 1/137$ gives $G_\beta = 1.1 \times 10^{-4}$ MeV fm^3. Although only approximate as g is not exactly $e/\sqrt{\varepsilon_0}$ and the $W^\pm$ particles are vector particles (i.e. have intrinsic spin 1) this estimate is quite close to the measured value of $G_\beta = 0.874 \times 10^{-4}$ MeV fm^3.

3.3 Fermi and Gamow–Teller beta transitions

3.3.1 Allowed transitions

The matrix element for a nuclear β^--decay involving a single neutron described by the wavefunction ψ_n is given by equation 3.2 above:

$$M_{if} = G_\beta \int_0^R \psi_e^*(r)\psi_{\bar\nu}^*(r)\psi_p^*(r)\psi_n(r)\, d^3r$$

where R is the radius of the nucleus. The lepton wavefunctions as a first approximation can be taken as plane waves, i.e. $\psi_e = L^{-3/2}\exp(i\mathbf{k}_e \cdot \mathbf{r})$ and $\psi_{\bar\nu} = L^{-3/2}\exp(i\mathbf{k}_{\bar\nu} \cdot \mathbf{r})$, in which case:

$$M_{if} = \frac{G_\beta}{L^3} \int_0^R \exp(-i\mathbf{q} \cdot \mathbf{r})\psi_p^*(r)\psi_n(r)\, d^3r$$

where $\mathbf{q} = \mathbf{k}_e + \mathbf{k}_{\bar\nu}$ is the momentum transfer in the β-decay. In a typical β-decay $\hbar qc \approx 1$ MeV so for $R = 5$ fm, $\mathbf{q} \cdot \mathbf{r}$ ($\lesssim qR$) will be $\lesssim \frac{1}{40}$. Therefore to a good approximation $L^{-3}\exp(-i\mathbf{q} \cdot \mathbf{r}) = L^{-3} = \psi_e^*(0)\psi_{\bar\nu}^*(0)$ over the volume of the nucleus. This corresponds to the leptons taking away no orbital angular momentum and the β-decay is then called an allowed decay. The matrix element M_{if} is given by:

$$M_{if} = G_\beta \int_0^R \psi_e^*(0)\psi_{\bar\nu}^*(0)\psi_p^*(r)\psi_n(r)\, d^3r \qquad (3.3)$$

which in the plane wave approximation becomes:

$$M_{if} = \frac{G_\beta}{L^3} \int_0^R \psi_p^*(r)\psi_n(r)\, d^3r$$

This is called an allowed Fermi β^--decay and the interaction just changes a neutron into a proton (or vice versa in β^+- or EC decay) without any affect on the space or spin part of the wavefunction of the nucleon. There is therefore no change in the l or s of the wavefunction of the nucleon involved in the decay. The selection rules for allowed Fermi transitions are therefore:

allowed Fermi: $\Delta J = 0$ no change of parity

and in such a transition the leptons take away no orbital or spin angular momentum, i.e. the two spin $\frac{1}{2}$ vectors of the e^- and $\tilde{\nu}$ (or e^+ and ν) couple to $J = 0$. An example is the decay

$$^{14}O(0^+) \rightarrow {}^{14}N^*(0^+) + e^+ + \nu$$

where the ground state of ^{14}O β^+-decays to its analogue state in ^{14}N, which is the first excited state of ^{14}N and the $T_z = 0$ member of the $T = 1$ isospin triplet.

It was soon realised after Fermi had proposed his theory that there occurred β-decays with comparable transition rates to allowed Fermi decays for which there was no change in parity but $\Delta J = 1$. This implies another component in the weak interaction which was originally proposed by Gamow and Teller, and as explained below this is an axial-vector weak current A_μ. This current gives rise in the non-relativistic limit (appropriate to nuclear β-transitions) to a matrix element for allowed transitions given by:

$$M_{if} = \lambda G_\beta \int_0^R \psi_e^*(0)\boldsymbol{\sigma}\psi_{\tilde{\nu}}^*(0)\psi_p^*(r)\boldsymbol{\sigma}\psi_n(r)\,d^3r$$

where λG_β is the Gamow–Teller coupling constant and $\boldsymbol{\sigma}$ is the spin $\frac{1}{2}$ operator. The operator $\boldsymbol{\sigma}$ can cause a change $\Delta s = 0$ or 1 in the spin part of the nucleon wavefunction but no change in l. However, the matrix element of $\boldsymbol{\sigma}$ is zero for $J = 0 \rightarrow 0$ transitions so the selection rules for allowed Gamow–Teller transitions are:

allowed Gamow–Teller: $\Delta J = 0$ or 1 but $0 \nrightarrow 0$ no change of parity

The spins of the leptons in such a decay couple to $J = 1$. An example is the decay:

$$^6He(0^+) \rightarrow {}^6Li(1^+) + e^- + \tilde{\nu}$$

of ^6He to the ground state of ^6Li.

There are also mixed allowed transitions in which both the vector (Fermi) and axial-vector (Gamow–Teller) parts of the weak current contribute such as

$$n \rightarrow p + e^- + \tilde{\nu}$$
$$^3H \rightarrow {}^3He + e^- + \tilde{\nu}$$

Because in an allowed Fermi transition a neutron just changes into a proton (or vice versa) without any affect on the rest of the wavefunction, these transitions are between analogue states, such as $^{14}O(0^+) \rightarrow {}^{14}N^*(0^+)$. The isospin selection rule is therefore:

$$\text{allowed Fermi: } \Delta T = 0$$

However, in an allowed Gamow–Teller transition the nucleon's wavefunction can alter so the isospin change is that due to a single nucleon's transition. As a nucleon has isospin $\frac{1}{2}$ the change is either 0 or 1 so the isospin selection rule is:

$$\text{allowed Gamow–Teller: } \Delta T = 0 \text{ or } 1$$

3.3.2 The shape of the β-spectrum

In a β^--decay a nucleus decays to another nucleus with the emission of an electron and an anti-neutrino which share the available decay energy E_0 ($\equiv Q$). The probability of a decay is given by Fermi's golden rule:

$$w = 2\pi/\hbar |M_{if}|^2 \rho_f$$

where w is the probability of a decay per unit time, M_{if} is the matrix element of the weak interaction and ρ_f is the density of final states. In an allowed β-decay M_{if} is a constant so the decay rate is determined by ρ_f. The momentum vectors and kinetic energies involved in a β-decay are shown in figure 3.4, which illustrates the decay:

$${}^{A}_{Z}M \rightarrow {}_{Z+1}{}^{A}M + e^- + \tilde{\nu}$$

Conservation of energy and momentum gives:

$$E_0 = E + E_{\tilde{\nu}} + E_R$$

$$\mathbf{P} + \mathbf{p} + \mathbf{q} = 0$$

As ${}^{A}_{Z}M \gg m_e$ then $E_R \ll E_0$ so to a very good approximation $E_0 = E + E_{\tilde{\nu}}$. The number of states per unit volume available to an electron with momentum lying between p and $p + \mathrm{d}p$ is $4\pi p^2\,\mathrm{d}p/h^3$, and likewise for an anti-neutrino with momentum lying between q and $q + \mathrm{d}q$ the number is

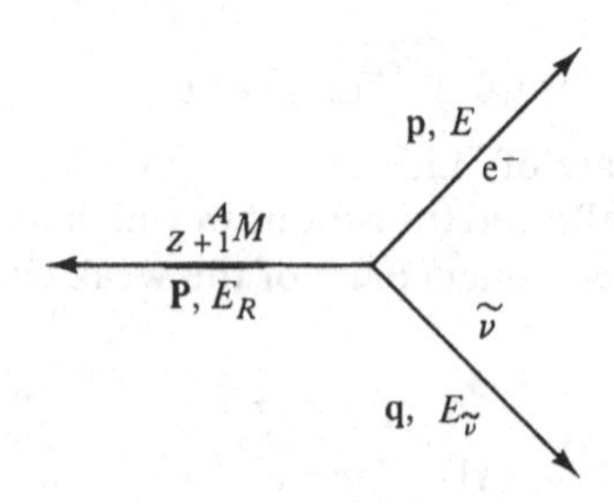

3.4. The momentum vectors in a β^--decay. E, E_R, $E_{\tilde{\nu}}$ denote kinetic energies.

$4\pi q^2\, \mathrm{d}q/h^3$. The total number of final states is therefore $\mathrm{d}N = 16\pi^2 p^2 q^2\, \mathrm{d}p\, \mathrm{d}q/h^6$. The density of final states $\rho_f \equiv \mathrm{d}N/\mathrm{d}E_0$ so $\mathrm{d}N$ is required as a function of $\mathrm{d}p$ and $\mathrm{d}E_0$. The relation:

$$\mathrm{d}x\mathrm{d}y = \begin{vmatrix} \frac{\partial x}{\partial s} & \frac{\partial x}{\partial t} \\ \frac{\partial y}{\partial s} & \frac{\partial y}{\partial t} \end{vmatrix} \mathrm{d}s\, \mathrm{d}t$$

where the determinant is the Jacobian for the transformation, enables $\mathrm{d}N/\mathrm{d}E_0$ to be evaluated as:

$$\frac{\mathrm{d}N}{\mathrm{d}E_0} \equiv \rho_f = \frac{16\pi^2}{h^6 c^3} p^2 (E_0 - E)^2\, \mathrm{d}p$$

where the mass $m_{\bar{\nu}}$ has been assumed to be zero and hence $qc = E_0 - E$. More simply, but less rigorously, the electron momentum and energy can be first considered fixed and then $\mathrm{d}q = \mathrm{d}E_0/c$ so $\mathrm{d}N_{\bar{\nu}} = 4\pi q^2\, \mathrm{d}E_0/h^3 c$, and then p and E be allowed to vary between p and $p + \mathrm{d}p$ yielding the same expression for ρ_f.

A final factor, the Coulomb correction factor $F(Z, p)$, must be included to account for the distortion of the electron's wavefunction from a plane wave caused by the interaction with the electrostatic field of the nucleus. For the same E_0 and M_{if} the effect of the Coulomb interaction is to shorten β^- while lengthening β^+ lifetimes as the phase space is larger for the β^--decay. This is because the initial maximum energy of an electron after decay is greater than E_0 as the electron loses energy as it passes out of the nuclear Coulomb field. Quantitatively the effect of the Coulomb field is to alter the magnitude of the electron's wavefunction at the origin $\psi_e^*(0)$, which in the plane wave approximation is taken as $\psi_e^*(0) = L^{-3/2}$. From equation 3.3 above, $|M_{if}|^2$ will be altered by:

$$F(Z, p) = L^3 |\psi_e(0)|^2 \approx 2\pi\eta/[1 - \exp(-2\pi\eta)]$$

where $\eta = \pm Ze^2/(4\pi\varepsilon_0 \hbar v)$ with the positive sign for β^-- and the negative sign for β^+-decay and v is the final velocity of the β-particles. The factor $F(Z, p)$ is called the Fermi function.

The probability for an electron with a momentum between p and $p + \mathrm{d}p$ being emitted is therefore given by:

$$w(p)\mathrm{d}p = \frac{64\pi^4}{h^7 c^3} |M_{if}|^2 p^2 (E_0 - E)^2 F(Z, p)\, \mathrm{d}p \tag{3.4}$$

If the function $K(Z, p)$ given by:

$$K(Z, p) = [w(p)/p^2 F(Z, p)]^{1/2} \equiv [w(E)/p(E + m_e c^2) F(Z, p)]^{1/2} \tag{3.5}$$

is plotted against E a straight line will be obtained intersecting the electron energy axis at the end-point E_0 if M_{if} is a constant. Such a plot is called a Kurie plot and an example is shown in figure 3.5.

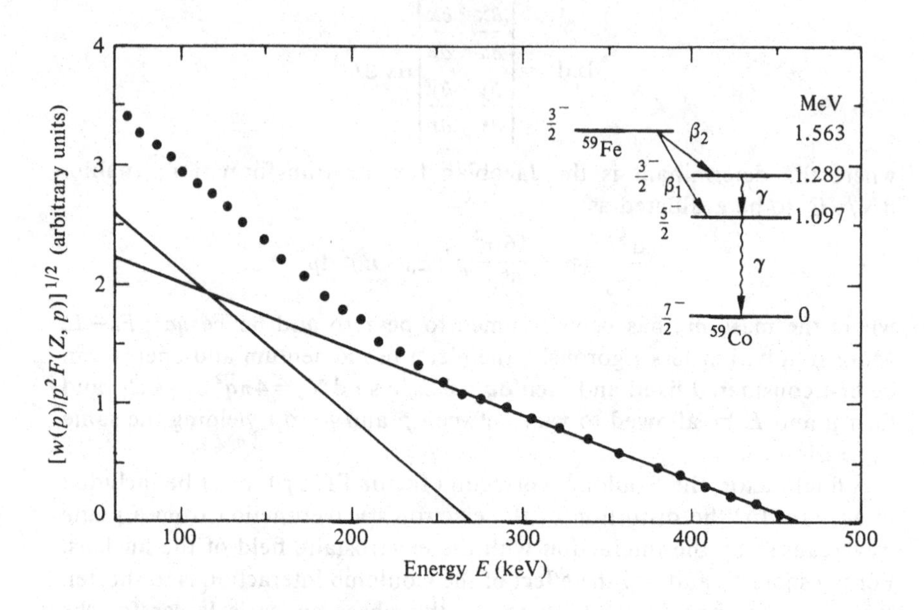

3.5. Kurie plot and its decomposition for the mixed β^--decay $^{59}Fe \xrightarrow{\beta^-} {}^{59}Co$ measured by Metzger, F. (From Marmier, P. and Sheldon, E., *Physics of Nuclei and Particles.* Academic Press (1970).)

3.3.3 β-decay lifetimes and ft values

As mentioned above there are transitions in which both the Fermi and Gamow–Teller interactions can contribute. For unpolarised nucleons these do not interfere and:

$$|M_{if}|^2 = G_\beta^2 |M_F|^2 + \lambda^2 G_\beta^2 |M_{GT}|^2 = G_\beta^2 |M|^2$$

where $|M|^2$ is the average squared matrix element. The total transition rate R for an allowed β-decay is obtained by integrating $w(p)\,dp$ over all electron momenta to give:

$$R = G_\beta^2 \frac{64\pi^4}{h^7 c^3} |M|^2 \int^{p\max} F(Z, p) N(p)\,dp$$

where $N(p)\,dp = p^2(E_0 - E)^2\,dp$. A useful approximation for estimating transition rates when $E_0 \gg m_e c^2$ is to ignore $F(Z, p)$ and take $pc = E + m_e c^2$, which gives Sargant's rule that $R \propto E_T^5$ where $E_T = E_0 + m_e c^2$.

If p and E are measured in units of $m_e c$ and $m_e c^2$, respectively, R becomes:

$$R = G_\beta^2 \frac{64\pi^4}{h^7 c^3} |M|^2 m_e^5 c^7 f$$

where f is dimensionless and is the value of the integral in the above expression for R. If t is the β-decay lifetime then $R = (\ln 2)/t$ so:

$$ft = \text{const.}/|M|^2 \text{ where const.} = \frac{h^7 \ln 2}{64 G_\beta^2 \pi^4 m_e^5 c^4} \tag{3.6}$$

The quantity ft is therefore a measure of the squared nuclear matrix element $|M|^2$. As the values of ft for β-transitions vary over such a large range it is customary to quote $\log_{10} ft$.

3.3.4 Super-allowed β-decays

An important group of β-decays are the pure Fermi $0^+ \to 0^+$ analogue transitions such as $^{14}O \to {}^{14}N$. These are called super-allowed or favoured transitions as the overlap between initial and final nuclear wavefunctions (which M_F (Fermi) is a measure of) is complete, neglecting the very small isospin mixing. In the ^{14}O β^+-decay there are two protons which can contribute to the transition (the two $p_{1/2}$ protons outside the ^{12}C core). In such a case the matrix element M_F becomes:

$$M_F \propto \sum_p \int \psi_n^*(r)\psi_p(r)\, d^3r$$

where the sum is over the two equivalent protons. (The other protons cannot decay as the final states are already occupied.) In ^{14}O the initial state is a 0^+ $T=1$ state. The probability amplitude that after a transition of $p \to n$ the final state has 0^+ $T=1$ is $\sqrt{2}$, as the final state has a wavefunction of the form $\{|p\rangle|n\rangle + |n\rangle|p\rangle\}/\sqrt{2}$. Therefore $|M_F|^2 = 2$. The observed ft value is 3127 s and on substitution in equation 3.6 this yields a value for G_β. The value derived from several β-decays is

$$G_\beta = 1.400 \times 10^{-62}\ \text{Jm}^3$$
$$G_\beta(\hbar c)^3 = 1.137 \times 10^{-11}\ \text{MeV}^{-2}$$

Another group of super-allowed transitions are the mixed transitions between mirror nuclei such as ${}^{13}_{7}N \to {}^{13}_{6}C$, ${}^3H \to {}^3He$ and ${}^{41}_{21}Sc \to {}^{41}_{20}Ca$. An important example is neutron decay as its lifetime enables the ratio of Gamow–Teller to Fermi coupling constants to be obtained. The matrix element for the Fermi part has the value $|M_F|^2 = 1$ and for the Gamow–Teller part $|M_{GT}|^2 = 3$ (the number 3 comes from summing the expectation value of $\boldsymbol{\sigma}$ over the spin 1/2 substates). The ratio of ft values for ^{14}O and the neutron therefore yields the ratio λ of Gamow–Teller to Fermi coupling constants:

$$(ft)^{14}\text{O}/(ft)\text{n} = 3127/1115 = (G_\beta^2 + 3\lambda^2 G_\beta^2)/2G_\beta^2$$
$$\therefore \quad |\lambda| = 1.24$$

The sign of λ is negative (see section on $(V-A)$ theory).

3.3.5 Forbidden β-decays

In allowed β-decays the leptons take away no orbital angular momentum and this corresponds to only considering the first and largest term in the expansion of the lepton wavefunction:

$$\exp(i\mathbf{q}\cdot\mathbf{r}) = 1 + i\mathbf{q}\cdot\mathbf{r} + \cdots$$

The second term $\mathbf{q}\cdot\mathbf{r}$ (and higher terms) $\ll 1$ over the nuclear volume and so will be insignificant unless the β-transition is such that an allowed decay is forbidden by the selection rules. An example is $^{39}\text{Ar}(\frac{7}{2}^-) \rightarrow {}^{39}\text{K}(\frac{3}{2}^+)$ for which $\Delta J = 2$ and there is a change of parity. For this decay the first term will give zero but the second term $\mathbf{q}\cdot\mathbf{r}$ in a Gamow–Teller transition will not, as the operator $\mathbf{r}$ changes l by one and causes a change of parity and the Gamow–Teller operator $\boldsymbol{\sigma}$ allows s to change by one allowing a $\Delta J = 2$ transition. Such a decay is called first-forbidden as the leptons carry away one unit of orbital angular momentum, and in general a β-transition is nth-forbidden if there are n units of orbital angular momentum carried away by the leptons.

This classification of β-transitions can be seen more formally by considering the expansion of a plane wave in angular momentum eigenstates:

$$\exp(i\mathbf{q}\cdot\mathbf{r}) = \exp(iqr\cos\theta) = \sum_l B_l(qr)\, Y_{l0}(\theta) \tag{3.7}$$

where $Y_{l0}(\theta)$ are the spherical harmonics with $m = 0$ and $B_l(qr)$ is related to the spherical Bessel function $j_l(qr)$ by

$$B_l(qr) = i^l[4\pi(2l+1)]^{1/2} j_l(qr)$$

The first two terms of this expansion correspond to those in the more familiar expansion

$$1 + iqr\cos\theta + \frac{(iqr\cos\theta)^2}{2!} + \cdots \tag{3.8}$$

The function $\exp(i\mathbf{q}\cdot\mathbf{r})$ can therefore be expressed in terms of orbital angular momentum eigenstates, the amplitude of each being $B_l(qr)$. The probability of each $|B_l(qr)|^2$ is shown as a function of qr in figure 3.6. For $qr \ll 1$ there is only any appreciable probability for $l = 0$ while in the limit of very high l (the classical limit) there is only appreciable amplitude for $qr = l$ (the correspondence principle).

From the approximate correspondence between equations 3.7 and 3.8, the probability that the leptons take away orbital angular momentum $l\hbar$ is

Table 3.1. *Some typical beta decays*

Decaying nucleus	Initial J^π	Product nucleus	Final J^π	$t_{1/2}$	E_{max} (MeV)	$\log_{10} ft$
n	$\frac{1}{2}^+$	H	$\frac{1}{2}^+$	622 ± 11 s	0.782	3.04
^{3}H	$\frac{1}{2}^+$	^{3}He	$\frac{1}{2}^+$	12.26 yr	0.0186	3.06
^{6}He	0^+	^{6}Li	1^+	0.81 s	3.51	2.91
^{14}O	0^+	^{14}N*	0^+	70.6 s	1.81	3.49
^{26}Al*	0^+	^{26}Mg	0^+	6.35 s	3.21	3.49
^{34}Cl	0^+	^{34}S	0^+	1.53 s	4.47	3.48
^{14}C	0^+	^{14}N	1^+	5730 yr	0.157	9.04
^{39}Ar	$\frac{7}{2}^-$	^{39}K	$\frac{3}{2}^+$	269 yr	0.565	9.92
^{38}Cl	2^-	^{38}Ar	0^+	37.2 min	4.92	7.43
^{22}Na	3^+	^{22}Ne	0^+	2.62 yr	1.82	12.7
^{10}Be	0^+	^{10}B	3^+	1.6×10^6 yr	0.56	13.4
^{40}K	4^-	^{40}Ca	0^+	1.28×10^9 yr	1.31	18.05
^{115}In	$\frac{9}{2}^+$	^{115}Sn	$\frac{1}{2}^+$	5.1×10^{14} yr	0.50	22.6

Source: From Segrè, E., *Nuclei and Particles.* Benjamin (1977).

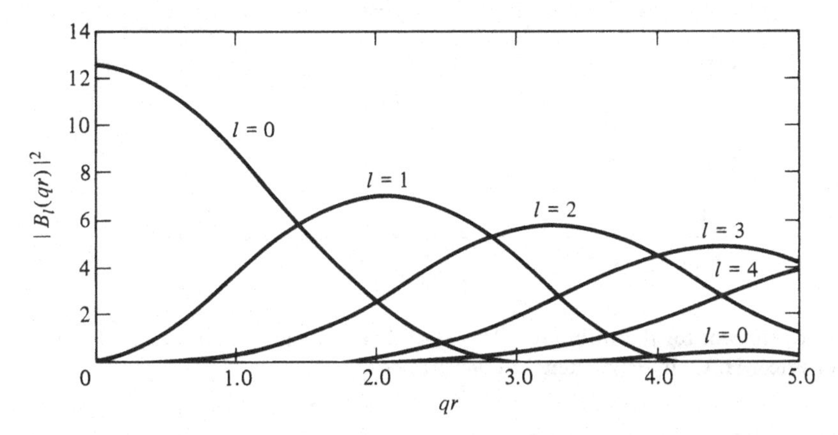

3.6. The functions $|B_l(qr)|^2$. (From Enge, H., *Introduction to Nuclear Physics.* Addison-Wesley (1966).)

expected to be of the order $(qR)^{2l}$ where R is the radius of the nucleus. Estimating $q \approx E_0/\hbar c$, where E_0 is the end-point energy, this predicts the matrix element for a first-forbidden transition with $E_0 = 1$ MeV and $R = 5$ fm to be $\sim\frac{1}{40}$ that of an allowed decay. The square of the matrix element for a transition is proportional to the ft value and this estimate of $\sim 10^{-3}$ is in rough agreement with what is found experimentally, as can be seen from the typical $\log_{10} ft$ values shown in Table 3.1.

The selection rules for first-forbidden β-decays are:

$$\Delta J = 0, 1 \text{ or } 2 \qquad \Delta T = 0 \text{ or } 1 \qquad \text{change of parity}$$

The matrix element for forbidden decays contains momentum dependent terms; for example, the $\mathbf{q} \cdot \mathbf{r}$ term in a first-forbidden transition gives rise to a q^2 correction to the shape of the β-spectrum and hence in general makes the Kurie plot non-linear. An example of a first-forbidden transition is the decay of ${}^{91}\mathrm{Y}(\frac{1}{2}^-) \to {}^{91}\mathrm{Zr}(\frac{5}{2}^+)$ and its Kurie plot is shown in figure 3.7. Besides the momentum dependent terms arising from the orbital angular momentum of the leptons, there are comparable terms arising from treating the nucleon motion relativistically. The matrix elements are as a result quite complex.

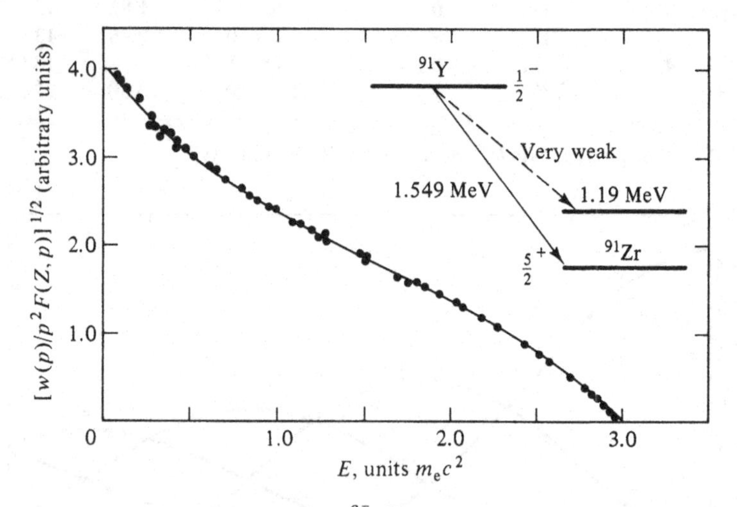

3.7. Kurie plot for the forbidden ${}^{91}\mathrm{Y} \xrightarrow{\beta^-} {}^{91}\mathrm{Zr}$ β^--decay. (From Langer, L. M. and Price, H. C. Jr, *Phys. Rev.* **75** (1949) 1109.)

3.3.6 Electron capture

In electron capture an atomic electron interacts weakly with a proton in the nucleus and initiates the reaction $e^- + p \to n + \nu$. Neglecting the intrinsic spin of the leptons the matrix element for an allowed Fermi transition is therefore:

$$M_{if} = G_\beta \int_0^R \psi_\nu^*(0)\psi_e(0)\psi_n^*(r)\psi_p(r)\, d^3r$$

Capture is most likely for a 1s-state electron as then $\psi_e(0)$ is a maximum and is given by:

$$\psi_e(0) = \pi^{-1/2}\left(\frac{Zm_e e^2}{4\pi\varepsilon_0\hbar^2}\right)^{3/2}$$

(for electrons in $l \neq 0$ states $\psi_e(0) = 0$). The matrix element is then:

$$|M_{if}|^2 = \frac{G_\beta^2}{L^3\pi}\left(\frac{Zm_e e^2}{4\pi\varepsilon_0\hbar^2}\right)^3 |M_F|^2$$

The capture rate w_K is given by the golden rule with the density of final states:

$$\rho_f = \frac{4\pi q^2}{h^3}\frac{dq}{dE_0}\cdot L^3$$

where q is the neutrino momentum. Neglecting nuclear recoil $q = E_0/c$ and $dq/dE_0 = 1/c$ so w_K is given by:

$$w_K = \frac{G_\beta^2 E_\nu^2}{\pi^2\hbar^4 c^3}|M_F|^2\left(\frac{Zm_e e^2}{4\pi\varepsilon_0\hbar^2}\right)^3 \quad (3.9)$$

3.3.7 Muon capture

When muons (μ^-) are slowed down in a material some are captured into high n hydrogenic states and subsequently cascade down into the $n = 1$ level emitting X-rays. The charge distribution of the nucleus must be taken into account in determining these X-ray energies. For light nuclei the energies are sensitive to the mean square charge distribution while for heavy nuclei they are also sensitive to higher moments.

Capture can occur for muons through the interaction:

$$\mu^- + p \rightarrow n + \nu_\mu, \qquad Q = 104.4\ \text{MeV}$$

and its rate is given approximately by:

$$w_K = \frac{G_\beta^2 E_\nu^2}{\pi^2\hbar^4 c^3}|M|^2\left(\frac{Zm_\mu e^2}{4\pi\varepsilon_0\hbar^2}\right)^3$$

which is the same as equation 3.9 above for electron capture but for $m_e \Rightarrow m_\mu$. This is an approximate formula because typically several protons near the Fermi level can contribute to the decay. If n contributed then the rate would be increased by a factor n.

3.4 Parity violation in β-decay

It had been assumed prior to 1956 that parity was conserved in the weak interaction like in the electromagnetic interaction. Lee and Yang pointed out in 1956 that this statement had not been experimentally tested and shortly afterwards it was demonstrated that parity was not conserved in the weak interaction. As discussed in chapter 1, conservation of a quantity follows from an invariance principle and parity conservation follows when the Hamiltonian is invariant with respect to the operation $\mathbf{r} \to -\mathbf{r}$. This operation is often described as a reflection but is not quite, as a reflection involves in addition a rotation; however, as long as the Hamiltonian is rotationally invariant then invariance under a reflection is an equivalent condition.

To develop the consequences let P represent the parity operator and $\psi(\mathbf{r})$ be an eigenfunction of P with eigenvalue p. Then $P\psi(\mathbf{r}) = p\psi(-\mathbf{r})$ so $P^2\psi(\mathbf{r}) = p^2\psi(\mathbf{r}) \equiv \psi(\mathbf{r})$, since two parity operations result in a return to the initial state. Therefore $p^2 = 1$ and the eigenvalues of P are $+1$ parity and -1 odd parity. The invariance of a Hamiltonian under a parity operation is equivalent to the commutator of P and H being zero (see p. 10), $[P, H] = 0$, since this means $\mathrm{d}\langle P\rangle/\mathrm{d}t = 0$, i.e. the rate of change of the expectation value of P with time is zero. If H is written as $H = T + V$ where T is the kinetic energy and V the potential energy operator then:

$$[P, H] = 0 \Rightarrow [P, V] = 0$$

This means that in a transition from an initial state ψ_i of parity p_i to a final state ψ_f of parity p_f caused by an interaction described by the potential V, then $p_f = p_i$ as can be seen from the following. The transition matrix element M_{if} is given by the volume integral:

$$\begin{aligned} M_{if} &= \int \psi_f^* V \psi_i \, \mathrm{d}\tau \\ &= \int \psi_f^* V P^2 \psi_i \, \mathrm{d}\tau \qquad P^2 = 1 \\ &= \int \psi_f^* P V P \psi_i \, \mathrm{d}\tau \qquad [P, V] = 0 \\ &= \int (P\psi_f)^* V P \psi_i \, \mathrm{d}\tau \qquad P \text{ Hermitian} \\ &= p_f p_i M_{if} \end{aligned}$$

so either $M_{if} = 0$ or $p_f = p_i$.

Since the final state after a transition brought about by a parity-conserving interaction has a definite parity ($p_f = p_i$) there is an important experimental consequence, which is that the expectation value in the final state of any pseudoscalar will be zero. A pseudoscalar quantity is one whose value changes sign under the parity operation. An example is $\mathbf{s} \cdot \mathbf{p}$ where $\mathbf{s}$ is the spin operator of the particle and $\mathbf{p}$ is the momentum of the particle. Under parity $\mathbf{p} \to -\mathbf{p}$ while $\mathbf{s} \to \mathbf{s}$ so $\mathbf{s} \cdot \mathbf{p} \to -\mathbf{s} \cdot \mathbf{p}$. The quantity $\mathbf{s} \cdot \mathbf{p}$ divided by $|\mathbf{p}||\mathbf{s}|$ is called the helicity H or handedness of a particle:

$$H = \mathbf{s} \cdot \mathbf{p}/|\mathbf{s}||\mathbf{p}|$$

For example, if an electron has its spin vector pointing along its direction of motion then $H=+1$ and the electron is called right-handed, and if $H=-1$ then it is called left-handed.

To show why the expectation value E of a pseudoscalar operator O is zero in a state ψ_f whose parity eigenvalue is p_f consider:

$$\begin{aligned} E &\equiv \int \psi_f^* O \psi_f \, d\tau \equiv \langle f|O|f\rangle \\ &= \langle f|OP^2|f\rangle \qquad P^2=1 \\ &= -\langle f|POP|f\rangle \qquad PO=-OP \\ &= -p_f^2\langle f|O|f\rangle \qquad P \text{ Hermitian} \\ &= -E \end{aligned}$$

so E is zero. The expectation value would not necessarily be zero if the state ψ_f did not have good parity. Therefore if in a transition from a state ψ_i of good parity p_i to a state ψ_f the expectation value of a pseudoscalar quantity in the final state is not zero then parity has been violated in the transition.

Another important experimental consequence concerns the angular distribution of particles emitted in a transition. Consider an initial state with a definite parity and a parity-conserving interaction which brings about a transition to a final state containing two or more particles, e.g. $^{16}O^* \to {}^{12}C+\alpha$, $^{24}Mg^* \to {}^{24}Mg+\gamma$ or $^{16}O^* \to {}^{12}C+\alpha+\gamma$, then parity conservation means that the parity of the final state is the same as that of the initial state. The parity of the final state is the product of the intrinsic parities of the particles and the parities of the wavefunctions describing the relative motions of the particles.

If the wavefunction describing one of the particles in the final state is $\phi(\mathbf{r})$ with $P\phi(\mathbf{r})=\phi(-\mathbf{r})=p\phi(\mathbf{r})$ then:

$$|\phi(-\mathbf{r})|^2 = p^2|\phi(\mathbf{r})|^2 = |\phi(\mathbf{r})|^2$$

so the probability of the particle being at $\mathbf{r}$ is the same as its being at $-\mathbf{r}$. This means that in a parity-conserving decay an emitted particle is as likely to be travelling in the direction $\mathbf{r}$ as $-\mathbf{r}$. Figure 3.8 shows the γ-ray angular distribution produced by a $1^- \to 0^+$ decay when the initial state is polarised along the z-axis, i.e. only the $m=1$ substate is populated. As can be seen any direction $\mathbf{r}$ is as equally likely as $-\mathbf{r}$, which is equivalent to saying the distribution is symmetric about any plane through the origin. Such an anisotropic distribution will not be seen if there is a random orientation of initial nuclear orientations as these will produce a random superposition of γ-ray intensity patterns which will give an isotropic intensity distribution.

Therefore if the distribution of an emitted particle from a polarised initial state of good parity is not symmetric about any plane through the origin

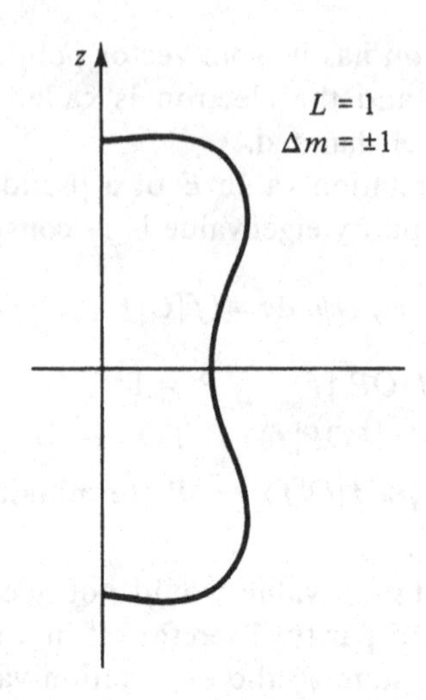

3.8. Polar representation of the dipole angular distribution for a 1^- $(m=1) \rightarrow$ 0^+-decay.

of the particle then the transition has been brought about by an interaction which violates parity. It was such an observation that provided the first evidence that parity is violated in the weak interaction.

3.4.1 Experimental evidence for parity violation

The first experiment that showed that parity was not conserved in the weak interaction was carried out by Wu *et al.* in 1957, who looked at the β-decay of ^{60}Co$(5^+) \rightarrow {}^{60}Ni(4^+)$, which is an allowed Gamow–Teller transition (see figure 3.9). They polarised the ^{60}Co nuclei by first cooling the ^{60}Co nuclei to 4.2 K using liquid ^{4}He and then to ~0.01 K by adiabatic demagnetisation. An external magnetic field was then applied which polarised the atomic magnetic moments. The resulting internal magnetic field polarised the nuclear magnetic moments and hence the spins of the ^{60}Co nuclei. The

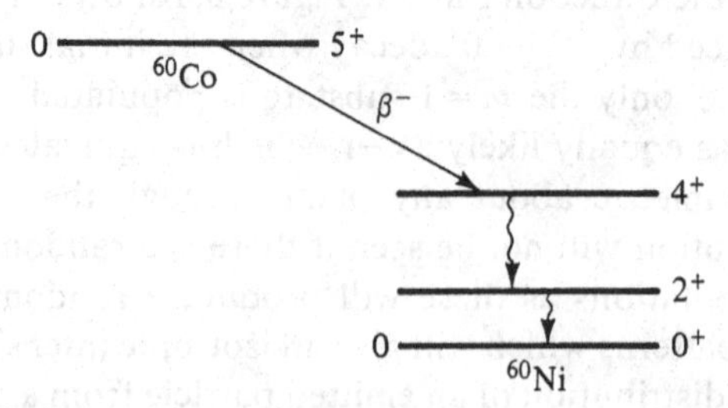

3.9. The $^{60}\text{Co} \xrightarrow{\beta^-} {}^{60}\text{Ni}$ β-decay scheme.

degree of polarisation was measured by looking at the anisotropy of the γ-rays from the decay of the 4^+ state in ^{60}Ni. A schematic diagram of the experimental arrangement is shown in figure 3.10.

The ratio of counts in the two γ-detectors is a measure of the degree of polarisation. Changing the direction of the magnetic field **B** and hence the direction of polarisation significantly altered the intensity of electrons in the beta detector, with more emitted in a direction anti-parallel to the spin of the ^{60}Co(5^+) state than parallel, i.e. more in a direction **r** than in −**r**. Since the initial state ^{60}Co(5^+) has a well-defined parity, this is a clear indication that parity is violated in the weak interaction.

The actual angular distribution of the electrons in the pure Gamow–Teller beta decay of ^{60}Co is given by the expression:

$$I(\theta) = 1 - \frac{1}{3}\frac{v}{c}\cos\theta$$

where v is the velocity of the electron. For such a distribution $I(\theta) \neq I(180-\theta)$ as the electrons are in a state of mixed parity, the state being a mixture of $l=0$ and $l=1$ wavefunctions.

In beta decay there is in general a non-zero expectation value for the electron helicity and this is present whether the source is polarised or not. It is interesting to note that a non-zero value for a beta particle's helicity was reported by Cox in 1928 but this was attributed by others at the time to an instrumental effect. Since Wu *et al.*'s experiment the helicity of electrons produced in β-decay has been measured and found to be $-v/c$ in agreement with $(V-A)$ theory (see below).

3.4.2 P and CP invariance

The consequences of parity conservation may also be expressed by the statement:

The mirror image of a process is also an observable process if parity is conserved.

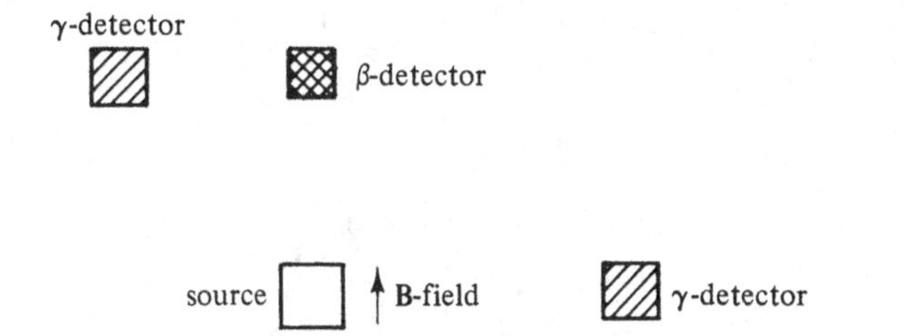

3.10. Schematic diagram of the experimental arrangement to look for parity violation in ^{60}Co β-decay.

This statement is illustrated in figure 3.11 for ^{60}Co β-decay. The electrons have on average negative helicity. The mirror image shows that if parity were conserved then as many electrons should have been emitted in the direction of the spin of ^{60}Co as opposed to it. Also as many electrons should have positive as negative helicity. It should be noted that although in a γ-ray transition photons of definite helicity, i.e. circularly polarised, are emitted there are as many right-hand as left-hand circularly polarised photons emitted in a transition since the electromagnetic interaction conserves parity. This is an example of the statement that a zero expectation value for a pseudoscalar quantity does not mean that every measurement must be zero, only that the average must be zero.

Although the weak interaction is not invariant under parity it is invariant (to a very good approximation) under the combined operation, called CP,

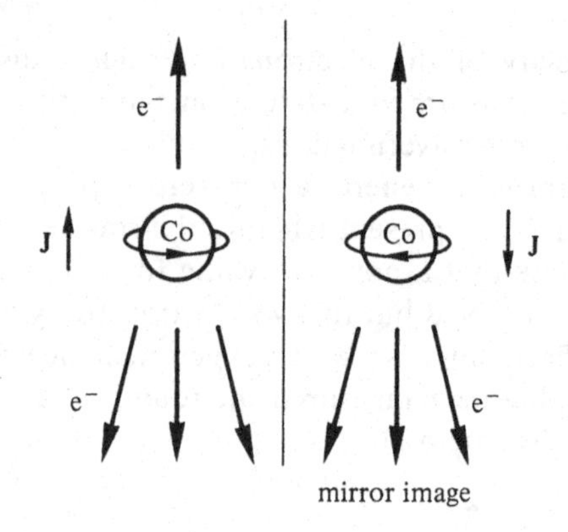

3.11. The violation of mirror symmetry in ^{60}Co β-decay.

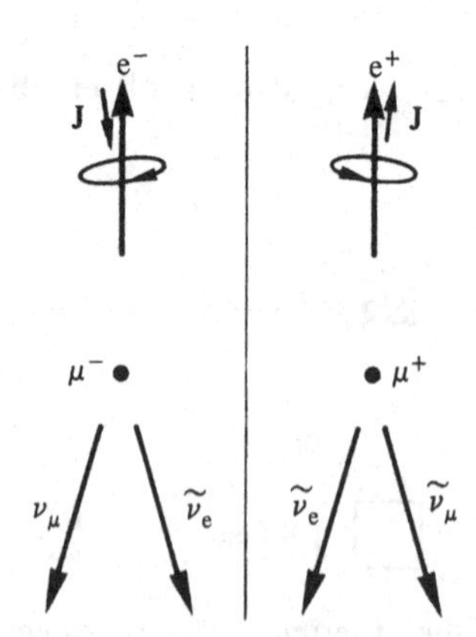

3.12. The charge-conjugate mirror symmetry of $\mu^{\pm}$ β-decay.

of charge conjugation and parity. Charge conjugation changes a particle into its anti-particle or vice versa so CP invariance is equivalent to saying that:

The charge conjugated mirror image of a process is also an observable process if CP is conserved.

This statement is illustrated in figure 3.12 for the weak decay $\mu^- \to e^- + \tilde{\nu}_e + \nu_\mu$ which predicts that in μ^+ decay positive helicity positrons will be emitted and this is what is found. There is, however, a very small violation of CP invariance observed in the weak decay of the strange particle K^0, the magnitude of which is still unexplained.

3.5 The neutrino

The anti-neutrino was originally proposed to account for the continuous energy spectrum of electrons produced in a β^--decay. Conservation of angular momentum alone would limit the anti-neutrino's spin to $\frac{1}{2}$ or $\frac{3}{2}$; however, some $0^+ \to 0^+$ β-transitions occur with ft values similar to that of neutron decay, which indicates that they are allowed decays and rules out $\frac{3}{2}$. Charge conservation requires the anti-neutrino to be electrically neutral and in β-decays the observed closeness of the maximum electron energy to E_0, the energy release in the β-decay, indicated early on that its rest mass was very small. The discovery of parity violation in the weak interaction indicated definite helicities for the anti-neutrino and neutrino and in 1958 the helicity of the neutrino was measured in a most elegant experiment by Goldhaber, Grodzins and Sunyar.

3.5.1 The helicity of the neutrino

The measurement used the β-decay of ${}^{152}_{63}Eu$. The nucleus ${}^{152}_{63}Eu$ decays part of the time by electron capture to an excited 1^- state in ${}^{152}_{62}Sm$ which has a lifetime of 3.5×10^{-14} s. A branch of the decay of this 1^- state is to the 0^+ ground state of ${}^{152}_{62}Sm$. The relevant portion of the decay scheme of ${}^{152}_{63}Eu$ is shown in figure 3.13. The experiment takes advantage of the near equality of the neutrino energy in the electron-capture ($E_\nu = 840$ keV) and the γ-ray energy of the $1^- \to 0^+$ decay ($E_\gamma = 960$ keV), to enable resonant scattering to be observed of the $1^- \to 0^+$ γ-ray when its helicity is the same as that of the neutrino. The spins and momenta involved in the experiment are illustrated in figure 3.14.

In figure 3.14*a* the ${}^{152}_{63}Eu$ captures an electron essentially with zero momentum. To conserve momentum the excited ${}^{152}_{62}Sm$ nucleus with $J^\pi = 1^-$ recoils with a momentum $\mathbf{R}$ equal to the momentum of the neutrino $\mathbf{q}$. In this figure the helicity of the neutrino has been assumed negative ($H = (\mathbf{s} \cdot \mathbf{p})/(s \cdot p)$), which is correct, so the direction of the spin of the 1^- state in ${}^{152}_{62}Sm$ is as

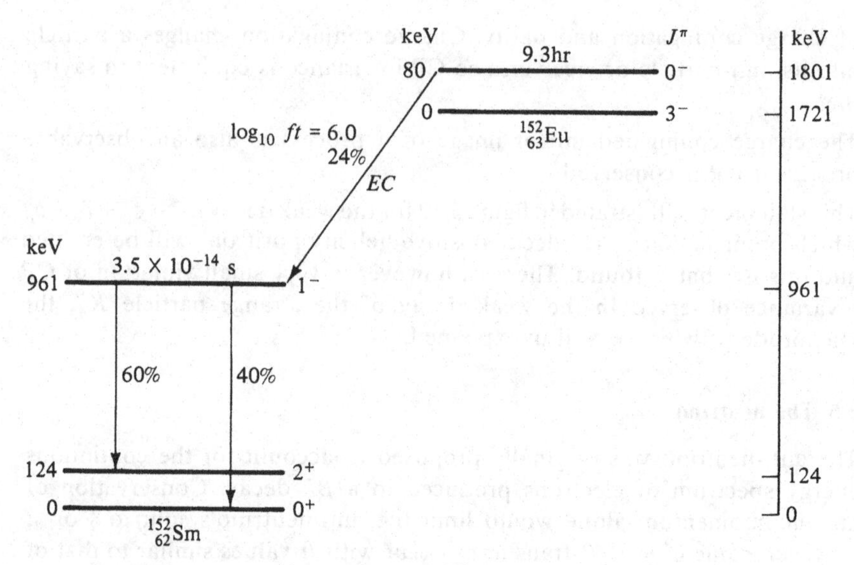

3.13. A portion of the β-decay scheme of ^{152}Eu.

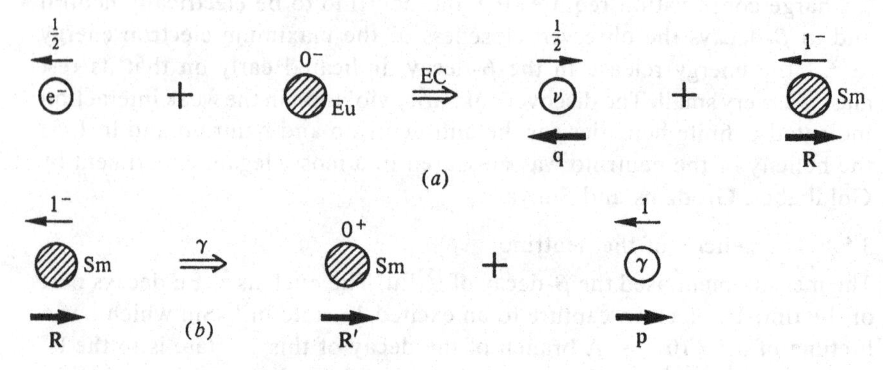

3.14. Representation of (a) the β-decay and (b) the γ-decay involved in the ^{152}Eu neutrino helicity experiment.

shown in order to conserve angular momentum. In figure 3.14b the γ-decay of the 1^- $^{152}_{62}$Sm state is shown with the γ-ray emitted in the same direction as the $^{152}_{62}$Sm nucleus is recoiling. In the rest frame of the recoiling $^{152}_{62}$Sm the energy E_0 of the $1^- \rightarrow 0^+$ γ-ray is given by:

$$E_0 = E_x\left(1 - \frac{E_x}{2Mc^2}\right) \qquad \text{(conservation of momentum)}$$

where M is the mass of the $^{152}_{62}$Sm nucleus and E_x is the excitation energy of the 1^- state. In the laboratory frame the energy of the γ-ray is Doppler-shifted because the $^{152}_{62}$Sm nucleus is moving, and its energy E_γ is related to E_0 by:

$$E_\gamma = E_0\left(1+\frac{v}{c}\right)$$
$$= E_0\left(1+\frac{E_\nu}{Mc^2}\right)$$

since $\mathbf{R}=\mathbf{q}$ so $v = q/M = E_\nu/Mc$. Therefore:

$$E_\gamma = E_x\left(1-\frac{E_x}{2Mc^2}\right)\left(1+\frac{E_\nu}{Mc^2}\right)$$
$$\approx E_x\left(1+\frac{E_x}{2Mc^2}\right) \quad \text{since } E_\nu \approx E_x \tag{3.10}$$

The γ-ray therefore has the required energy for absorption by $^{152}_{62}$Sm from the 0^+ ground state to the 1^- excited state and this is the condition for resonant scattering of the γ-ray by a piece of $^{152}_{62}$Sm. The energy E_γ does not have to exactly satisfy equation 3.10 because of the thermal motion of the $^{152}_{62}$Sm nuclei. The experimental arrangement used is shown in figure 3.15.

To measure the polarisation of the γ-rays an analysing magnet was used to magnetise an iron absorber as shown, which changes the absorption coefficient for the γ-rays depending on their polarisation. Measuring the effect of reversing the magnetic field enabled the polarisation of the γ-rays to be measured. The result was that the neutrino has a negative helicity and was consistent with complete polarisation (i.e. $H_\nu = -1$).

3.5.2 The mass of the neutrino

Early measurements comparing the maximum electron energy to E_0, the energy released in the beta decay, indicated that the electron anti-neutrino rest mass $m_{\bar{\nu}}$ was very small. Some of the most precise experiments that have been carried out have looked at the beta decay of tritium, which is a super-allowed decay with a low E_0 of 18.6 keV. A low E_0 is useful as the percentage instrumental electron resolution required to see the effect of a finite $m_{\bar{\nu}}$ is smaller the larger the value of E_0. The shape of a beta decay spectrum near the end point is altered if $m_{\bar{\nu}}$ is not zero. The probability for an electron with a momentum between p and $p+\mathrm{d}p$ being emitted is given by:

$$w(p)\,\mathrm{d}p = \frac{64\pi^4}{h^7c^3}|M_{if}|^2 p^2(E_0-E)[(E_0-E)^2 - m_{\bar{\nu}}^2c^4]^{1/2}F(Z,p)\,\mathrm{d}p$$

The symbols are as used in equation 3.4 above. If the mass $m_{\bar{\nu}}$ is not zero then the Kurie plot (a plot of $K(Z,p)$ (equation 3.5) versus the electron energy E) is no longer straight but is curved near the end point. The

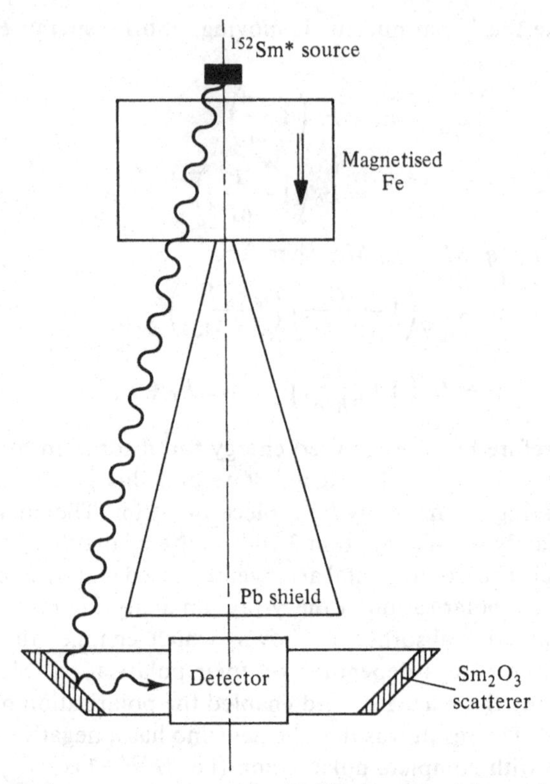

3.15. Experimental arrangement used by Goldhaber, Grodzins and Sunyar to determine the helicity of the neutrino. (From Marmier, P. and Sheldon, E., *Physics of Nuclei and Particles*. Academic Press (1970).)

calculated shape for tritium beta decay for a mass $m_{\bar{\nu}}$ equal to 30 eV is shown in figure 3.16 and the effect of the finite neutrino mass is seen to be most significant near the end point.

In order to detect or put a limit on the effect of a finite mass $m_{\bar{\nu}}$, a strong tritium source is required as only a very small fraction (10^{-9}) of the emitted electrons are within 20 eV of the end point. However if the source is too thick then the energy loss of the electrons within the source will mask the effect of a finite anti-neutrino mass. In 1970 Bergkvist devised an elegant solution to this problem which is illustrated in figure 3.17. By applying a potential difference along the extended thin source all the emitted electrons of energy E can be detected. The improvement in statistics enabled Bergkvist to put a limit of $m_{\bar{\nu}} < 60\ \text{eV}/c^2$ (part of the data is shown in figure 3.18).

Since then a Russian group of experimentalists have carried out another experiment on tritium beta decay and have concluded that the electron anti-neutrino mass is $\sim 30\ \text{eV}/c^2$, though this is in conflict with the result

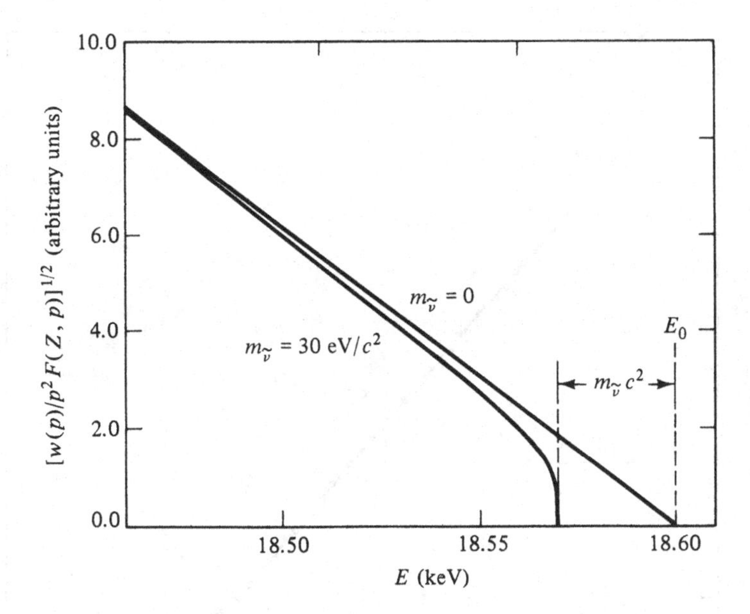

3.16. Calculated Kurie plots for tritium β-decay with $m_{\tilde{\nu}} = 0$ and $m_{\tilde{\nu}} = 30$ eV/c^2.

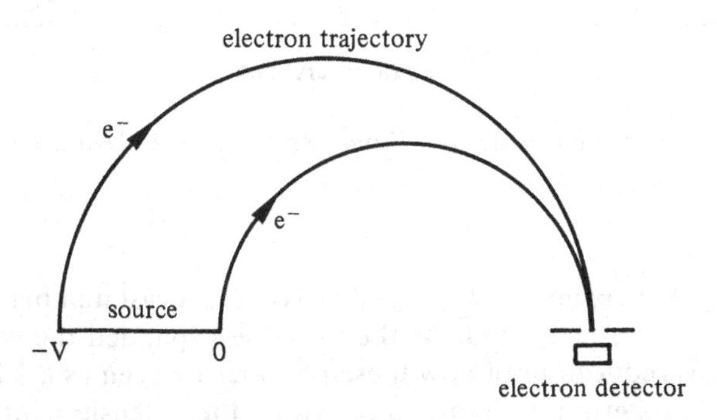

3.17. Schematic diagram illustrating Bergkvist's extended β-source. The potential gradient along the source matches the dispersion of the β-spectrometer. The analysing magnetic field is normal to the diagram.

from a Swiss group of less than 18 eV/c^2. Currently several groups are trying to resolve this conflict. If the neutrino should have a mass of approximately 30 eV/c^2 this would have very interesting cosmological implications.

The universe appears to have evolved from a 'Big Bang' some 15 thousand million years ago. The principal evidence for this is the observed recession (from the red shift of their spectra) of galaxies from each other with a velocity which is proportional to their separation (Hubble's law). In this

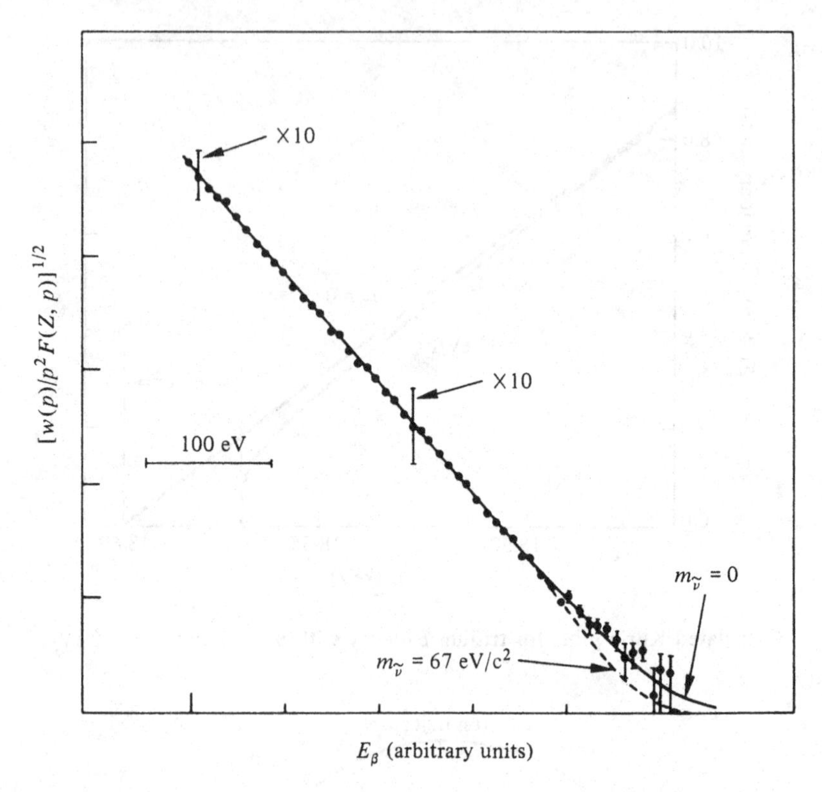

3.18. Kurie plot of tritium β-decay. (From Bergkvist, H. E., *Nucl. Phys.* **B39** (1972) 317.)

'Big Bang' the light elements were synthesised and a vast number of photons and neutrinos were generated. As the universe expanded the wavelengths of the photons reduced until now these photons are seen as a 2.7 K cosmic microwave background radiation with the density of photons ~400 photons cm^{-3}. The neutrinos which were produced in the 'Big Bang' also cooled and their temperature is now approximately 2.3 K corresponding to ~100 neutrinos of each type per cm^3 (there are three types of neutrino known: electron, muon and tau).

There is considerable evidence, in particular from the motions of stars and galaxies, that the amount of material in the universe is an order of magnitude greater than what we observe; the dark matter or missing matter problem; and it had been speculated prior to the Russian result that the dark matter might be neutrinos. If the missing matter were entirely made up of one type of neutrino then its mass would be approximately 50 eV/c^2, hence the excitement when the Russian result of approximately 30 eV/c^2 was announced. However, there are difficulties, particularly in understanding galaxy formation, if neutrinos are the dark matter and the question of what the dark matter consists of is still unresolved.

3.5.3 Inverse β-decay and neutrino detection

Inverse β-decay is the process in which a neutrino or anti-neutrino is captured by a nucleus and an electron or positron of definite energy is emitted:

$$\nu + {}^A_Z M \rightarrow {}^A_{Z+1}M + e^- \quad \text{or} \quad \tilde{\nu} + {}^A_Z M \rightarrow {}^A_{Z-1}M + e^+$$

The probability for neutrino capture is directly related to that of ordinary β-decay and is given by Fermi's golden rule:

$$w_c = \frac{2\pi}{\hbar} G_\beta^2 |M|^2 \rho_f / \Omega^2$$

where Ω is the normalisation volume. The density of final states $\rho_f \equiv dN/dE_0 = 4\pi p^2 \Omega (dp/dE_0)/h^3$, where p is the electron momentum. As ${}^A_Z M \gg m_e$ then $E_0 = E$ (the electron kinetic energy) to a very good approximation, and $dp/dE_0 = (E_0 + m_e c^2)/p = E_T/p$ so:

$$w_c = \frac{2\pi}{\hbar} G_\beta^2 |M|^2 \frac{4\pi p E_T}{h^3 c^2 \Omega} \tag{3.11}$$

where $|M|^2$ is the same as for the decay:

$${}^A_Z M \rightarrow {}^A_{Z+1} M + e^- + \tilde{\nu}$$

and is given in terms of the ft value (equation 3.6) for the decay by:

$$G_\beta^2 |M|^2 = h^7 c^3 \ln 2 / (64\pi^4 m_e^5 c^7 ft)$$

The probability w_c can also be expressed in terms of an effective cross-sectional area σ_c within which a neutrino will be captured. If a normalisation of one neutrino per volume Ω is used then the relation between w_c and σ_c is

$$\Omega w_c = \sigma_c c \tag{3.12}$$

where c is the velocity of the neutrino (m_ν assumed zero). Combining equations 3.6, 3.11 and 3.12 gives the following expression for σ_c:

$$\sigma_c = 2\pi^2 (\hbar c)^3 (\ln 2) p E_T / [(m_e c^2)^5 ft]$$

For the capture reaction $\tilde{\nu}_e + p \rightarrow n + e^+$ the threshold is $E_{\tilde{\nu}} = 1.8$ MeV. The ft value for neutron decay is $\sim 10^3$, so for $E_{\tilde{\nu}} = 2.8$ MeV incident anti-neutrinos, $E_e = 1$ MeV and the capture cross-section $\sigma_c \approx 10^{-47}$ m^2 (using $p \approx E_T/c$, $c = 3 \times 10^{23}$ fm s^{-1}, $\hbar c = 200$ MeV fm and $m_e c^2 = 0.5$ MeV). The relation between cross-section and mean free path is $l = 1/n\sigma_c$ where n is the number of nuclei per m^3. For protons in water n is 3×10^{28} giving $l \approx 3 \times 10^{18}$ m or ~ 300 light years.

This reaction was first observed by Reines and Cowan in 1959 using a 1000 MW reactor as the source of anti-neutrinos. The anti-neutrinos with a flux of $\sim 10^{17}$ m^{-2} s^{-1} passed through a target of water containing $\sim 10^{28}$ protons in which some $CdCl_2$ was dissolved. A positron produced in the reaction $\tilde{\nu}+p\rightarrow n+e^+$ quickly annihilates with an electron in $\sim 10^{-9}$ s to produce two 511 keV γ-rays travelling in opposite directions. The recoiling neutron slows down through collisions with protons, and is then captured by a cadmium nucleus Cd(n, γ)Cd* producing γ-rays of total energy 9 MeV (see figure 3.19). This capture process takes $\sim 10^{-6}$ s, so a characteristic signature of two simultaneous 511 keV γ-rays, followed by γ-rays of total energy $\sim$9 MeV, a few microseconds later, indicates a $\tilde{\nu}+p\rightarrow n+e^+$ reaction. The expected event-rate is $FN\sigma_c\varepsilon$ where F is the $\tilde{\nu}$ flux, N is the number of protons, σ_c the cross-section and ε the efficiency of the detector, which was $\sim 3\times 10^{-2}$. Substituting gives a rate of $\sim$1 per hour, which was what was observed.

3.5.4 Solar neutrinos

As an example of neutrino capture consider the process:

$$\nu+{}^{37}\text{Cl}\rightarrow{}^{37}\text{Ar}+e^- \qquad Q=-0.814\text{ MeV}$$

For neutrinos with sufficient energy the analogue state in ^{37}Ar is preferentially populated as this is a super-allowed transition. The ^{37}Ar is unstable and β-decays by EC to ^{37}Cl with a half-life of 35 days. This EC process produces initially an excited ^{37}Cl, which has a K-shell vacancy, and its subsequent de-excitation gives rise to both X-rays and electrons (Auger electrons).

This neutrino capture reaction has been used to detect some of the neutrinos emitted by the nuclear processes taking place in the sun (solar neutrinos). One of these is the reaction and subsequent decay:

$${}^{7}_{4}\text{Be}+p\rightarrow{}^{8}_{5}\text{B}+\gamma \quad \text{and} \quad {}^{8}_{5}\text{B}\rightarrow{}^{8}_{4}\text{Be}^*(2^+)+e^++\nu+14.02\text{ MeV}$$

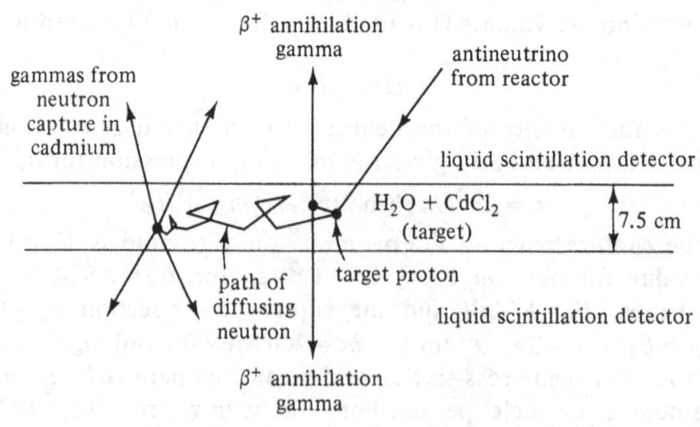

3.19. Schematic diagram of the detector arrangement used by Reines and Cowan to identify the $\tilde{\nu}+p\rightarrow n+e^+$ reaction. (From Segrè, E., *Nuclei and Particles*. Benjamin (1977).)

which only produces a small fraction of the total number of solar neutrinos but generates neutrinos with sufficient energy to initiate the $\nu + {}^{37}Cl \rightarrow {}^{37}Ar + e^-$ reaction. The experiment used a very large tank holding 615 tons of a detergent (C_2Cl_4) containing chlorine. This tank was ~1 mile below ground to reduce production of ^{37}Ar through the reaction $^{37}Cl(p, n)^{37}Ar$ initiated by cosmic rays. Helium gas was bubbled through the detergent every 80 days to sweep up any ^{37}Ar atoms produced by solar neutrinos, whose flux is unaffected by passing through the earth as their mean free path is many light years. The ^{37}Ar atoms were counted by detecting the Auger electrons. Even though the calculated flux of solar neutrinos from the ^8B reaction is very large ($6 \times 10^{10}\ m^{-2}\ s^{-1}$) and there are a large number of ^{37}Cl nuclei (2×10^{30}) in the tank, the expected count-rate is exceedingly small because of the very low cross-section σ_c ($10^{-46}\ m^2$).

The surprising result of this experiment by Davis *et al.* is that the observed flux of solar neutrinos is approximately a quarter of the calculated flux. This calculated value depends on a model of the solar interior but this is believed to be understood sufficiently well that there is a very significant discrepancy, which is now known as the solar neutrino problem.

There is currently speculation that the reduced flux of solar neutrinos might arise through the phenomenon of neutrino oscillations in which the flux of electron neutrinos is reduced to about one third of their initial value while the remaining flux becomes made up of muon and tau neutrinos. Two experiments which are presently being planned to measure the solar neutrino flux are neutrino capture by heavy water and neutrino capture by gallium producing germanium. This latter reaction is sensitive to the main flux of solar neutrinos made in the $p + p \rightarrow {}^2_1H + e^+ + \nu$ reaction which is the main energy source within the sun.

3.6 (V − A) theory

Fermi assumed that the weak interaction was like the electromagnetic interaction and was described by a Lorentz-invariant scalar matrix element formed by the dot product of two vector operators, i.e. a $V \cdot V$ interaction as described above. This interaction is not complete, however, as it does not account for Gamow–Teller transitions and in fact there are five possible combinations that give a Lorentz-invariant scalar matrix element: $S \cdot S$, $P \cdot P$, $V \cdot V$, $A \cdot A$ and $T \cdot T$, where S, P, V, A and T stand for scalar, pseudoscalar, vector, axial-vector and tensor operators, respectively. The $A \cdot A$ and $T \cdot T$ combinations would both give rise to Gamow–Teller transitions.

The discovery of parity violation showed that the matrix element $|M_{if}|$ must be a mixture of scalar and pseudoscalar terms so it would not be invariant under the parity operation. Feynman and Gell-Mann postulated that the interaction was an equal mixture of V and A operators. They considered the following wave equation for a spin 1/2 massive particle:

$$\left(\frac{i\hbar}{c}\frac{\partial}{\partial t}-\boldsymbol{\sigma}\cdot\mathbf{p}\right)\left(\frac{i\hbar}{c}\frac{\partial}{\partial t}+\boldsymbol{\sigma}\cdot\mathbf{p}\right)\phi=m^2c^2\phi \tag{3.13}$$

where ϕ is a two-component wavefunction and $\boldsymbol{\sigma}$ is the Pauli spin operator. This can be deduced from the energy-momentum relation

$$\frac{E^2}{c^2}-p^2=m^2c^2$$

substituting the operator identity $(\boldsymbol{\sigma}\cdot\mathbf{p})^2=\mathbf{p}^2$. For a massless particle then either $E\phi_R/c=\boldsymbol{\sigma}\cdot\mathbf{p}\phi_R$ or $E\phi_L/c=-\boldsymbol{\sigma}\cdot\mathbf{p}\phi_L$ (these equations were first proposed by Weyl in 1929), i.e. there are just two solutions corresponding to left- (L) and right- (R) handed helicity and these describe a massless neutrino (ϕ_L) and anti-neutrino (ϕ_R). The wavefunctions ϕ_L and ϕ_R are not eigenstates of parity as under parity $\phi_L\rightarrow\phi_R$ and $\phi_R\rightarrow\phi_L$; however they are eigenstates under the combined operation of charge conjugation and parity (CP) (see section 3.4.2 above).

If the mass is not zero then equation 3.13 is equivalent to two equations:

$$\left.\begin{aligned}\left(\frac{i\hbar}{c}\frac{\partial}{\partial t}+\boldsymbol{\sigma}\cdot\mathbf{p}\right)\phi_L=mc\phi_R\\ \left(\frac{i\hbar}{c}\frac{\partial}{\partial t}-\boldsymbol{\sigma}\cdot\mathbf{p}\right)\phi_R=mc\phi_L\end{aligned}\right\} \tag{3.14}$$

Now in β-decay only left-handed neutrinos are involved, i.e. ϕ_L. Feynman and Gell-Mann assumed that the electron must also be described by only ϕ_L, with ϕ_L satisfying equations 3.14 above. With this assumption the only combination of S, P, V, A and T operators which gives a non-zero matrix element is $(V-A)\cdot(V-A)$. (If the neutrino had right-handed helicity then it would be $(V+A)\cdot(V+A)$.) The angular distributions of the leptons emitted in a β-decay predicted by this interaction are shown in figure 3.20. Also shown are those for a scalar $(S\cdot S)$ and tensor $(T\cdot T)$ interaction, which would also give rise to Fermi and Gamow–Teller transitions, respectively. Experimentally the electron-neutrino correlation can be inferred from measuring both the electron momentum and the nuclear recoil, e.g. in the decay ${}^6He(0^+)\rightarrow{}^6Li(1^+)+e^-+\tilde{\nu}$. The pseudoscalar interaction P is insignificant in nuclear β-decay as the nucleon velocities are much less than c. The theory also predicts the helicities of the emitted leptons: $e^-=-v/c$, $e^+=+v/c$, $\tilde{\nu}=+1$, $\nu=-1$.

The electron helicity can be measured by observing the circular polarisation of γ-rays produced when the electrons are stopped in an absorber. The

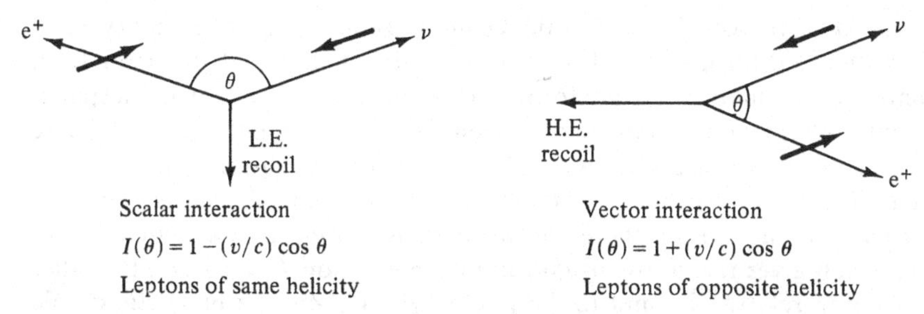

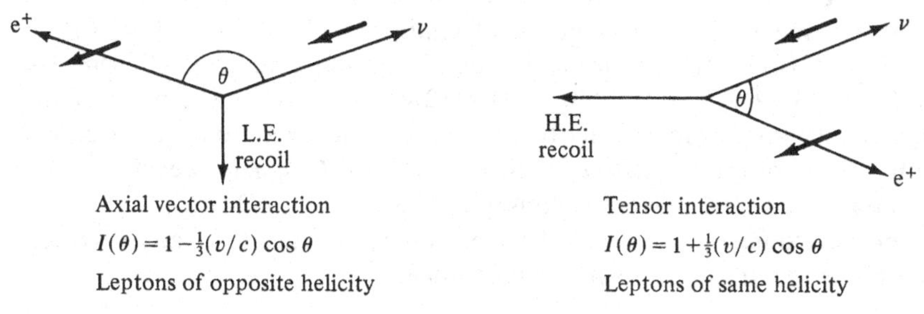

3.20. Angular distribution, $I(\theta)$, in β-decay predicted by various types of interaction. (From Perkins, D. H., *Introduction to High-Energy Physics.* Addison-Wesley (1982).)

sign of the helicity of the forward γ-rays is the same as that of the electrons and can be determined by measuring the γ-ray absorption in magnetised iron as in the experiment on the helicity of the neutrino.

Incidentally, the observed asymmetry in the electron distribution from the decay of polarised ^{60}Co nuclei can be seen as a direct consequence of the negative helicity of the electron and the leptons' spins coupling to spin 1 in a Gamow–Teller transition. The transition is from a $J = 5$ state in ^{60}Co to a $J = 4$ state in ^{60}Ni so to conserve angular momentum the electron must tend to be emitted in the opposite direction to the initial orientation of the $J = 5$ ^{60}Co nuclei.

3.6.1 The conserved-vector current hypothesis

For a purely leptonic decay process, such as:

$$\mu^- \to e^- + \tilde{\nu}_e + \nu_\mu$$

the weak interaction matrix element is $G_F(V - A) \cdot (V - A)$ where G_F is called the Fermi coupling constant. However, nuclear β-decay is a semi-leptonic weak interaction and the nucleon weak current is proportional to

(V$-\lambda$A) where λ is the ratio of the Gamow–Teller (axial) to Fermi (vector) coupling constants, $\lambda = 1.24$, and is the ratio of the axial to vector form factors for q^2 small. There are form factors because the nucleon has a spatial extent and the conserved-vector current hypothesis relates some of these form factors to those of the nucleon's electromagnetic current. The nucleons have an anomalous magnetic moment (i.e. not that of a Dirac spin 1/2 particle) and as a result there are two parts to the electromagnetic current: one which gives rise to the usual form of interaction $F_1(q^2)J_\mu A_\mu$, the other in the non-relativistic limit to $F_2(q^2)(\boldsymbol{\sigma}\times\mathbf{q})\cdot\mathbf{A}$. This is a magnetic dipole interaction as $\mathbf{q}\equiv -\mathrm{i}\nabla$ so the interaction is proportional to $(\boldsymbol{\sigma}\times\nabla)\cdot\mathbf{A}=\boldsymbol{\sigma}\cdot(\nabla\times\mathbf{A})=\boldsymbol{\sigma}\cdot\mathbf{B}$. $F_1(q^2)$ depends on the distribution of charge and for $q=0$ is a measure of the amount of charge so $F_1^{\mathrm{p}}(0)=1$ and $F_1^{\mathrm{n}}(0)=0$. Similarly $F_2(0)$ is equal to the magnitude of the anomalous magnetic moment so $F_2^{\mathrm{p}}(0)=1.79/2M$ and $F_2^{\mathrm{n}}(0)=-1.91/2M$ where M is the mass of the nucleon. By Lorentz invariance there could be a third part to the electromagnetic current proportional to q_μ with a form factor $F_3(q^2)$. However, because the electromagnetic current is conserved, i.e. $\partial J_\mu/\partial x_\mu=0$, $F_3(q^2)=0$.

The electromagnetic current of the proton J_{p} or neutron J_{n} can be written as a nucleon current J_{N} using isospin since:

$$\begin{aligned} J_{\mathrm{N}} &= \tfrac{1}{2}(1-\tau_3)J_{\mathrm{p}}+\tfrac{1}{2}(1+\tau_3)J_{\mathrm{n}} \\ &= \tfrac{1}{2}(J_{\mathrm{p}}+J_{\mathrm{n}})+\tfrac{1}{2}\tau_3(J_{\mathrm{n}}-J_{\mathrm{p}}) \\ &= J_{\mathrm{N}}^{S}+\tau_3 J_{\mathrm{N}}^{V} \end{aligned}$$

where $\tau_3/2$ is the nucleon's third component of isospin. There are therefore isoscalar J_{N}^S and isovector J_{N}^V contributions to the electromagnetic current. Likewise the weak nucleon current can be expressed as:

$$J_{\mathrm{W}}=\tau^{\pm}(V_{\mathrm{N}}-\lambda A_{\mathrm{N}})$$

where $\tau^{\pm}$ cause a change in τ_3 of $+1$ or -1 and so correspond to p$\to$n or n$\to$p, respectively. This is an isovector current and the term giving rise to Fermi transitions is $\tau^{\pm}V_{\mathrm{N}}$, which is an isovector weak vector current. By Lorentz invariance there could be three parts to this current with associated form factors $f_1(q^2)$, $f_2(q^2)$ and $f_3(q^2)$, just like there could be to the electromagnetic vector current. In allowed Fermi transitions only the first part of this current contributes and the matrix element is multiplied by the form factor $f_1(q^2)$, i.e. the Fermi constant G_β includes $f_1(q^2)$ while the muon weak coupling constant G_{F} does not as the muon behaves as a point-like particle.

The constants G_β and G_{F} are very similar in magnitude and this similarity led Feynman and Gell-Mann to propose that the isovector weak vector current and the isovector electromagnetic vector current are members of an isotriplet of currents all of which are conserved. This idea is called the conserved-vector current (CVC) hypothesis and it means that $f_1(q^2) = F_{V1}(q^2)$, $f_2(q^2) = F_{V2}(q^2)$ and $f_3(q^2) = 0$, where:

$$F_V(q^2) = [F^n(q^2) - F^p(q^2)]/2 \text{ as } J_N^V = (J_n - J_p)/2.$$

The prediction that $f_1(q^2) = F_{V1}(q^2)$ would also predict that $G_\beta = G_F$ as $F_{V1}(q^2)$ is essentially 1 at the q^2 found in nuclear β-decay. The lack of exact equality between G_β and G_F is not accounted for by radiative corrections and is now understood in terms of Cabbibo's hypothesis that the weak interaction is shared between $\Delta S = 0$ and $\Delta S = 1$ (S = strangeness) transitions such that the coupling strength in nuclear β-decay ($\Delta S = 0$) is $G_F \cos\theta_C$ and in $\Delta S = 1$ transitions is $G_F \sin\theta_C$. This means that $G_\beta = G_F \cos\theta_C$ and the experimental value for $\cos\theta_C = 0.975$.

A striking test of the CVC hypothesis is given in a comparison of the β-decays: ${}^{12}_{5}\text{B}(1^+) \xrightarrow{\beta^-} {}^{12}_{6}\text{C}(0^+)$ and ${}^{12}_{7}\text{N}(1^+) \xrightarrow{\beta^+} {}^{12}_{6}\text{C}(0^+)$, which are allowed Gamow–Teller transitions. The dominant nuclear term is $\lambda\langle\boldsymbol{\sigma}\rangle$ but the term involving $f_2(q^2)$ in the vector current can also contribute and its effect is called weak magnetism as the analogous term in the electromagnetic current gives rise to a magnetic dipole interaction. In the non-relativistic limit the weak magnetic interaction is of the form $(\boldsymbol{\sigma}\times\mathbf{q})\cdot\mathbf{L}$ (cf. $(\boldsymbol{\sigma}\times\mathbf{q})\cdot\mathbf{A}$) where $\mathbf{L}$ is the lepton current so its contribution to the nuclear part of the matrix element is $f_2(q^2)\langle\boldsymbol{\sigma}\rangle\times\mathbf{q}$. The CVC hypothesis is that $f_2(q^2) = F_{V2}(q^2)$, which for nuclear β-decay means $f_2(q^2) = F_{V2}(0) = [F_2^p(0) - F_2^n(0)]/2 = 3.70/2M$. Allowing for this term alters the shape of the spectrum of emitted positrons and electrons in different ways and the agreement with experiment is good.

3.7 Summary

Fermi's theory, in which an electron and neutrino are created by an interaction which has a zero range, accounts for the observed electron energy spectra and spread of β-decay lifetimes. The modern theory of the weak interaction, as arising through the exchange of heavy (~ 80 GeV/c^2) intermediate vector bosons, predicts a very short range for the weak interaction of $\sim 10^{-3}$ fm. The observation of parity violation showed that the Fermi theory was incomplete and that the weak matrix element must contain a mixture of scalar and pseudoscalar terms. This is provided by the (V – A) theory, which accounts for the observed helicities of the emitted neutrinos and electrons.

3.8 Questions

1.* Why is it possible for ${}^{40}_{19}\text{K}$ to exhibit β^-, β^+ and EC types of decay?

2.* The nucleus ${}^{21}_{11}\text{Na}$ decays to ${}^{21}_{10}\text{Ne}$ with the emission of a positron. The radius of the nuclei of mass number 21 is 3.6 fm; estimate the maximum energy of the positron. (A spherical uniform distribution of charge Q has a Coulomb potential energy of $(3/5)Q^2/4\pi\varepsilon_0 R$.)

3. Calculate a value for G_β from the observed ft value for ^{14}O β-decay of 3127 s.

4.* The nucleus ^{87}Rb decays into the ground state of ^{87}Sr, with a half-life of 4.7×10^{10} years and a maximum β^- energy of 272 keV. Discuss briefly the difficulties one might encounter in attempting to measure this half-life.
Five different samples of chondritic meteorites are found to have the following proportions of ^{87}Rb, ^{87}Sr and ^{86}Sr:

Meteorites	^{87}Rb/^{86}Sr	^{87}Sr/^{86}Sr
Modoc	0.86	0.757
Homestead	0.8	0.751
Bruderheim	0.72	0.747
Kyushu	0.6	0.739
Buth Furnace	0.09	0.706

Given that the nucleus ^{86}Sr is not a daughter product of any long-lived radioactive nucleus, show that these data are consistent with a common primordial ratio ^{87}Sr/^{86}Sr and a common age for all these meteorites and find that age.

5.* Classify the following β-decays as allowed, first forbidden, Fermi or Gamow–Teller transitions: (*a*) $^3\text{H}\to{}^3\text{He}$, (*b*) $^{14}\text{O}_{\text{gs}}(0^+)\to{}^{14}\text{N}^*(0^+)$, (*c*) $^{47}_{21}\text{Sc}(\frac{7}{2}^-)\to{}^{47}_{22}\text{Ti}^*(\frac{7}{2}^-)$, (*d*) $^{36}_{16}\text{S}(0^+)\to{}^{36}_{17}\text{Cl}(2^-)$, (*e*) $^{152}_{63}\text{Eu}^*(0^-)\to{}^{152}_{62}\text{Sm}(0^+)$.

6.* Explain qualitatively why all of the following processes can be attributed to the weak force despite the very different lifetimes involved.

(*a*) $\mu^+\to e^+ + \nu_e + \bar{\nu}_\mu$	$t_{1/2}=1.5\times10^{-6}$ s	$Q=105$ MeV
(*b*) $n\to p+e^-+\bar{\nu}_e$	$t_{1/2}=900$ s	$Q=0.782$ MeV
(*c*) $^7\text{Be}+e^-\to{}^7\text{Li}+\nu_e$	$t_{1/2}=4.6\times10^6$ s	$Q=0.86$ MeV
(*d*) $^{12}\text{C}+\mu^-\to{}^{12}\text{B}+\nu_\mu$	$t_{1/2}=1.4\times10^6$ s	$Q=92$ MeV
(*e*) $^6\text{He}\to{}^6\text{Li}+e^-+\bar{\nu}_e$	$t_{1/2}=1.2$ s	$Q=3.54$ MeV
(*f*) $^{10}\text{Be}(0^+)\to{}^{10}\text{B}(3^+)$	$t_{1/2}=5\times10^{13}$ s	$Q=0.55$ MeV

7.* The three leptons e, μ and τ may be assumed to have the same weak interaction coupling constant. The τ-lepton (mass 1784 MeV/c^2) and the muon (mass 105 MeV/c^2) have decay modes and branching values as follows:

$$\mu\to e+\nu+\bar{\nu} \quad (100\%)$$
$$\tau\to e+\nu+\bar{\nu} \quad (17\%)$$

Given the mean lifetime of the muon is 2.2×10^{-6} s make an estimate of the mean lifetime of the τ-lepton.
If τ leptons with momentum 5 GeV/c are produced in an e^+e^- collider, calculate the mean flight path before decay in the laboratory system.

8. $^{57}_{27}$Co decays by electron capture to $^{57}_{26}$Fe. Will X-rays associated with a $Z=26$ or $Z=27$ atom be seen following a $^{57}_{27}$Co decay?
9.* Indicate how the momentum spectrum in β^- decay would be changed if the universe were to contain a degenerate gas of (*a*) neutrinos, (*b*) anti-neutrinos with Fermi energy $\varepsilon \leq E_0$.
10.* Parity violating effects in β-decay are observed in the form of helicities and asymmetries with respect to the nuclear spin. However, in allowed decays the parities of initial and final nuclear states are always the same. Why is this?
11. Calculate the fraction of electrons emitted within 100 eV of the end-point in tritium β-decay, assuming the mass of the neutrino is zero. ($^3\text{H} \to {}^3\text{He} + \text{e}^- + \bar{\nu}_\text{e} + 18.60$ keV.)
12. If in β^--decay the probability that an anti-neutrino was emitted of zero mass was 0.99 and of 17 keV mass was 0.01, sketch the Kurie plot for a β^- decay with a Q-value of 65 keV.
13.* The possibility exists that neutrinos have mass and that there is an interaction between the neutrinos ν_e and ν_μ produced in weak-interaction processes, so that the neutrinos which are mass eigenstates ν_1 and ν_2 are in fact linear combinations of ν_e and ν_μ:

$$\nu_1 = \nu_\text{e} \cos\theta + \nu_\mu \sin\theta$$

$$\nu_2 = \nu_\mu \cos\theta - \nu_\text{e} \sin\theta$$

If at time $t=0$ electron neutrinos are produced in a weak interaction show that at time t the probability P of detecting a muon neutrino is given by:

$$P = \sin^2 2\theta \sin^2(1.27\Delta m^2 L/pc)$$

where $\Delta m^2 = (m_1^2 - m_2^2)c^4$ is in (eV)^2, L is the distance from the source in metres, and p is the neutrino beam momentum in MeV/c. (The masses m_1 and m_2 may be taken to be so small that $E_i = pc + m_i^2c^3/2p$, $i = 1, 2$.)
14. Verify that the capture cross-section $\sigma_\text{c} \approx 10^{-47}$ m^2 for 2.8 MeV incident $\tilde{\nu}$ on protons.
15.* The positron decay of ^{8}B to ^{8}Be has an end-point energy of 14.09 MeV. What fraction of the neutrinos produced in the decay of ^{8}B have sufficient energy to induce the reaction $\nu + {}^{37}\text{Cl} \to {}^{37}\text{A} + \text{e}^-$? ($Q = -0.81$ MeV.)
16. By considering the probability that the spin of a spin $\frac{1}{2}$ particle is at an orientation of θ to the quantisation axis, deduce the form of the ν-e$^+$ angular distribution (figure 3.20) for a vector interaction in the limit $v \to c$: $I(\theta) = 1 + \cos\theta$. Explain qualitatively why $\pi \to \mu + \nu$ rather than e$+\nu$ even though the decay $\pi \to \text{e} + \nu$ has a much larger Q-value.

4 Gamma decay

4.1 The theory of γ-decay

Excited states if bound to particle emission nearly always decay through the electromagnetic interaction usually with the emission of a γ-ray, though if the nucleus has a large Z and the excitation is low then decay by internal conversion in which an electron is ejected is more likely. Very occasionally when the angular momentum change in a low-energy γ-transition is very large, β-decay competes with γ-decay. For unbound states well above threshold, particle decay predominates. Near threshold, charged particle decay is hindered by the Coulomb barrier and neutron decay by the transmission across the potential discontinuity at the nuclear surface, with the result that γ-decay can be comparable to particle decay for excited states near threshold.

4.1.1 Spontaneous decay

A bound excited state of a nucleus (or of an atom or molecule) would be stable with respect to γ-decay were it not for the fluctuations in the electromagnetic field intensity of the vacuum. These fluctuations, which give rise to a non-zero mean-square value for the electric and magnetic field intensities, arise from quantising the electromagnetic field. They perturb an excited state and induce spontaneous decays.

Classically the energy of an electromagnetic field, U, is given by:

$$U = \tfrac{1}{2}\int (\mathbf{D}\cdot\mathbf{E}+\mathbf{B}\cdot\mathbf{H})\,d\tau$$

If there are no free charges anywhere then **E** and **B** can be represented by:

$$\mathbf{E} = -\frac{\partial \mathbf{A}}{\partial t} \quad \text{and} \quad \mathbf{B} = \text{curl}\,\mathbf{A}$$

The vector potential **A** can be expanded in plane waves:

$$\mathbf{A} = \sum_k [\mathbf{A}_k \exp(i\mathbf{k}\cdot\mathbf{r}) + \mathbf{A}_k^* \exp(-i\mathbf{k}\cdot\mathbf{r})]$$

and the energy of the field in unit volume equals:

$$U=\sum_k\left(\tfrac{1}{2}\varepsilon_0|\dot{\mathbf{A}}_k|^2+\tfrac{1}{2}\frac{k^2}{\mu_0}|\mathbf{A}_k|^2\right)$$

This is like the energy of a set of harmonic oscillators with mass $m\equiv\varepsilon_0$, force constants $\alpha_k\equiv k^2/\mu_0$ and frequencies $\omega_k=\sqrt{\alpha_k/m}=kc$ $(c^2=1/\varepsilon_0\mu_0)$.

In quantum mechanics the energy of this set of harmonic oscillators is:

$$U=\sum_k(n_k+\tfrac{1}{2})\hbar\omega_k$$

where n_k is the number of photons with wave vector **k**. In the ground state, the vacuum, $U=\sum_k\frac{1}{2}\hbar\omega_k$, so $|\mathbf{A}_k|^2_{\text{vac}}$ (and hence $\mathbf{E}^2$ and $\mathbf{B}^2$), has a non-zero value given by:

$$\tfrac{1}{2}\hbar\omega_k=\tfrac{1}{2}\left(\varepsilon_0\omega_k^2+\frac{k^2}{\mu_0}\right)|\mathbf{A}_k|^2_{\text{vac}}$$

$$\therefore\quad |\mathbf{A}_k|^2_{\text{vac}}=\frac{\mu_0\hbar c^2}{2\omega_k}$$

The $\mathbf{A}_k$ are operators and have matrix elements between photon states which differ in their number of photons by one (cf. the matrix elements of x between eigenfunctions of a one-dimensional harmonic oscillator). Therefore:

$$\begin{aligned}|\mathbf{A}_k|^2_{\text{vac}}&\equiv\langle 0||\mathbf{A}_k|^2|0\rangle\\&=\sum_n\langle 0|\mathbf{A}_k|n\rangle\langle n|\mathbf{A}_k^*|0\rangle\qquad\left(\sum_n|n\rangle\langle n|=1\right)\\&=\sum_n|\langle 0|\mathbf{A}_k|n\rangle|^2\\&=|\langle 0|\mathbf{A}_k|1\rangle|^2=|(\mathbf{A}_k)_{01}|^2\end{aligned}$$

since only for $n=1$ are the matrix elements of $\mathbf{A}_k$ non-zero. The matrix element $(\mathbf{A}_k)_{01}$ which represents the transition from the vacuum to a state with one photon present is therefore:

$$(\mathbf{A}_k)_{01}=\sqrt{\frac{\mu_0\hbar c^2}{2\omega_k}}$$

The interaction of a charged particle with an electromagnetic field is obtained by replacing the momentum **p** of the particle by $(\mathbf{p}-e\mathbf{A})$. The term $p^2/2m$ in the Hamiltonian becomes:

$$\frac{p^2}{2m}-\frac{e}{m}\mathbf{p}\cdot\mathbf{A}+\frac{e^2}{2m}\mathbf{A}^2$$

and the interaction is given by the $\mathbf{p}\cdot\mathbf{A}$ and $\mathbf{A}^2$ terms. The $\mathbf{A}^2$ term is small and corresponds to a two-photon event and for one-photon processes the interaction is $-(e/m)\mathbf{p}\cdot\mathbf{A}$. The matrix element M_{21} for the transition of a charged particle between single particle states ψ_1 and state ψ_2 with the emission of a photon is therefore

$$-\langle\psi_2|\frac{e}{m}\mathbf{p}\cdot\mathbf{A}|\psi_1\rangle = -\frac{e}{m}\langle\psi_2|\mathbf{p}\cdot\boldsymbol{\varepsilon}\exp(\mathrm{i}\mathbf{k}\cdot\mathbf{r})|\psi_1\rangle\times\sqrt{\frac{\mu_0\hbar c^2}{2\omega_k}} \tag{4.1}$$

where the unit vector $\boldsymbol{\varepsilon}$, which is in the direction $\mathbf{A}_k$, represents the polarisation of the photon.

4.1.2 Multipole fields

The vector function $\mathbf{A}_k\exp(\mathrm{i}\mathbf{k}\cdot\mathbf{r})$, which represents the photon, can be expanded in a multipole expansion, each term of which corresponds to a definite angular momentum. The usefulness of this is that only a few terms which depend on the spin and parity of ψ_1 and ψ_2 contribute significantly to the matrix element for a transition. In this expansion the function $\exp(\mathrm{i}\mathbf{k}\cdot\mathbf{r})$ is expanded in terms of spherical harmonics, $Y_{lm}(\theta,\psi)$, each of which has an associated orbital angular momentum $\mathbf{l}$. Scalar functions describe spin zero particles in quantum mechanics and vector functions describe particles with an intrinsic spin $\mathbf{s}$ of one and a negative intrinsic parity ($\mathbf{A}_k\to-\mathbf{A}_k$ under parity). The $\mathbf{l}$ and $\mathbf{s}$ couple to give a total angular momentum conventionally labelled $\mathbf{L}$ (not to be confused with the orbital angular momentum $\mathbf{l}$) with $l=L$ or $l=L\pm1$. The spin $\mathbf{s}$ of the photon is either along or opposed to its direction of motion, i.e. $m_s=\pm1$ but $m_s\neq0$. Its orbital angular momentum is perpendicular to its direction of motion ($\mathbf{l}=\mathbf{r}\times\mathbf{p}$) so $m_l=0$. Therefore the spin $\mathbf{s}$ and $\mathbf{l}$ cannot couple up to $L=0$. The terms with $l=L$ correspond to magnetic radiation and with $l=L\pm1$ to electric radiation. For example, for $l=0$ then $L=1$ and this corresponds to electric dipole (E1) radiation. The parity of the radiation is the product of the intrinsic parity of the photon and that of the spherical harmonic $(-1)^l$ and is therefore equal to $(-1)^{l+1}$, which for E1 is negative.

4.2 Transition rates

4.2.1 Single-particle estimates

In a typical nuclear γ-decay $E_\gamma=1$ MeV, which means that for a nucleus with a radius $R=5$ fm, $\mathbf{k}\cdot\mathbf{r}\leq kR=\frac{1}{40}$ so to a reasonable first approximation $\exp(\mathrm{i}\mathbf{k}\cdot\mathbf{r})=1$ in the transition matrix element, which corresponds to taking just the $l=0$ term of the multipole expansion. Taking the photon polarisation

$\boldsymbol{\varepsilon}$ to be along the x-axis then the matrix element M_{21} (equation 4.1) for the transition of a single proton is given by:

$$\begin{aligned} M_{21} &\propto \langle\psi_2|\frac{ep_x}{m}|\psi_1\rangle \\ &= \frac{\mathrm{i}}{\hbar}\langle\psi_2|e[H, x]|\psi_1\rangle \\ &= \mathrm{i}\omega\langle\psi_2|ex|\psi_1\rangle \end{aligned}$$

i.e. to the electric dipole matrix element D_{21} between ψ_1 and ψ_2.

The transition rate is given by Fermi's golden rule:

$$\begin{aligned} w &= \frac{2\pi}{\hbar}|M_{21}|^2\frac{4\pi p^2}{h^3}\frac{\mathrm{d}p}{\mathrm{d}E} \\ &= \frac{\omega^2}{\pi\hbar^2 c^3}|M_{21}|^2 \\ &= \frac{\mu_0}{2\pi\hbar}\frac{\omega^3}{c}|D_{21}|^2 \end{aligned} \tag{4.2}$$

This refers to a transition between two substates and ignores the spin of the proton, so it is not the average rate. However, estimating D_{21} assuming pure single-particle states is in general a very poor approximation, so refining the above expression to correct for these omissions (which give a factor of order one) is not important. Estimating $|D_{21}|^2 = e^2R^2$ for a single-particle transition then for a 1 MeV decay, $w_{E1} \approx 10^{16}\,\mathrm{s}^{-1}$ for $R \approx 5$ fm. Excited states can only decay by an E1 transition if there is a change of parity, otherwise $D_{21} = 0$, and if $|J_i - J_f| \leq 1$ but not $J_i = 0 \rightarrow J_f = 0$. The restriction on angular momentum change comes about because in an E1 decay the photon carries away one unit of angular momentum and angular momentum must be conserved in the decay.

For transitions involving larger changes in angular momentum then higher terms corresponding to $l > 0$ in the expansion of $\exp(\mathrm{i}\mathbf{k}\cdot\mathbf{r})$ must be considered:

$$\exp(\mathrm{i}\mathbf{k}\cdot\mathbf{r}) = 1 + \mathrm{i}\mathbf{k}\cdot\mathbf{r} + \cdots$$

The matrix element M_{21} for the $\mathrm{i}\mathbf{k}\cdot\mathbf{r}$ term is given by:

$$M_{21} \propto \langle\psi_2|\frac{e}{m}\mathbf{p}\cdot\boldsymbol{\varepsilon}\mathbf{k}\cdot\mathbf{r}|\psi_1\rangle$$

Since $\boldsymbol{\varepsilon}$ is in the direction of $\mathbf{A}_k$ and $\mathbf{B} = \nabla \times \mathbf{A} = \mathrm{i}\mathbf{k} \times \mathbf{A}$ then $\boldsymbol{\varepsilon}$ and $\mathbf{k}$ are perpendicular. If $\boldsymbol{\varepsilon}$ is in the direction of $\mathbf{x}$ then $\mathbf{k}$ is in the direction of $\mathbf{z}$

and:

$$M_{21} \propto k\langle\psi_2|\frac{e}{m}p_x z|\psi_1\rangle$$

$$=\frac{e\omega}{2mc}\langle\psi_2|(p_x z - p_z x)+(p_x z + p_z x)|\psi_1\rangle$$

$$=\frac{e\omega}{2mc}\langle\psi_2|l_y|\psi_1\rangle+\frac{ie\omega^2}{2c}\langle\psi_2|zx|\psi_1\rangle \qquad \left(p_x=\frac{im}{\hbar}[H,x]\right)$$

$$=\frac{\omega}{c}(\mu_y)_{21}+\frac{i\omega^2}{2c}(Q_{zx})_{21}$$

where μ_y is the orbital component of the magnetic moment operator and contributes to M1 transitions and Q_{zx} is a component of the quadrupole operator and gives rise to E2 transitions. (There is also an intrinsic magnetic moment contribution to M1 transitions.)

Substituting into the expression for the transition rate above (equation 4.2) gives:

$$w_{\mathrm{M1}}=\frac{\mu_0}{2\pi\hbar}\frac{\omega^3}{c^3}|\mu_{21}|^2$$

and

$$w_{E2}=\frac{\mu_0}{2\pi\hbar}\frac{\omega^5}{4c^3}|Q_{21}|^2$$

If μ_{21} is estimated as one nuclear Bohr magnetron, i.e. $\mu_{21}=e\hbar/2m$, then the ratio of $w_{\mathrm{M1}}/w_{\mathrm{E1}}$ is given by:

$$w_{\mathrm{M1}}/w_{\mathrm{E1}}=\hbar^2/4m^2c^2R^2$$

Since by the uncertainty principle $mvR \approx \hbar$ then:

$$w_{\mathrm{M1}}/w_{\mathrm{E1}}=\frac{v^2}{4c^2}$$

which for a nucleon at the Fermi level with $T \approx 30$ MeV equals $\sim 1.5\times10^{-2}$. For an E2 transition the single-particle matrix element can be estimated by $Q_{21}=eR^2$, which gives for $E_\gamma=1$ MeV and $R=5$ fm:

$$w_{\mathrm{E2}}/w_{\mathrm{E1}}=\frac{k^2}{4}R^2 \approx 1.5\times10^{-4}$$

The selection rules for M1 and E2 transitions are: no change of parity, otherwise μ_{21} and Q_{21} are zero, and for M1: $|J_i - J_f| \leq 1$ but not $0 \to 0$ and for E2: $|J_i - J_f| \leq 2$ but not $0 \to 0$ or $\frac{1}{2} \to \frac{1}{2}$. These angular momentum restrictions come about because the photon carries away one (M1) or two (E2) units of angular momentum. For higher multipolarity transitions, corresponding to higher terms in the expansion of $\exp(i\mathbf{k}\cdot\mathbf{r})$, the selection rules are:

$$\begin{array}{lll} \mathrm{M}L & \Delta\pi=(-1)^{L+1} & \mathbf{J}_i+\mathbf{J}_f=\mathbf{L} \\ \mathrm{E}L & \Delta\pi=(-1)^{L} & \mathbf{J}_i+\mathbf{J}_f=\mathbf{L} \end{array}$$

The single-particle rates are down for higher multipole transitions of the same type (E or M) by a factor of order $(kR)^2$ between one multipole and the next higher one while the ratio of (ML/EL) transitions of the same energy is of the order of 1%. The energy dependence of a multipole transition of order L is predicted to be $(E_\gamma)^{2L+1}$.

When discussing transition rates an averaged value for the square of the transition matrix element, called the reduced transition probability, $B(\mathrm{E}L)$ or $B(\mathrm{M}L)$, is generally quoted. These transition matrix elements are proportional to the matrix elements of the appropriate multipole operator, which for EL transitions is proportional to $r^L P_L(\cos\theta)$. Their connection with the transition rate w is:

$$\left.\begin{aligned} w_{\mathrm{E1}} &= 1.59\times 10^{15} E_\gamma^3 B(\mathrm{E1}) \\ w_{\mathrm{M1}} &= 1.76\times 10^{13} E_\gamma^3 B(\mathrm{M1}) \\ w_{\mathrm{E2}} &= 1.23\times 10^{9} E_\gamma^5 B(\mathrm{E2}) \end{aligned}\right\} \quad (4.3)$$

where $B(\mathrm{E}L)$ are in units of e^2 (fermis)^{2L}, $B(\mathrm{M1})$ is in units of μ_{N}^2, E_γ is in MeV and w is in (seconds)$^{-1}$. Also single-particle estimates similar to those deduced above, called Weisskopf estimates, are often used when comparing γ-ray rates. For E1, M1, E2 and E3 transitions the Weisskopf units (W.u.) for the gamma widths are:

$$\left.\begin{aligned} \Gamma_{\mathrm{W}}(\mathrm{E1}) &= 6.8\times 10^{-2} A^{2/3} E_\gamma^3 \\ \Gamma_{\mathrm{W}}(\mathrm{M1}) &= 2.1\times 10^{-2} E_\gamma^3 \\ \Gamma_{\mathrm{W}}(\mathrm{E2}) &= 4.9\times 10^{-8} A^{4/3} E_\gamma^5 \\ \Gamma_{\mathrm{W}}(\mathrm{E3}) &= 2.3\times 10^{-14} A^2 E_\gamma^7 \end{aligned}\right\} \quad (4.4)$$

where Γ_γ is in eV and E_γ is in MeV. Generally in a decay more than one type of transition is possible, e.g. for a $3^+ \rightarrow 2^+$ decay M1, E2, M3, E4 and M5 transitions are all possible. But because of the strong dependence on multipole order only the lowest or the two lowest multipolarities are usually important, i.e. for a $3^+ \rightarrow 2^+$ transition M1 and E2.

4.2.2 Enhanced transition rates

In the collective model enhanced electric multipole transition rates arise through the collective motion of many protons. In a microscopic description, although gamma decay only involves the transition of a single nucleon, as a photon only interacts with one nucleon, a transition matrix element can be considerably larger than the single-particle estimate through a superposition of terms in the matrix element. For example, consider the E2 transition between the $\frac{5}{2}^-$ 0.32 MeV first excited and $\frac{7}{2}^-$ ground state of ${}^{51}_{23}\mathrm{V}_{28}$. In the shell model both states are described by the configuration $(\pi \mathrm{f}_{7/2})^3$ coupled to $\frac{5}{2}^-$ and $\frac{7}{2}^-$, respectively. There are three equivalent $\mathrm{f}_{7/2}$ protons in the initial state, any one of which may make the transition to the ground state. The E2 matrix element Q_{fi} between initial and final states is therefore given by:

$$Q_{fi} = \langle (f_{7/2})^3_{7/2} | \sum_{i=1}^{3} Q_i | (f_{7/2})^3_{5/2} \rangle$$

where the sum is over the three equivalent protons and Q_i is the single particle E2 transition operator. The sum gives the possibility of an enhanced matrix element and for this particular transition $\Gamma(\text{E2})$, which is proportional to $(Q_{fi})^2$, is $\sim 5\Gamma_W(\text{E2})$, where $\Gamma_W(\text{E2})$ is the single-particle Weisskopf estimate. The maximum possible Γ_γ width for N equivalent protons is

$$\Gamma_\gamma^{\max} = N^2 \Gamma_{\text{sp}}$$

where Γ_{sp} is the appropriate average single-particle width, which may be estimated by Γ_W. So an enhancement of 100 over the single-particle estimate would correspond to the equivalent of ten protons acting coherently.

In general the initial and final wavefunctions are much more complex and have many components:

$$\psi_i = \sum_n a_n \psi_n, \qquad \psi_f = \sum_m a_m \psi_m$$

so

$$\langle \psi_f | Q | \psi_i \rangle = \sum_{nm} a_n a_m^* Q_{mn}$$

The amplitudes a_n and a_m^* can be of different sign and there can be a lot of cancellation in the sum with the result that $Q_{fi} \ll Q_{\text{sp}}$. If however ψ_i and ψ_f are the wavefunctions describing effectively N equivalent protons then there is an enhancement in the sum and $Q_{fi} = N^2 Q_{\text{sp}}$. An example of this is seen in the transition rates for E2 decays between members of a rotational band where the intrinsic deformed wavefunctions of the levels are the same.

Another example is found in the giant dipole resonance, which in the single-particle shell model is described by a superposition of 1p–1h states (e.g. in ^{16}O one such p–h state would be $[(1d_{5/2})(1p_{3/2})^{-1}]_1$). The electric dipole matrix element between the 0^+ ground state and the 1^- giant dipole state is a coherent superposition of matrix elements of the single-particle dipole operator between the ground state wavefunction and the $(1\text{p}-1\text{h})^{-1}$ wavefunctions. For a self-conjugate nucleus the resultant increase in Γ_γ is given by $\Gamma_\gamma = Z\Gamma_{\text{sp}}$.

4.2.3 Isospin selection rules

In a gamma-ray transition only a single nucleon is involved as a photon only interacts with a single nucleon. As the isospin of a nucleon is a half then there is the general selection rule:

$$\Delta T = 0 \text{ or } 1$$

for a γ-decay between two pure isospin states. An example of this rule is the absence of a $\Delta T = 2$ transition in the decay of the lowest $T = 2$ 0^+ state in ^{28}Si at 15.22 MeV excitation to the 2^+ first excited state at 1.78 MeV excitation. As the Coulomb interaction (and to a very much lesser extent the weak interaction) mixes states of different isospin with the same J^π then violations of this selection rule would be expected at the level of about 1%.

More restrictive selection rules apply in certain cases. Consider first E1 transitions. Qualitatively, to create an electric dipole moment in a nucleus there must be a separation of the charge and the centre of mass of the nucleus: this involves differentiating between neutrons and protons and so a $\Delta T = 0$ E1 transition in a self-conjugate nucleus ($N = Z$) in which there is no such differentiation will be forbidden.

This selection rule follows more formally from considering the correct operator $\mathbf{O}_{E1}$ for an E1 transition:

$$\mathbf{O}_{E1} = \sum_{\text{protons}} e(\mathbf{r}_p - \mathbf{R})$$

where $\mathbf{R}$ is the centre of mass of the nucleus. (The single-particle operator $e\mathbf{r}$ discussed above neglects the effect of the centre of mass.) This can be rewritten as a sum over all nucleons:

$$\mathbf{O}_{E1} = \tfrac{1}{2} \sum_i (1 - 2t_{z_i}) e(\mathbf{r}_i - \mathbf{R})$$

$$= -\sum_i e t_{z_i} (\mathbf{r}_i - \mathbf{R}) \qquad (4.5)$$

since $(1 - 2t_{z_i}) = 0$ for neutrons and $\mathbf{R} = (\sum_i \mathbf{r}_i)/A$. The operator $\mathbf{O}_{E1}$ is therefore a pure isovector operator as it only involves t_{z_i} so $T = 0 \rightarrow T = 0$ E1 transitions are forbidden (cf. the operator $\sum_i e z_i$, the z-component of the single-particle dipole operator, which is a vector operator for which the selection rule $J = 0 \nleftrightarrow J = 0$ holds). It is only in self-conjugate nuclei that $T = 0$ levels are found as $T \geq T_z = (N - Z)/2$.

For M1 transitions (and for higher multipole transitions) there is little centre of mass effect as no differentiation between charge and centre of mass is required. The single-particle operator $\mathbf{O}_{M1}$ in units of nuclear magnetons using the isospin notation is:

$$\mathbf{O}_{M1} = \tfrac{1}{2} \sum_i (1 - 2t_{z_i}) \mathbf{l}_i + \tfrac{1}{2} \sum_i (1 - 2t_{z_i}) g_p \mathbf{s}_i + \tfrac{1}{2} \sum_i (1 + 2t_{z_i}) g_n \mathbf{s}_i$$

$$= \tfrac{1}{2} \sum_i [\mathbf{l}_i + (g_p + g_n)\mathbf{s}_i] + \sum_i t_{z_i} [(g_n - g_p)\mathbf{s}_i - \mathbf{l}_i]$$

$$= \tfrac{1}{2}\mathbf{J} + \tfrac{1}{2} \sum_i (g_p + g_n - 1)\mathbf{s}_i + \sum_i t_{z_i} [(g_n - g_p)\mathbf{s}_i - \mathbf{l}_i]$$

The first term $\frac{1}{2}\mathbf{J}$ does not contribute as $\langle f|\mathbf{J}|i\rangle = 0$. As $g_p = 5.59$ and $g_n = -3.83$ the isovector term is enhanced while the isoscalar term is reduced. Neglecting the orbital contribution to the isovector term for an order of magnitude estimate, the ratio of isovector to isoscalar matrix elements will be approximately $(g_p + g_n - 1):(g_n - g_p) = 1:12$. In a self-conjugate nucleus $T = 0 \rightarrow T = 0$ M1 transitions, which are pure isoscalar transitions, would therefore be expected to be inhibited in comparison with isovector M1 transitions by roughly two orders of magnitude.

There is no distinction between neutrons and protons for collectively enhanced E2 transitions between members of a rotational band as their intrinsic states are the same, nor between vibrational levels associated with different numbers of quadrupole vibrational quanta (phonons). Therefore enhanced E2 transitions have $\Delta T = 0$. Likewise enhanced E3 transitions corresponding to octupole vibrations have $\Delta T = 0$.

4.2.4 Analogue transitions

Since the wavefunctions of members of an isotopic multiplet are the same except for an exchange of $n \leftrightarrow p$, i.e. they only differ in T_z, there are predicted to be similarities in the decay schemes of analogue states. An example is found in the decays of the lowest $\frac{3}{2}^-$ levels in the mirror nuclei, ${}^{19}_{10}Ne$ and ${}^{19}_{9}F$ (see figure 4.1).

4.2.5 Experimental transition rates

The wavefunctions of nuclear states are generally much more complex than single-particle shell model wavefunctions and involve many components.

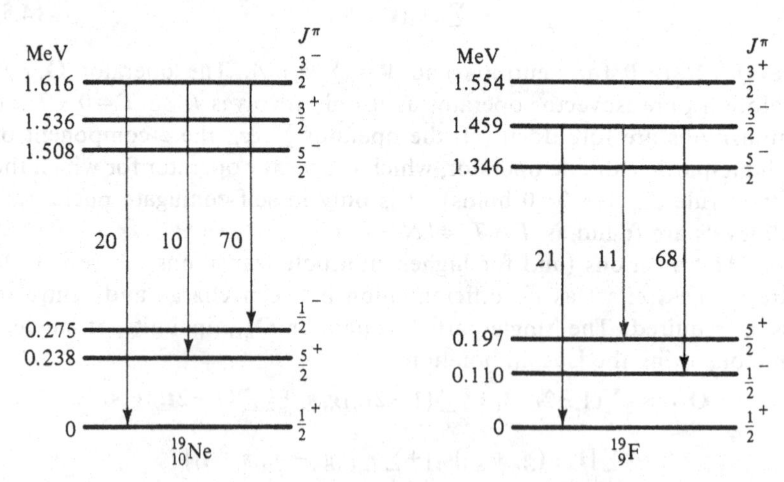

4.1. Diagram showing the equality of branching ratios (within experimental errors) of the γ-transitions from the lowest $\frac{3}{2}^-$ analogue states in ${}^{19}_{10}Ne$ and ${}^{19}_{9}F$.

This would suggest that unless there was some coherence, as in collective transitions, or inhibition, as in $T=0 \to T=0$ E1 transitions, transition widths would be a small fraction of Γ_{sp}.

(i) E1 transitions

Generally E1 transitions are found to be strongly inhibited compared with Γ_{sp}. This is principally because for low-lying states which could decay by an E1 transition ($\mathbf{J}_i+\mathbf{J}_f=\mathbf{1}, \Delta\pi=(-1)$) there is no change in isospin and a tendency for the neutrons and protons to move in phase which strongly reduces the E1 matrix element. In the case of self-conjugate nuclei, as discussed above, $\Delta T=0$ E1 transitions are forbidden while $\Delta T=1$ are allowed. An example of this is seen in ^{16}O (figure 4.2).

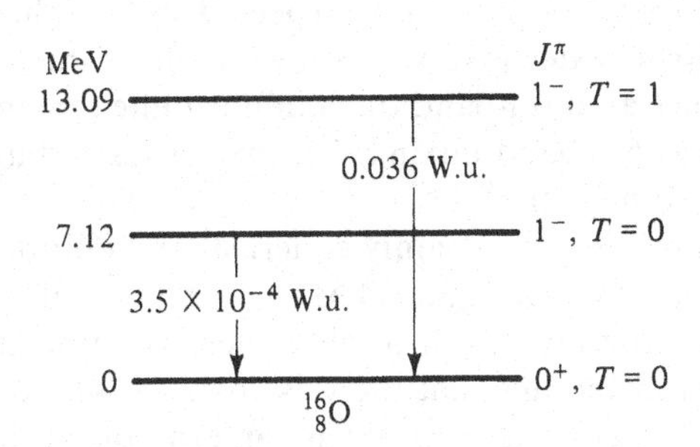

4.2. Illustration of the strong inhibition of $\Delta T=0$ E1 transitions in self-conjugate nuclei. The 13.09 MeV $\Delta T=1$ E1 transition is allowed and is ~100 times stronger than the 7.12 MeV $\Delta T=0$ E1 transition. (The giant dipole 1^- $T=1$ state is at ~22 MeV excitation in ^{16}O). The strengths are in single-particle units.

Where large E1 matrix elements are found is between the ground state and the giant dipole resonance, which in a self-conjugate nucleus has $T=1$. It corresponds to a collective oscillation of neutrons against protons or, in shell model terms, a superposition of one-particle one-hole 1^- states. These E1 matrix elements have values typically an order of magnitude larger than Γ_{sp}.

(ii) M1 transitions

Because the low-lying states in many nuclei have the same parity as the ground state M1, together with E2 or mixed M1/E2, transitions are the most common in nuclei. The M1 strength is typically a few percent of the single-particle estimate, reflecting the overlap with single-particle wavefunctions. As discussed above, in self-conjugate nuclei a pure isoscalar $T=0\rightarrow T=0$ M1 transition is expected to be strongly suppressed compared with an isovector M1 transition; an example of this is seen in figure 4.3.

(iii) E2 transitions

The single-particle estimates Γ_{sp} predict E1 transitions to be $\sim 10^2$ stronger than M1 and $\sim 10^4$ stronger than E2 transitions for $E_\gamma \sim 1$ MeV. Experimentally E1 transitions are generally strongly inhibited while E2 transitions are often enhanced by an order of magnitude with the result that M1 and E2 rates are often comparable.

As discussed above this enhancement reflects a collective motion of the nucleons and enhanced transitions are found between members of a rotational band, called interband transitions. Transitions from one band to another, intra-band transitions, are not expected to be enhanced because of the different intrinsic states involved. An example of this is seen in ^{40}Ca (see figure 4.4). Where an intra-band transition is quite strong this suggests that there has been some band-mixing, i.e. the intrinsic states of the two bands are mixed to some extent.

Mixing is also found between mainly spherical and deformed states. An example is found in ^{16}O (see figure 4.5). The rotational band built on 6.05 MeV 0^+ states is mainly 4p–4h in character. The transition from the 6.92 2^+ state to ground implies that there is some 4p–4h configuration in the ground state of ^{16}O. The two 0^+ states are an example of the coexistence of mainly spherical and significantly deformed states within the same nucleus.

Enhanced transitions are also found between vibrational levels, an example of which is found in ^{76}Se (see figure 4.6). The E2 transitions between the triplet of levels just above 1 MeV excitation and the 0.56 MeV 2^+ level

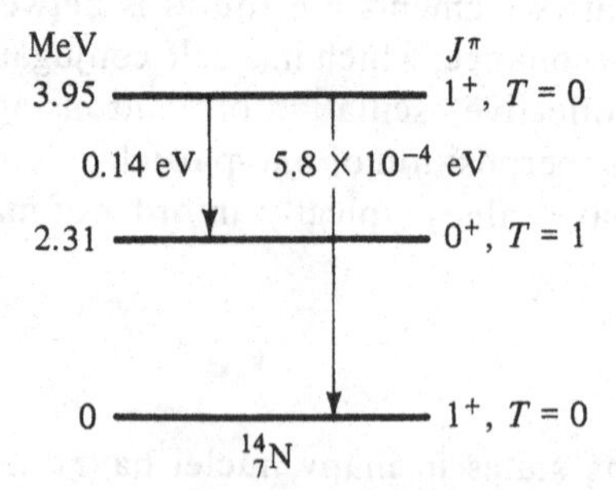

4.3. An example of the strong suppression of $\Delta T=0$ M1 transitions compared with $\Delta T=1$ M1 transitions in self-conjugate nuclei.

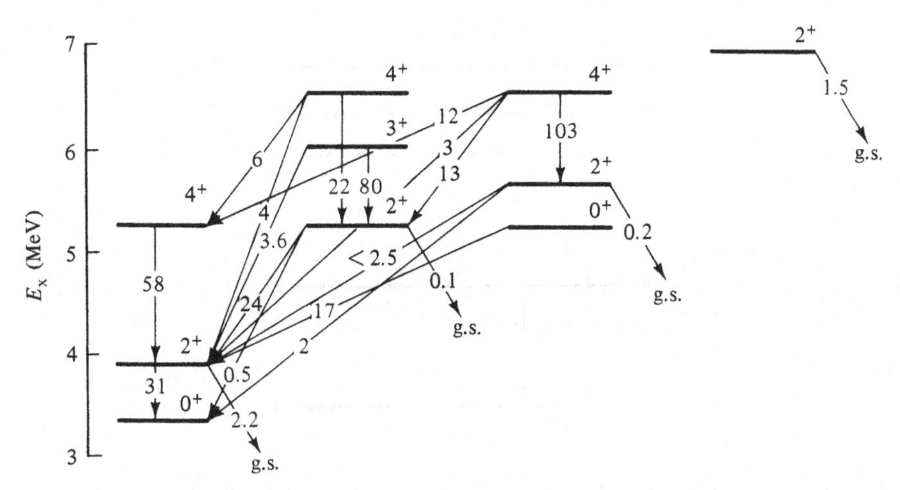

4.4. A diagram indicating the strengths (in single-particle units) of some interband and intraband E2 transitions in ^{40}Ca. (After Wilkinson, D. H., *Nuclear Physics Lecture Notes.* Oxford Nuclear Physics Laboratory.)

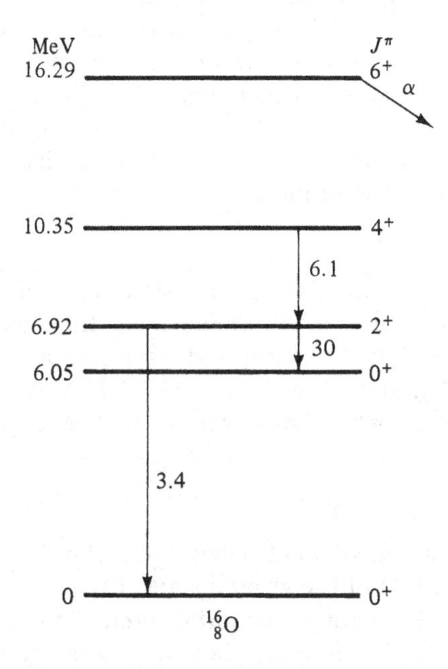

4.5. A partial energy level scheme of $^{16}_{8}$O showing the strengths (in single-particle units) of some E2 transitions.

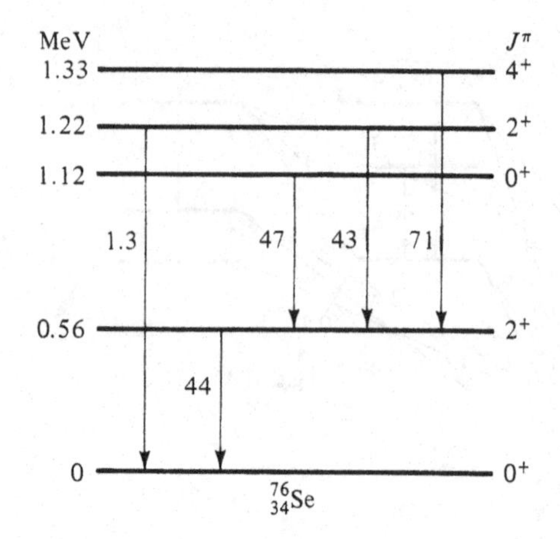

4.6. Diagram showing the strengths of E2 transitions (in single-particle units) between vibrational levels in ^{76}Se.

are all enhanced as is the 0.56 MeV $2^+ \to 0^+$ g.s. transition. The 1.22 $2^+ \to 0^+$ ground state transition, however, is not enhanced. It involves a transition from a two-phonon to a zero-phonon state, which for pure states would be forbidden as a photon can only cause a change of one or zero phonons in a transition. That a transition is seen suggests some small mixing of the one-phonon and two-phonon 2^+ states or of the two-phonon and zero-phonon 0^+ states. Generally mixing will be more likely the closer the excitation energies of levels of the same J^π.

(iv) E3 transitions

Besides the collective dipole 1^- vibrational states in even–even nuclei there are also collective octupole 3^- vibrational states. Enhanced octupole transitions to the ground state are seen and examples are the decays from ^{20}Ne (3^-, 5.63 MeV), 10.6 W.u.; ^{40}Ca (3^-, 3.73 MeV), 26 W.u.; ^{208}Pb (3^-, 2.61 MeV), 39 W.u. (where W.u. stands for Weisskopf units, equation 4.4).

(v) Isomeric transitions

In some nuclei there are low-lying excited states whose angular momentum is very different from that of the ground state. This is easily understood to be a consequence of the strong spin–orbit interaction in the simple shell model (see figure 2.9). Such states can only γ-decay by a low-energy high-order multipole transition and their decay rates can be very slow, as

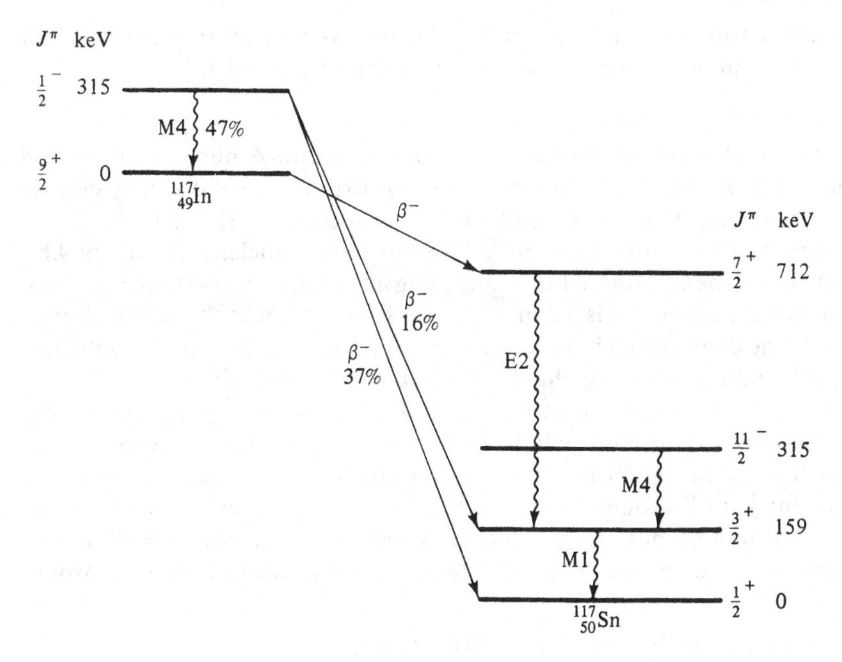

4.7. Diagram showing some of the energy levels in $^{117}_{49}$In and $^{117}_{50}$Sn. Multipolarities of observed γ-transitions are indicated. The 315 keV excited state of ^{117}In has a mean life of 1.93 hr and decays by γ-emission or by first-forbidden β^--emission with branching ratios as shown. The ground state of $^{117}_{49}$In has a mean life of 0.7 hr and decays by allowed β^--emission to the 712 keV state of $^{117}_{50}$Sn with a maximum electron energy of 740 keV.

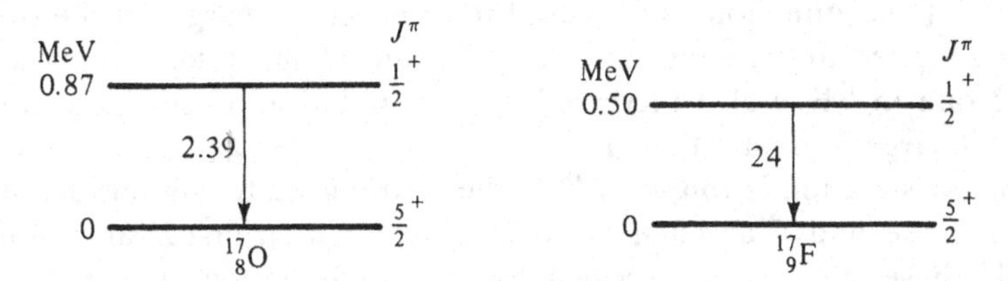

4.8. A comparison of the E2 transition strengths in $^{17}_{8}$O and $^{17}_{9}$F from the $\frac{1}{2}^+$ first excited to $\frac{5}{2}^+$ ground states (in single-particle units).

expected from the single-particle estimates. As a result competition with β-decay can occur and an example is shown in figure 4.7.

4.2.6 Effective charges

In the single-particle picture excited states in odd-A nuclei with an odd number of neutrons, e.g. $^{17}_{8}O$, would not be expected to γ-decay by electric multipole transitions as the odd nucleon is uncharged. However this is not observed, e.g., in transitions in $^{17}_{8}O$ and its mirror nucleus $^{17}_{9}F$ (figure 4.8), and this can be accounted for by assigning an effective charge to the neutron. For E1 transitions this is a consequence of the motion of the centre of mass and in the expression above (equation 4.5) protons and neutrons contribute equally with an effective charge of $+\frac{1}{2}e$ and $-\frac{1}{2}e$ respectively.

For E2 transitions the centre of mass effect is small but there are important admixtures, called core polarisation, to the single-particle wavefunctions which result in significant E2 transition probabilities. As an example consider the $\frac{1}{2}^+$ to $\frac{5}{2}^+$ ground state transition in ^{17}O. The wavefunctions for these states cannot be pure single-particle wavefunctions as then the E2 matrix element would be zero. A possible pair of wavefunctions, however, would be:

$$\psi(\tfrac{1}{2}^+) = \alpha\psi(^{16}O_{gs})\phi(2s_{1/2}) + \beta\psi(^{16}O_{2+})\phi(1d_{5/2})$$

$$\psi(\tfrac{5}{2}^+) = \gamma\psi(^{16}O_{gs})\phi(1d_{5/2}) + \delta\psi(^{16}O_{2+})\phi(2s_{1/2}) + \varepsilon\psi(^{16}O_{2+})\phi(1d_{5/2})$$

where α and γ are close to 1 and $\psi(^{16}O_{2+})$ is the wavefunction of a 2^+ excited state in ^{16}O. An E2 transition is now possible between the $\frac{1}{2}^+ \to \frac{5}{2}^+$ with a strength Γ estimated by:

$$\Gamma = (\gamma\beta + \alpha\delta)^2\Gamma(^{16}O_{2+} \to {}^{16}O_{gs})$$

The observed strength is $\sim\Gamma_{sp}$ implying $\Gamma(^{16}O_{2+} \to {}^{16}O_{gs}) \gg \Gamma_{sp}$ as $(\gamma\beta + \alpha\delta)^2 \ll 1$, i.e. that the $^{16}O_{2+}$ state is a collective 2^+ state. There is such a collective state, which is a vibrational state corresponding to a superposition of 1p–1h states with the particle promoted through two oscillator shells (to have positive parity), called the giant quadrupole resonance. The simple shell model wavefunction can be used if an effective charge (of the order of 1) is assigned to the neutron to take account of small core admixtures. The E2 rate in ^{17}F is also increased by core polarisation so the proton's effective charge is greater than 1.

A similar situation is found in ^{18}O where strong E2 transitions are seen between the second 0^+ and first 2^+ state and between the first 2^+ and ground state. If these states are described by the configuration $[(sd)^4p^{-2}]$, i.e. promoting two nucleons from the ^{16}O core, then reasonable agreement with experiment is found using an effective neutron charge of $0.5e$ and an effective proton charge of $1.5e$.

In terms of the model of an effective charge to account for the enhancement of transition rates the electric multipole single-particle operator is proportional to $\beta\mathbf{O}$ for neutrons and to $(1+\beta)\mathbf{O}$ for protons where β is the effective charge (in units of e) and $\mathbf{O}$ is the electromagnetic transition operator for a single particle of charge e. The total transition operator $\mathbf{O}_{\mathrm{T}}$ is given by:

$$\begin{aligned}\mathbf{O}_{\mathrm{T}} &= \sum_{i}^{N} \beta\mathbf{O}_i + \sum_{j}^{Z} (1+\beta)\mathbf{O}_j \\ &= \sum_{k}^{A} [\tfrac{1}{2}(1+2t_{z_k})\beta\mathbf{O}_k + \tfrac{1}{2}(1-2t_{z_k})(1+\beta)\mathbf{O}_k] \\ &= \sum_{k}^{A} \tfrac{1}{2}(1+2\beta)\mathbf{O}_k - \sum_{k}^{A} \mathbf{O}_k t_{z_k}\end{aligned}$$

So for $\Delta T=1$ transitions $\mathbf{O}_{\mathrm{T}} = -\sum_k^A \mathbf{O}_k t_{z_k}$, which is the same as if $\beta=0$, i.e. $\Delta T=1$ EL transitions are not expected to be enhanced.

4.3 Measurement of lifetimes

For relatively long lifetimes down to $\tau \sim 10^{-11}$ s the technique of delayed coincidence can be used. In this method a signal is generated by a particle or photon, which is emitted promptly (in comparison with τ), and another signal by the photons from the state of interest, which will depend on time as $\exp(-t/\tau)$. By looking at the number of coincidences within a small interval as a function of the delay time of the prompt signal the lifetime can be inferred.

Shorter lifetimes down to $\tau \sim 10^{-14}$ s can be measured using the Doppler effect. In a reaction nuclei are often produced moving at a high velocity of the order of 1% of the velocity of light. For a thin enough target these nuclei will escape with little energy loss and travel in vacuum. If they are stopped after a known time $t \approx \tau$ then the number of nuclei decaying in flight compared to the number decaying at rest will determine the lifetime of the excited state. The Doppler effect can be used to differentiate between those decaying in flight and those at rest, as illustrated in figure 4.9. This technique, called the recoil distance technique, can be used to measure lifetimes down to $\sim 10^{-12}$ s.

A similar technique which uses the time of slowing down in a medium to provide the clock can be used for lifetimes down to $\sim 10^{-14}$ s. In this method, called the Doppler shift attenuation method (DSAM), the magnitude of the Doppler shift measures the velocity of the nucleus when it decays. The time taken for the nucleus to slow down to a particular velocity is calculated using measurements of the stopping power, which is the loss

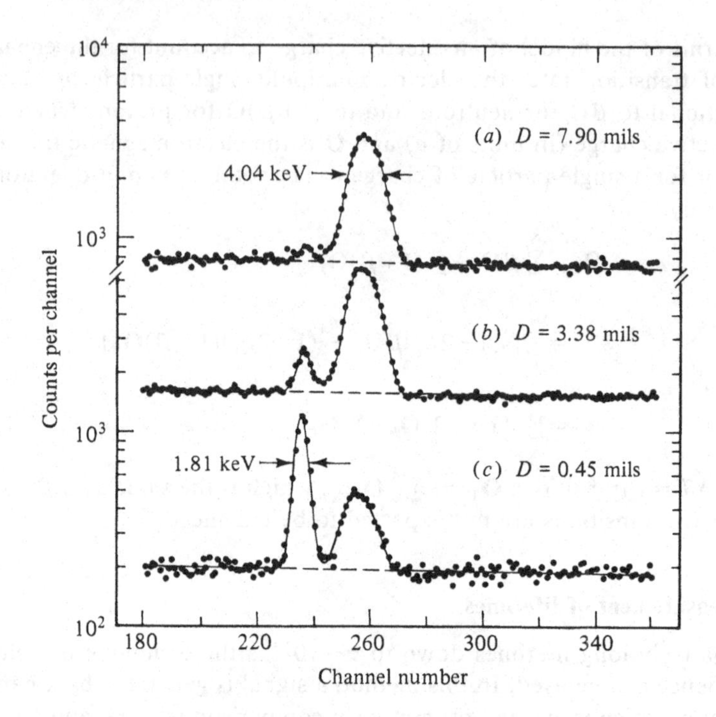

4.9. The stopped and Doppler-shifted peaks from the ^{22}Na 891 to 0 keV transition observed with a Ge(Li) detector at 0° to the beam in the ^{19}F(α, n)^{22}Na reaction at $E_\alpha = 5.5$ MeV (1 mil = 25.4 μm). D is the distance of the surface stopping the recoils (called the plunger) from the target foil. (From Jones, K. W., Schwartzschild, A. Z. and Warburton, E. K., *Phys. Rev.* **178** (1969) 1773.)

of energy per unit thickness, for the medium in which the nucleus is moving. An example of a spectrum obtained is shown in figure 4.10.

Other methods which can be used for even shorter lifetimes are Coulomb excitation, inelastic electron scattering and resonance reactions, all of which measure the width of the state through its effect on the cross-section for a reaction. This technique is also used in resonance absorption studies.

4.3.1 Resonance absorption

For a stationary free nucleus the energy E_γ of a γ-ray emitted in the decay of a state x at an excitation energy E to the ground state is given to a good approximation by:

$$E_\gamma = E(1 - E/2Mc^2) = E - R \tag{4.6}$$

where M is the mass of the nucleus and the reduction in energy from E is due to the recoil energy R of the nucleus. If the atom containing the nucleus

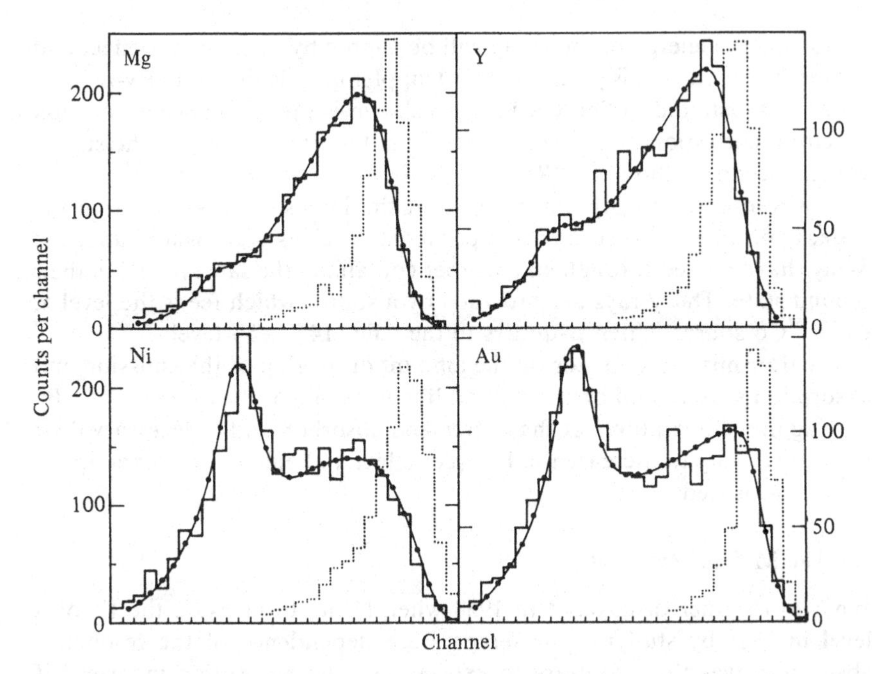

4.10. A comparison of computed (solid curve) and observed line shapes for the gamma-ray from the 2.24 to 0 MeV transition in ^{30}Si excited in the reaction ^{27}Al $(\alpha, p)^{30}Si$. Bombarding energies were in the range 4.5 to 5.0 MeV and results are shown for four different backings. The dotted histogram is the line shape observed with the target unbacked. (From Currie, W. M., *Nucl. Inst. and Methods* **73** (1969) 173.) ($\tau = 360$ fs).

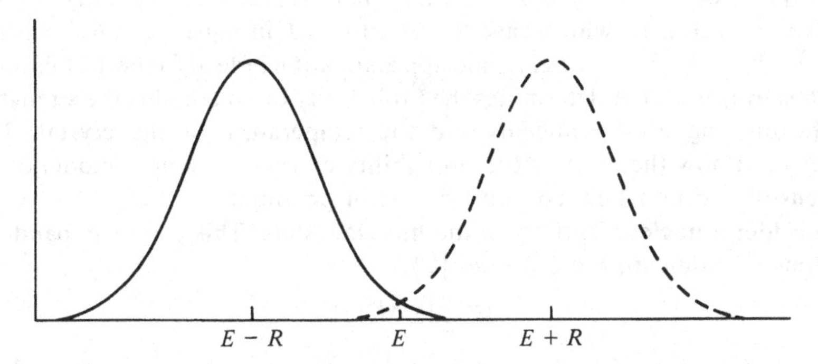

4.11. The effect of recoil and Doppler broadening on the γ-ray emission spectrum (solid line) and absorption cross-section (dotted line) for transitions between a state at an excitation energy E and the ground state.

is moving, the energy of the γ-ray will be altered by the Doppler effect and for thermal motion this gives rise to a Doppler broadening of the γ-ray line shape as illustrated in figure 4.11. Also shown is the resonance absorption spectrum corresponding to transitions from the ground state to the state x which is centred about $E+R$.

In a resonance absorption experiment the intensity of γ-rays from the γ-decay of an excited state x in a particular nucleus is measured after the γ-rays have passed through an absorber containing the same nuclei in their ground state. The γ-rays are provided by a source which feeds the level x, e.g. a ^{57}Co source which β-decays to the ^{57}Fe (14.4 keV) level.

The transmission depends on the amount of overlap of the emission and absorption spectra and on the natural line width Γ_γ of the excited state. By varying the temperatures of the source and absorber, and if necessary their relative velocity to increase the Doppler effect and enhance the overlap, Γ_γ can be measured.

4.4 The Mossbauer effect

What Mossbauer discovered in 1958 when trying to measure the Γ_γ of a level in ^{191}Ir by studying the temperature dependence of the resonance absorption was that, contrary to expectation, the absorption increased if the temperatures of both the source and absorber were lowered to liquid nitrogen temperatures. The reason is that although the Doppler broadening is reduced at these temperatures, the effect of this is more than offset by the absorption arising from the recoilless emission and subsequent recoilless absorption of γ-rays from the source caused by the nuclei being bound and not free.

Recoilless emission and absorption can take place if the nucleus is bound in a crystal for then the possibility exists that the recoil can be taken up by the whole lattice, in which case the effective M in equation 4.6 is vastly increased and $E_\gamma = E$ to a very good approximation indeed for both emission and absorption. What determines the probability of no recoil is the strength of the binding of the nucleus and the temperature of the crystal. To understand how these affect the probability consider a simple model of a nucleus bound in a 1-D harmonic oscillator potential.

Consider a nucleus initially in the nth eigenstate. This can be expanded in terms of momentum eigenstates $|k'\rangle$:

$$|n\rangle = \sum_{k'} |k'\rangle\langle k'|n\rangle$$

After the emission of a photon with momentum k each momentum component $|k'\rangle$ is shifted by $-k$ by conservation of momentum so:

$$|k'\rangle \Rightarrow |k'-k\rangle = \mathrm{e}^{-ikx}|k'\rangle$$

Therefore the wavefunction $|f\rangle$ describing the nucleus after the γ-decay is given by:

$$|f\rangle = \sum_{k'} e^{-ikx}|k'\rangle\langle k'|n\rangle$$
$$= e^{-ikx}|n\rangle$$

This is not an eigenstate of the 1-D harmonic oscillator and the probability of recoilless emission is given by:

$$P = |\langle n|e^{-ikx}|n\rangle|^2$$
$$= \left|\int \psi^*(x)\, e^{-ikx}\psi(x)\, dx\right|^2$$
$$= \left|\int \rho(x)\, e^{-ikx}\, dx\right|^2 \qquad (4.7)$$

i.e. P equals the square of the Fourier transform of the probability density $\rho(x)$ for finding the nucleus at the point x. If k is such that $kx \ll 1$ for all values of x that $\rho(x)$ is appreciable then P will be close to 1. For a nucleus in the ground state $\rho(x)$ has the form of a gaussian, i.e. $\rho_0(x) \propto \exp(-\alpha x^2)$, and in an actual crystal a gaussian approximation for $\rho(x)$ is generally a very good approximation for excited states in which case the probability P is given by:

$$P = \exp(-k^2\langle x^2\rangle)$$

where $\langle x^2\rangle$ is the mean square displacement of the nuclei. The probability will therefore increase with decreasing temperature because $\langle x^2\rangle$ will become smaller and for nuclei in the ground state:

$$P = \exp(-k^2\langle x^2\rangle_0) = \exp(-R/\hbar\omega)$$

where R is the recoil energy and $\hbar\omega$ is the energy spacing of the harmonic oscillator potential well. The frequency $\omega = \sqrt{k/m}$ and for an appreciable probability $\hbar\omega \gtrsim R$, i.e. the binding must be sufficiently strong. In the Debye model of a crystal the strength of the binding is reflected in the Debye temperature θ_D and in this model an appreciable probability P of recoilless emission requires $k\theta_D \gtrsim R$. If P is very small then the nucleus behaves as a free nucleus, i.e. the γ-ray line is Doppler broadened as shown in figure 4.11.

In Mossbauer's original experiment the probability of recoilless emission, and hence recoilless absorption, is very small at room temperature; however, it is appreciable at liquid nitrogen temperature. It is therefore only when both source and absorber are lowered in temperature that an increased absorption is expected and this is what Mossbauer discovered.

Similar phenomena were known in other branches of physics but the similarity was not appreciated. In X-ray diffraction there are scattered X-rays with the same wavelength as the incident X-rays but going in a different

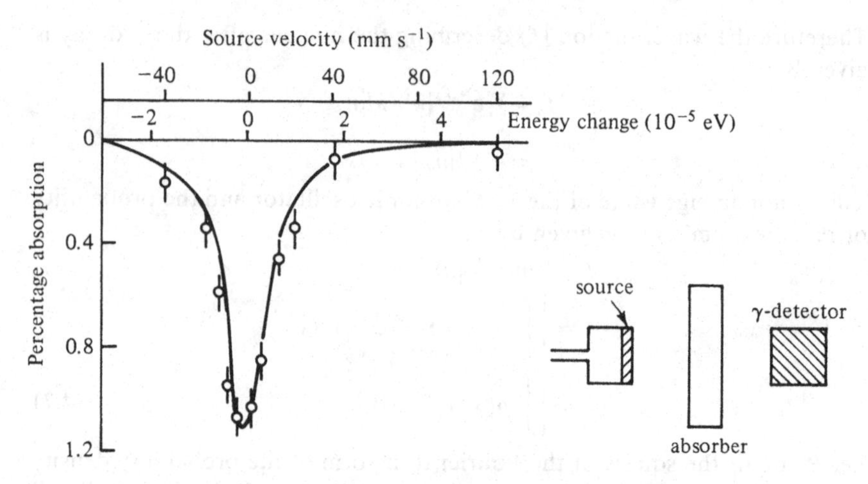

4.12. Determination of the natural line width of the 129 keV excited state in ^{191}Ir in a resonance absorption experiment, under conditions of significant recoilless emission and absorption. (From Mossbauer, R. L., *Zeits. für Physik* **151** (1958) 124.)

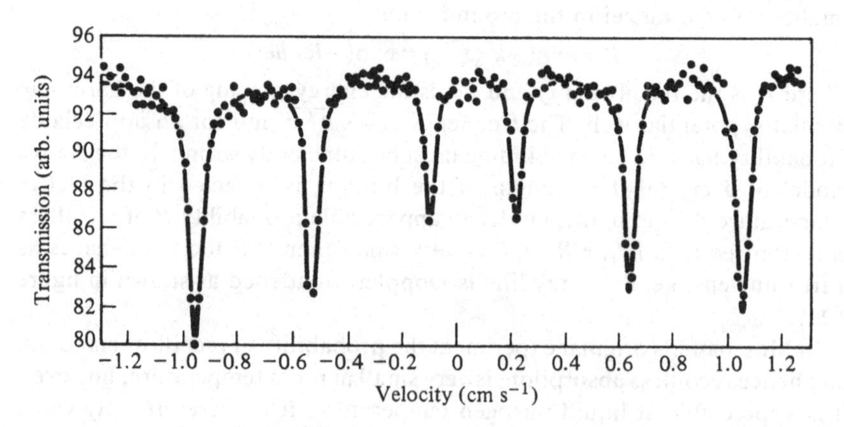

4.13. The effect of the hyperfine interactions within an iron absorber on the transmission of 14.4 keV γ-rays from the decay of the $J=\frac{3}{2}$ 14.4 keV excited to $J=\frac{1}{2}$ ground state of ^{57}Fe. (The difference in environment of emitting and absorbing nuclei accounts for the transmission pattern not being centred on zero speed.) (From Wertheim, G. K., *Mossbauer Effect.* Academic Press (1964).)

direction; there is therefore a momentum transfer with no loss of energy and the same thing occurs in neutron scattering in crystals. The expression for the probability that this scattering occurs is very similar to that for P above (equation 4.6) and in X-ray scattering $\rho(x)$ is the electron density and the probability P is known as the Debye–Waller factor.

Because the resonant absorption at liquid nitrogen temperature is due to the complete overlap of the recoilless emission and absorption natural line shapes, a very small relative velocity of source and absorber reduces the overlap because of the Doppler shift. By plotting the transmission as a function of relative velocity the natural line width of the line can be measured and Mossbauer's results for the $^{191}Ir^*$ (129 keV) level are shown in figure 4.12. As can be seen, the Mossbauer effect can also be used to measure very small ($\sim 10^{-5}$ eV in the above example) energy shifts such as arise through hyperfine interactions. An example of this is shown in figure 4.13.

4.5 Internal conversion

Internal conversion is the electromagnetic process in which an excited nuclear state de-excites with the ejection of an atomic electron. The interaction is the electromagnetic interaction between the nucleus and the bound electron. For an electric transition the magnetic interaction between the electron and nucleon currents can be neglected and just the Coulomb interaction considered, in which case the matrix element for a 1s electron is:

$$M_{if} \propto \sum_i \iint \exp(-i\mathbf{k}_e \cdot \mathbf{r}_e)\psi_f^* \frac{e^2}{|\mathbf{r}_e - \mathbf{r}_i|} \psi_i \frac{1}{a^{3/2}} \exp\left(\frac{-r_e}{a}\right) d^3r_i \, d^3r_e$$

where the sum is over the protons within the nucleus, $\mathbf{k}_e$ is the final electron wavevector, ψ_i and ψ_f the initial and final nuclear states, and a is the Bohr radius of the 1s electron. The term involving $1/|\mathbf{r}_e - \mathbf{r}_i|$ can be expanded for $r_e > r_i$ as:

$$\frac{1}{|\mathbf{r}_e - \mathbf{r}_i|} = \frac{1}{r_e} \sum_i \left(\frac{r_i}{r_e}\right)^l P_l(\cos\theta)$$

where θ is the angle between $\mathbf{r}_e$ and $\mathbf{r}_i$. The matrix element therefore becomes:

$$M_{if} \propto \frac{ek_e^{l-2}}{a^{3/2}} \int \exp(-i\mathbf{k}_e \cdot \mathbf{r}_e) \left(\frac{1}{k_e r_e}\right)^{l+1} \exp\left(\frac{-r_e}{a}\right) d^3(k_e r_e) \cdot [B_{if}(\mathrm{E}L)]^{1/2}$$

where the integral over r_i is proportional to the square root of the reduced transition probability B_{if} (EL) mentioned above (equation 4.3) and the integral over the electron coordinate r_e has been written in dimensionless form. The internal conversion rate R is proportional to $|M_{if}|^2$ times the density of final states, which for non-relativistic electrons is proportional to k_e. The rate R therefore depends on:

$$R \propto e^2 Z^3 k_e^{2l-3} B_{if}(EL)$$

as the Bohr radius $a \propto 1/Z$.

The ratio of internal conversion to γ-decay, the conversion coefficient, is therefore independent of the nuclear wavefunctions and can be used to determine the multipolarity of a transition. Since the γ-decay rate is proportional to $k_\gamma^{2l+1} B_{if}(EL)$, the K-shell conversion coefficient, α_{K_l}, for electric transitions is given by:

$$\alpha_{K_l} \propto Z^3 k_e^{2l-3} / k_\gamma^{2l+1} \propto Z^3 k_\gamma^{-l-5/2}$$

since $k_e \propto k_\gamma^{1/2}$ for non-relativistic electrons. The conversion coefficients for magnetic transitions are different but a general conclusion is that internal conversion is most likely for low-energy low-multipolarity transitions in heavy nuclei.

4.5.1 $0 \to 0$ transitions

While single-photon decay is strictly forbidden for $0 \to 0$ transitions, two-photon decay is possible but very unlikely and the most common form of decay for such transitions is by internal conversion of an s-state electron. In the above treatment only the field in the region $r_e > r_i$ was considered. For $0 \to 0$ transitions this is not varying and there is no contribution to the matrix element for $r_e > r_i$ corresponding in the above expression to $B_{if}(E0)$ being zero as ψ_i and ψ_f are orthogonal. An s-state electron, however, has a finite probability for being within the nucleus and having $r_e < r_i$, in which case:

$$\frac{1}{|\mathbf{r}_e - \mathbf{r}_i|} = \frac{1}{r_i} \sum_l \left(\frac{r_e}{r_i}\right)^l P_l(\cos\theta)$$

which for $l = 0$ becomes $1/r_i$ so the matrix element is given by:

$$M_{if} \propto \sum_i \int_0^R \psi_f^* \psi_i \left(\int_0^{r_i} \phi_f^* \frac{1}{r_i} \phi_i \, d^3 r_e \right) d^3 r_i$$

where R is the radius of the nucleus, ϕ_i is the initial s-state electron wavefunction, and ϕ_f is the final electron wavefunction which is also an s-state by conservation of angular momentum. To a good approximation $\phi_f^*(r_e) = \phi_f^*(0)$ and $\phi_i(r_e) = \phi_i(0)$ so the integral I over $d^3 r_e$ becomes

$$I = \phi_f^*(0) \phi_i(0) \int_0^{r_i} \frac{1}{r_i} 4\pi r_e^2 \, dr_e$$

so the matrix element is given by:

$$M_{if} \propto \sum_i \int \psi_f^* r_i^2 \psi_i \, d^3 r_i$$

and is therefore non-zero.

Another way a $0 \to 0$ can occur is by the creation of an electron–positron pair if the transition energy exceeds $2m_ec^2 = 1.022$ MeV. As three particles are involved in the final state the density of final states increases more quickly with transition energy than for internal conversion, and for the decay of the 6.06 MeV first excited 0^+ state in ^{16}O the process of internal pair creation is dominant with a lifetime of 7×10^{-11} s.

4.6 Angular correlations

Although the multipolarities of some transitions can be determined by measuring their internal conversion coefficients the most general method is by studying the angular correlations of the γ-rays emitted in the decay of an excited state. As an example consider the decay sequence $A(0^+) \to B(1^-) \to C(0^+)$. The decay of the initial state A will be isotropic. Let the direction of the photon in a particular decay of $A \to B$ define the z-axis. If the photon was right-circularly polarised then by conservation of angular momentum the state B will be formed in the $m_J = -1$ substate; likewise if left-circularly polarised then $m_J = +1$. However B is not formed in the substate $m_J = 0$ so the substates of B are not equally populated and in the subsequent decay of $B(1^-)$ to $C(0^+)$ the angular distribution of the emitted γ-ray relative to the first γ-ray is not isotropic. The decays of the $m_J = \pm 1$ substates have the same angular distribution (see below) equal to $\frac{1}{2}(1 + \cos^2 \theta)$, where θ is the angle of the second γ-ray with respect to the first (the z-axis), so the angular distribution is $\propto (1 + \cos^2 \theta)$.

If the spin and parity of A and C but not B had been known then the observation of a $(1 + \cos^2 \theta)$ correlation would have implied that B had spin 1 and that both γ-rays had unit multipolarity, but it does not determine the parity of B as two M1 decays would have the same correlation. The polarisation of the two γ-rays however would be different and if measured can be used to infer the parity of B.

In general if all the substates of a state are equally populated the decay of the state will be isotropic as no direction is defined. Further, the angular distributions from the decay of the (Jm) and $(J-m)$ substates of a state of spin J are equal. This follows from parity conservation. The wavefunction ϕ_γ describing the photon must have good parity, so $P\phi_\gamma(\mathbf{r}) = \pm\phi_\gamma(\mathbf{r})$, so $|\phi_\gamma(\mathbf{r})|^2 = |\phi_\gamma(-\mathbf{r})|^2$, i.e. the intensity of γ-rays in the direction $\mathbf{r}$ is the same as in the direction $-\mathbf{r}$. The angular distribution is therefore unaffected by changing $z \to -z$, which is equivalent to changing $m \to -m$, so the distributions for (Jm) and $(J-m)$ are the same. For a state of spin $\frac{1}{2}$, since $m = +\frac{1}{2}$ and $m = -\frac{1}{2}$ have the same distributions and equal amounts of both are isotropic then each substate has an isotropic distribution.

For a general cascade the calculation of the correlations is complicated and mixing of multipolarities, such as M1/E2, must be allowed for. Combining the results of correlation measurements with other information such as lifetimes is a powerful way of determining the spins and parities of nuclear levels.

4.7 Summary

The multipole expansion of the transition matrix element is shown to be important as only one or two terms contribute significantly in a gamma decay. The Weisskopf estimates of these based on a simple single-particle model provide a useful measure of transition strength. Isospin selection rules are derived and found helpful when discussing decay schemes.

The existence of enhanced transition rates corresponding to the collective motion of many nucleons is understood in the shell model as arising from the coherent superposition of many single-particle terms in the transition matrix elements. Examples are seen in the giant dipole resonances and the enhanced E2 transition rates down a rotational band. The use of effective charges, to account for core-polarisation, in a description of enhanced transitions is described.

Techniques for the measurement of lifetimes are briefly mentioned and the importance of the Mossbauer effect is explained. Internal conversion and $0 \rightarrow 0$ transitions are also described and the study of angular correlations is shown to provide information on the spins and parities of nuclear levels. Finally, from a discussion of several examples, it is seen that a considerable amount of information can be gained about the wavefunctions of nuclear states from the study of γ-decay.

4.8 Questions

1. Show that the matrix element $(\mathbf{A}_k)_{n,n+1}$, which represents the transition from a state with n photons to a state with $n+1$ photons, equals $[(n+1)\mu_0 \hbar c^2/2\omega_k]^{1/2}$.
2.* The energies (in MeV), spins and parities of the ground and first four excited states of ${}^{207}_{82}\mathrm{Pb}$ are as follows: $0.0\,\frac{1}{2}^-$; $0.57\,\frac{5}{2}^-$; $0.90\,\frac{3}{2}^-$; $1.63\,\frac{13}{2}^+$; $2.33\,\frac{7}{2}^-$. Discuss the shell model description of these states. The 0.57 MeV state has a lifetime of 10^{-10} s, while the lifetime of the 1.63 MeV state is 0.8 s. Account for this difference and predict the multipolarity of the γ-decay of the 1.63 MeV state. Estimate the lifetime of the 2.33 MeV state.
3.* Explain why radiative transitions of multipole order higher than electric dipole are relatively weak in atoms but often strong in nuclei.

4.* It is believed that, to a good approximation, parity is conserved in nuclear γ-transitions. How then is it possible for the nuclear parity sometimes to change in such a transition and sometimes not? Suggest an experimental test of parity conservation in nuclear γ-decay. Give reasons to expect a definite but small violation of parity conservation in nuclear γ-transitions, and suggest an order of magnitude for the effect.

5. Estimate the γ-decay lifetime of the $\frac{1}{2}^-$ level at 315 keV excitation in ${}^{117}_{49}\mathrm{In}$ shown in figure 4.7.

6. Deduce the expression (equation 4.6) for the energy E_γ of a γ-ray emitted in the decay of a state at an excitation energy E to the ground state.

7. In a Mossbauer experiment the source, which emits photons with frequency ν, is raised to a height l above the absorber and detector. By equating the photon's inertial mass $h\nu/c^2$ with its gravitational mass (the equivalence principle), show that the frequency ν' of the photon at the absorber is given by $\nu' = \nu(1+gl/c^2)$ where g is the acceleration due to gravity. For $l = 30$ m calculate the fractional change in frequency and the source velocity required to give resonance.

8. In the γ-decay $X(1^-) \rightarrow Y(0^+)$, the angular distributions from the $m_J = \pm 1$ substates of X are both equal to $\frac{1}{2}(1+\cos^2\theta)$. What would the angular distribution from the $m_J = 0$ substate be equal to?

9.* The particles $\Sigma^0(1192\ \mathrm{MeV}/c^2)$ and $\Lambda^0(1116\ \mathrm{MeV}/c^2)$, both have $J^\pi = \frac{1}{2}^+$ and the particle $\Delta^0(1236\ \mathrm{MeV}/c^2)$ has $J^\pi = \frac{3}{2}^+$. The main decay mode of Σ^0 is $\Sigma^0 \rightarrow \Lambda^0 + \gamma$. Give a reasoned estimate of the order of magnitude of the mean lifetime of Σ^0, given that 0.6% of all Δ^0 particles decay by $\Delta^0 \rightarrow \mathrm{n} + \gamma$ and the mean lifetime of Δ^0 is 0.6×10^{-23} s.

10. A state in a nucleus with $A = 160$ decays by a 1 MeV transition which is either of M1 or E2 multipolarity. The lifetime of the state is measured to be 10^{-12} s. What multipole is the transition most likely to be?

11. In a recoil distance measurement γ-rays are observed from the decay of excited states in ${}^{16}\mathrm{N}$ produced in the reaction ${}^{2}\mathrm{H}({}^{18}\mathrm{O}, \alpha){}^{16}\mathrm{N}^*$ at $E({}^{18}\mathrm{O}) = 37$ MeV. At this energy a significant proportion of the ${}^{16}\mathrm{N}^*$ nuclei leave the target foil as hydrogen-like ions, i.e. as 7^+ ions. The intensity of 276 keV γ-rays from the decay of stopped ${}^{16}\mathrm{N}^*(1^-)$ nuclei shows an oscillation about an exponential decay, as the target to stopper foil distance is varied, with a period of 4.2 ps. Explain this oscillation and deduce the magnetic moment of the 1^- excited state in ${}^{16}\mathrm{N}$.

5 Alpha decay, fission and thermonuclear fusion

Alpha decay, fission and thermonuclear fusion all illustrate in different ways the importance of the phenomenon of quantum-mechanical tunnelling. In α-decay, an α-particle is contained by a barrier, called the Coulomb barrier, formed by the strong attractive potential and the electrostatic repulsive $1/r$ potential between the α-particle and the rest of the nucleus. The strong dependence of the probability of tunnelling through this barrier on the α-particle's energy is shown to account for the observed systematics of α-decay. In fission there is also a barrier arising from the nuclear and Coulomb forces and the observed spontaneous fission lifetimes are also explained by tunnelling.

The shape of the fission barrier is shown to depend significantly on shell corrections to the liquid drop model. These corrections explain the occurrence of fission isomers. The possibility of a chain reaction following neutron-induced fission and its application in nuclear reactors is then described. Finally the process of thermonuclear fusion of hydrogen within the sun and the possibility of making fusion reactors are discussed. It is shown how the luminosity of the sun can be understood as arising from the weak process: $p+p \rightarrow d+e^{+}+\nu$, which is highly inhibited within the sun by the Coulomb barrier between the protons.

5.1. Alpha decay

From the semi-empirical mass formula most nuclei with $A > 150$ are unstable to decay by α-emission. This is not readily observed until nuclei with $A > 200$ because of the Coulomb barrier that the α-particles must penetrate by quantum-mechanical tunnelling. The probability of decay depends both on the probability that the α-particle is preformed within the initial nucleus and on the probability of tunnelling. To calculate the probability of tunnelling first consider a simple 1-D barrier defined by:

$$V = -V_0, x \leq 0; \qquad V = V_B, 0 < x \leq a; \qquad V = 0, x > a$$

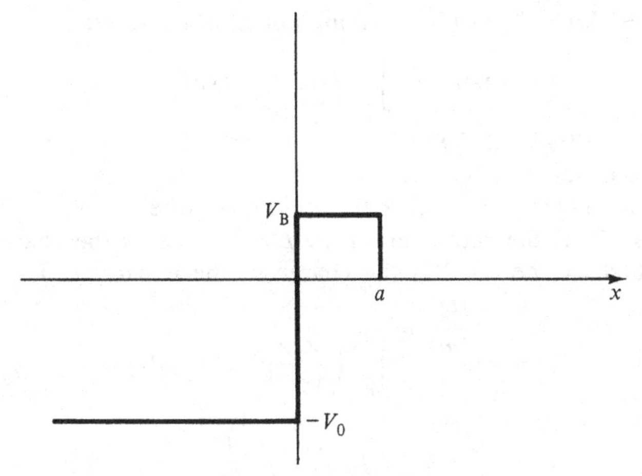

5.1. A simple one-dimensional barrier.

see figure 5.1. The wavefunctions describing the particle in the three regions are:

$$\begin{aligned}\psi &= \exp(\mathrm{i}k_0x) + R\exp(-\mathrm{i}k_0x) \qquad x \le 0\\ &= A\exp(-Kx) + B\exp(+Kx) \qquad 0 < x \le a\\ &= C\exp(\mathrm{i}kx) \qquad x > a\end{aligned}$$

If $Ka \gg 1$ where $\hbar^2K^2/2m = V_0 - E$ then $B \ll A$ and matching the boundary conditions (ψ and $\partial\psi/\partial x$) at $x=0$ gives $A \approx 2k_0/(k_0+\mathrm{i}K)$. The boundary conditions at $x=a$ then give:

$$C = \exp(-\mathrm{i}ka)\frac{4k_0K}{(K-\mathrm{i}k)(k_0+\mathrm{i}K)}\exp(-Ka)$$

The transmission coefficient T is the ratio of transmitted to incident flux, $|C|^2k/k_0$, so:

$$T = \frac{16kk_0K^2}{(k^2+K^2)(k_0^2+K^2)}\cdot\exp(-2Ka)$$

For a 3-D potential the transmission coefficient for $l=0$ particles through the potential barrier is the same, as the Schrödinger equation in 3-D is:

$$-\frac{\hbar^2}{2m}\frac{\mathrm{d}^2u}{\mathrm{d}r^2} + [V(r) + l(l+1)\hbar^2/2mr^2]u = Eu$$

where $\psi = r^{-1}u(r)Y_{lm}(\theta, \phi)$. For an $l=0$ spherical wave $u = \exp(\mathrm{i}kr)$ so the equation satisfied by u is the same as in the 1-D potential barrier problem. For $l \neq 0$ the effective barrier is increased by the centrifugal barrier $l(l+1)$ $\hbar^2/2mr^2$. The transmission coefficient is dominated by the exponential factor $\exp(-2Ka)$, which for a varying potential becomes:

$$T \approx \exp\left[-2\int K(r)\mathrm{d}r\right] \equiv \exp(-2\gamma) \tag{5.1}$$

where $K^2(r) = (2m/\hbar^2)[V(r) + l(l+1)\hbar^2/2mr^2 - E]$. This expression for T is called the Gamow factor.

In α-decay $V(r) \approx Zze^2/4\pi\varepsilon_0 r$ for $R \leq r < R_c$ where R is the radius of the nucleus, R_c is the radius at which $E = V(r)$, Z is the charge of the daughter nucleus and $z = 2$ is the charge of the α-particle. For an $l = 0$ α-particle:

$$\gamma = \frac{(2m)^{1/2}}{\hbar}\int_R^{R_c}\left(\frac{Zze^2}{4\pi\varepsilon_0 r} - E\right)^{1/2}\mathrm{d}r$$

$$= \frac{(2mE)^{1/2}}{\hbar}\int_R^{R_c}\left(\frac{R_c}{r} - 1\right)^{1/2}\mathrm{d}r$$

since $R_c = zZe^2/4\pi\varepsilon_0 E$. If $E = \frac{1}{2}mv^2$ is much less than the maximum barrier height $(zZe^2/4\pi\varepsilon_0 R)$ then $R_c \gg R$ and:

$$\gamma \approx zZe^2/4\varepsilon_0\hbar v \tag{5.2}$$

The rate of α-decay R will be the product of the number of times per second f an α-particle strikes the barrier times the transmission coefficient. The factor f is the product of the probability p of finding an α-particle moving with velocity v_0 ($mv_0 = \hbar k_0$) in the nucleus and the frequency of its striking the barrier $v_0/2R$. This latter factor is $\sim 10^{22}\ \mathrm{s}^{-1}$ but p is rather uncertain. However, as a first approximation f can be taken as the same for even–even nuclei of approximately the same mass, which assumes that p is roughly constant as v_0 is principally determined by the depth of the potential well V_0. The α-decay rate is therefore:

$$R = fT = f\exp(-2\gamma)$$

$$\therefore \quad \log_{10} R = a - 1.71ZE^{-1/2} \tag{5.3}$$

where E is in MeV and a is a constant. A plot of the log of the lifetime of a number of nuclei versus $ZE^{-1/2}$ is shown in figure 5.2 and this simple relation is borne out well. Such a plot is called a Geiger–Nuttall type of plot.

A rough estimate of the dependence of lifetime on energy from figure 5.2 is a factor of 10^5 per MeV change in α-energy. Besides this very strong energy dependence, there is a slight angular momentum dependence because of the additional centrifugal barrier. Calculations of this give approximate factors of 1, 0.37, 0.04, 1×10^{-3} for $l = 0, 2, 4, 6$ respectively. The ratio of α-decay rates for decays to members of a rotational band, such as in $^{238}_{94}\mathrm{Pu} \rightarrow {}^{234}_{92}\mathrm{U} + \alpha$, are to a large extent accounted for by the variation in the transmission coefficient indicating, as expected, little variation in preformation probability as the intrinsic wavefunctions are the same for members

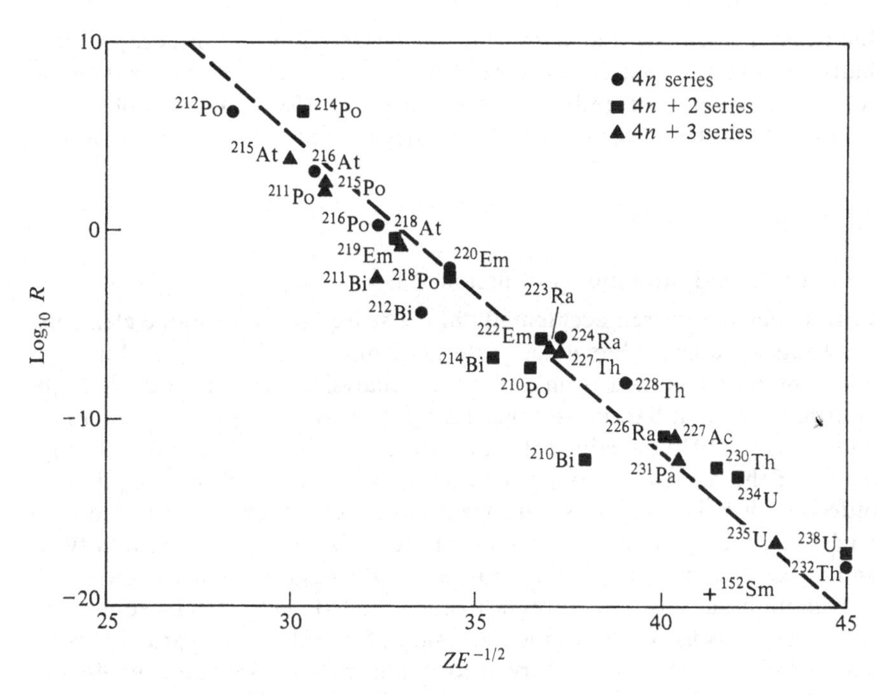

5.2. Plot of the logarithm of the decay rate R versus $ZE^{-1/2}$ for a number of α-decaying nuclei. The dashed line has the slope of -1.71 predicted by equation 5.3. (From Leighton, R. B. *Principles of Modern Physics*, McGraw-Hill (1959).)

of a rotational band. As the barrier height $B = Zze^2/4\pi\varepsilon_0 R$ then $\gamma \propto B/v$ (equation 5.2), so for decays with similar energy release, an exponential dependence of the lifetime on barrier height is expected. This is found in fission lifetimes (see below).

In the radioactive decay of a heavy element, a sequence of α- and β-decays is generally observed. A β-decay can result in an excited state of a nucleus being formed which then decays by α- as well as by γ-decay. These beta-delayed α-emitters give rise to high energy α-particles (long-range α's), e.g. ${}^{212}_{84}\mathrm{Po}^*$ (1.8 MeV) which gives rise to 10.55 MeV α-particles. Likewise the β-decay of a proton-rich nucleus can give rise to beta-delayed proton emission and of a neutron-rich nucleus, such as is found in fission, to beta-delayed neutron emission.

Proton radioactivity is possible and was first observed in 1970. An excited state of ${}^{53}\mathrm{Co}$ with a probable $J^\pi = \frac{19}{2}^-$, produced in the heavy-ion reaction ${}^{40}\mathrm{Ca}({}^{16}\mathrm{O}, 2\mathrm{np}){}^{53}\mathrm{Co}$, was found to proton-decay with a half-life of 245 ms. In 1982 proton radioactivity was first seen from the ground state of a nucleus:

the heavy proton-rich nucleus $^{151}_{71}$Lu was observed to proton-decay with a half-life of 85 ms. Radioactivity with the emission of nuclei larger than an α-particle has also recently been discovered. In 1984 ^{223}Ra was observed to decay with the emission of ^{14}C. This decay mode is a form of fission.

5.2 Fission

5.2.1 The liquid drop model of fission

Fission was discovered accidentally in the search for transuranic elements by neutron bombardment of natural uranium. In 1939, Hahn and Strassemann concluded from a chemical analysis of the products of such an irradiation that barium isotopes ($A \approx 140$) had been produced. Meitner and Frisch in 1939 explained this phenomenon in terms of the liquid drop model of the nucleus. They pointed out that heavy nuclei only required a little energy to become very deformed and unstable, breaking up into two nuclei of roughly equal size, similar to the division of a droplet into two, and suggested this energy was supplied in the capture of the neutron by the uranium nucleus. This form of fission is called neutron-induced fission.

Spontaneous fission, which is the fission of a nucleus in its ground state, also occurs and was first observed for $^{240}_{94}$Pu in 1940. Spontaneous fission into two equal mass fragments is energetically possible for nuclei with $A \gtrsim 90$ (for $A/Z = 2.3$) as can be shown from the semi-empirical mass formula:

$$M(A, Z) = Zm_{\mathrm{p}} + (A - Z)m_{\mathrm{n}} - a_{\mathrm{v}}A + a_{\mathrm{s}}A^{2/3} + a_{\mathrm{c}}Z^2A^{-1/3} + a_{\mathrm{a}}(A - 2Z)^2A^{-1} + \delta \qquad (5.4)$$

the terms of which are discussed in chapter 2, p. 18.

The reason that spontaneous fission is not observed for nuclei with $A \lesssim 240$ is that such nuclei are stable with respect to a small deformation from their equilibrium shape. This is because such a change causes an increase in surface area and a consequent loss in binding energy which is not offset by the decrease in Coulomb energy arising from the increased separation of charge. The result is a fission barrier of height E_{f}, of a form indicated in figure 5.3.

5.2.2 The fission barrier

The energy release in fission, E_{R} in figure 5.3, can be estimated from equation 5.4 and for nuclei with $A \approx 240$ it is around 200 MeV. After fission the two fission fragments are neutron-rich compared with stable nuclei of their mass with the result that a small number (~ 2.5) of neutrons are emitted immedi-

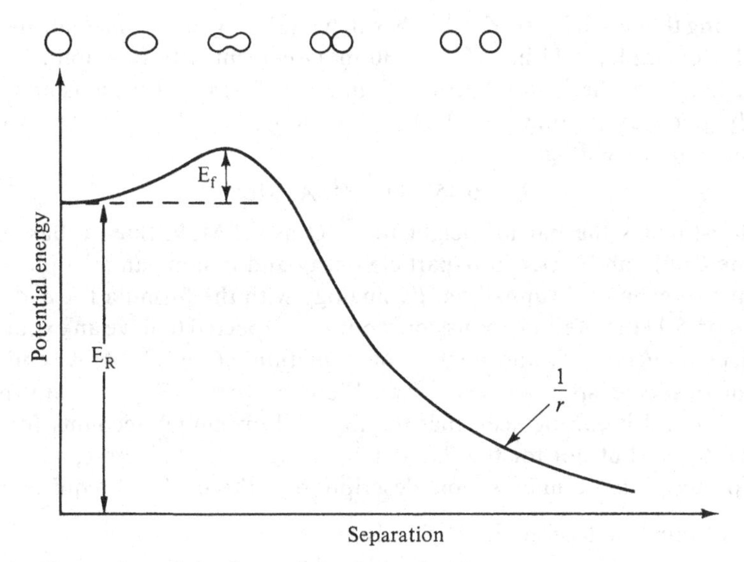

5.3. An illustration of the form of the fission barrier.

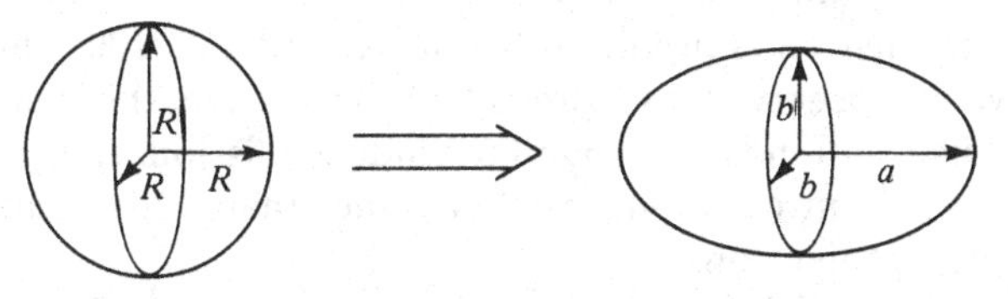

5.4. Deformation of sphere to ellipsoid of same volume: $a = R(1+\varepsilon)$; $b = R/(1+\varepsilon)^{1/2}$. For this prolate deformation ε is positive.

ately. These prompt neutrons can induce further fission and give rise to a chain reaction. This is discussed in the section below on neutron-induced fission.

The mass at which nuclei become unstable with respect to fission, i.e. when $E_f \approx 0$, can be estimated by considering the change in binding energy when a spherical nucleus undergoes a small deformation to an ellipsoid of the same volume, as illustrated in figure 5.4. To second order in ε, the surface area of the nucleus increases by $(1+2\varepsilon^2/5)$ while the Coulomb energy, assuming a uniform charge distribution, decreases by $(1-\varepsilon^2/5)$. Substituting these factors into the semi-empirical mass formula, equation 5.4, gives the change in binding energy $-\Delta E_B$ as:

$$\Delta E_B = 0.14A^{2/3}(49 - Z^2/A)\varepsilon^2 \text{ MeV} \qquad (5.5)$$

predicting that nuclei with $Z^2/A \leqslant 49$ will be stable against immediate fission and, for example, ^{238}U has $Z^2/A = 36$ in agreement with this limit.

The height of the fission barrier (E_f in figure 5.3) for $A \approx 240$ is approximately given by putting $|\varepsilon| = 0.29$ in equation 5.5 and noting $A^{2/3}$ varies slowly with A yielding:

$$E_f \approx 0.455(49 - Z^2/A)\ \text{MeV} \tag{5.6}$$

which estimates the barrier height for ^{238}U as 6.1 MeV. Such a barrier is like the Coulomb barrier in α-particle decay and fission can take place by quantum-mechanical tunnelling. By analogy with the formula for α-decay (equation 5.1) the lifetime for fission would be expected to have an exponential dependence on E_f and hence from equation 5.6 on Z^2/A. A semi-log plot of observed spontaneous fission lifetimes versus Z^2/A is shown in figure 5.5 and it can be seen that the liquid drop model accounts for the general trend, but not for the fine details of the data. For these, as would be expected, a more microscopic description of the nuclei is required.

5.2.3 Asymmetric fission and fission isomers

In fission, whether spontaneous or induced, the nucleus divides into two nuclei of roughly equal size. The liquid drop model can predict whether or not these nuclei should be exactly equal, i.e. symmetric fission. The result is that symmetric fission is predicted for nuclei with $A \gtrsim 200$. This is quite contrary to what is observed, as figure 5.6 illustrates, and this characteristic two-humped distribution is also found for induced fission. In figure 5.6 the curve for ^{252}Cf is centred about $A \approx 124$ rather than 126 because of the emission of prompt neutrons.

Another feature which the liquid drop model fails to describe is the occurrence of fission isomers. These were discovered in 1962 by Polikanov *et al.* and are isomeric states found in certain nuclei around $A \approx 240$ at 2-3 MeV excitation with lifetimes in the range 10^{-2} to 10^{-9} s which decay by fission rather than γ-decay. In the liquid drop model a 2-3 MeV excited state would have a fission lifetime in that range because of the reduced barrier height E_f, but such a state would be expected to γ-decay unless it were a very high spin state (cf. isomeric γ-decaying states, chapter 4, p. 126). However, measurements of the isomer to ground state ratio in reactions where different amounts of angular momentum were brought in indicated that the isomers did not have very high spins.

5.2.4 Shell corrections to the liquid drop model

The explanation of both the observed asymmetric fission and the occurrence of fission isomers lies in taking account of shell corrections to the liquid drop model. In a shell model description of a nucleus the nucleons fill the energy levels up to the Fermi level and the ground state energy of the

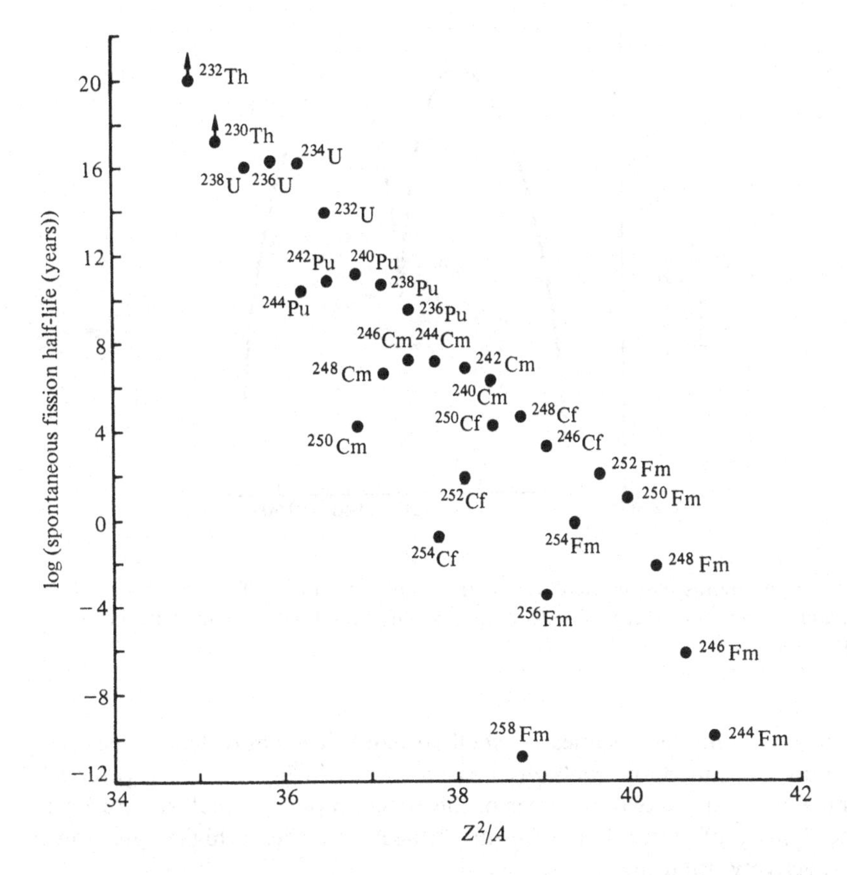

5.5. Plot of the logarithm of spontaneous fission lifetimes versus Z^2/A (data from Vandenbosch, R. and Huizenga, J. R., *Nuclear Fission.* Academic Press (1973)). (The significant decrease in lifetimes for isotopes with $N > 152$ is related to a subshell closure at $N = 152$.)

nucleus is determined by the sum of the energies of the occupied levels. As described in chapter 2, this procedure can be used to calculate the equilibrium deformation of a nucleus by using the Nilsson model to give the energy levels and then minimising the sum over occupied levels as a function of deformation. While this gives good agreement with observed ground state deformations (figure 2.17), for large deformations as occur in fission this technique fails. This is thought to be because the calculation is not a self-consistent one in that the matter distribution does not have to follow the potential distribution.

Strutinsky developed a technique in 1968 which avoids this problem by considering the difference between the liquid drop and shell model. As the

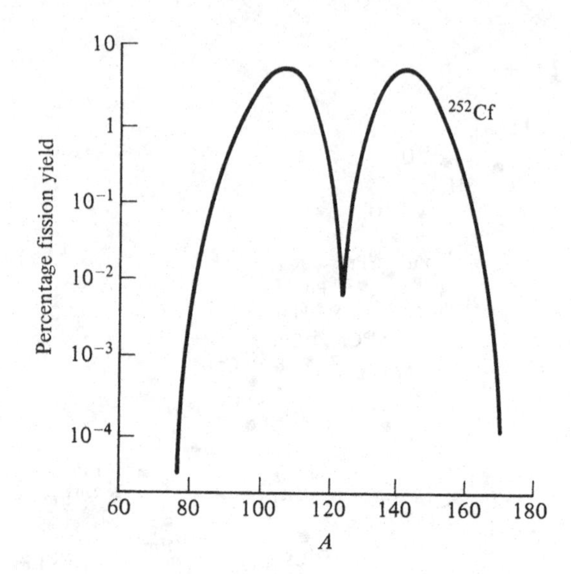

5.6. Spontaneous fission yield versus fragment mass of ^{252}Cf. (From Irvine, J. M., *Heavy Nuclei, Superheavy Nuclei and Neutron Stars.* Oxford University Press (1975).)

liquid drop model assumes no shell structure it is equivalent to assuming a uniform density of levels $\bar{\rho}(\varepsilon)$ as illustrated in figure 5.7*a*. Figures 5.7*b* and 5.7*c* illustrate two extremes of the effect on $\rho(\varepsilon)$ of shell structure with the density of states at the Fermi surface $\rho_F(\varepsilon)$ being higher and lower, respectively, than $\bar{\rho}(\varepsilon)$.

The sum over occupied levels (assuming two nucleons per level) is shown below each figure and it can be seen that a nucleus is more bound when $\rho_F(\varepsilon)$ is less than the liquid drop value $\bar{\rho}(\varepsilon)$. The increase in binding due to this shell effect is proportional to $[\bar{\rho}(\varepsilon)-\rho_F(\varepsilon)]$ and a correction to the liquid drop value can be made by evaluating this quantity. The Nilsson model can be used to find $\rho(\varepsilon)$ for a particular deformation and $\bar{\rho}(\varepsilon)$ and $\rho_F(\varepsilon)$ are then calculated by averaging $\rho(\varepsilon)$ over an energy range larger or smaller, respectively, than the spacing between major shells at the Fermi surface. Typically, values are over 5–10 MeV for $\bar{\rho}(\varepsilon)$ and 1–2 MeV for $\rho_F(\varepsilon)$.

As the level density $\rho(\varepsilon)$ varies with deformation the resulting correction to the liquid drop value will typically change sign as a function of deformation. This can be seen in the schematic diagram of the Nilsson level scheme for an odd proton above $Z=82$ shown in figure 5.8. The magic numbers of 82 and 114 for zero deformation are where $\rho(\varepsilon)$ is lower than average and nuclei are correspondingly more bound. As the deformation increases

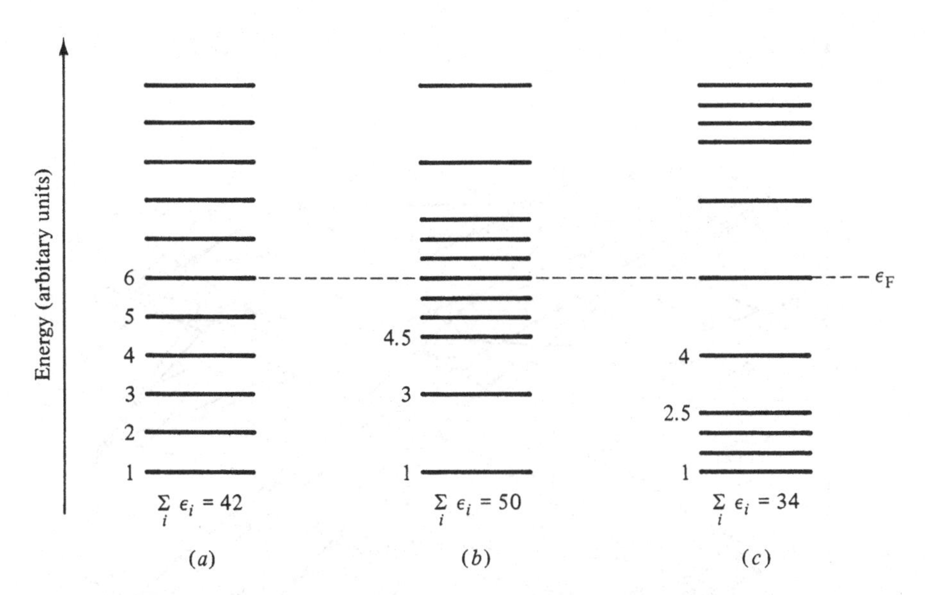

5.7. Level densities at the Fermi surface $\rho_F(\varepsilon)$: (*a*) $\rho_F(\varepsilon) = \bar{\rho}(\varepsilon)$; (*b*) $\rho_F(\varepsilon) > \bar{\rho}(\varepsilon)$; (*c*) $\rho_F(\varepsilon) < \bar{\rho}(\varepsilon)$. The sum over occupied levels (assuming two nucleons per level) is shown below each set of levels.

these magic numbers change and around $\delta \sim 0.3$ ($\delta \approx 3\varepsilon/2$) nuclei with 100 protons become magic.

In fission as the deformation increases the nucleus becomes like two intersecting spheroids of generally unequal size. To account for this a two-centre Nilsson model can be used and for each separation the shell correction must be calculated for different fragment mass ratios. Besides the shell correction, the effect of pairing must also be taken into account and this tends to reduce the magnitude of the correction. This is because the binding due to pairing increases as $\rho(\varepsilon)$ increases, as discussed in chapter 8, but the main effect on the shape of the fission barrier is due to the shell correction.

The resultant change to the liquid drop fission barrier, by including both shell and pairing corrections for $^{240}_{94}$Pu, is shown in figure 5.9 with the liquid drop curve shown as a dotted line. The most dramatic feature is the occurrence of a double hump in the fission barrier reflecting a change in sign of the shell correction as $\rho(\varepsilon)$ varies with deformation and separation. The calculations for ^{236}U (see figure 5.10) shows that fission starts off symmetrically through the first barrier but then becomes asymmetric favouring a fragment mass ratio of ~1.4 in good agreement with observation; cf. the fission of ^{252}Cf shown in figure 5.6.

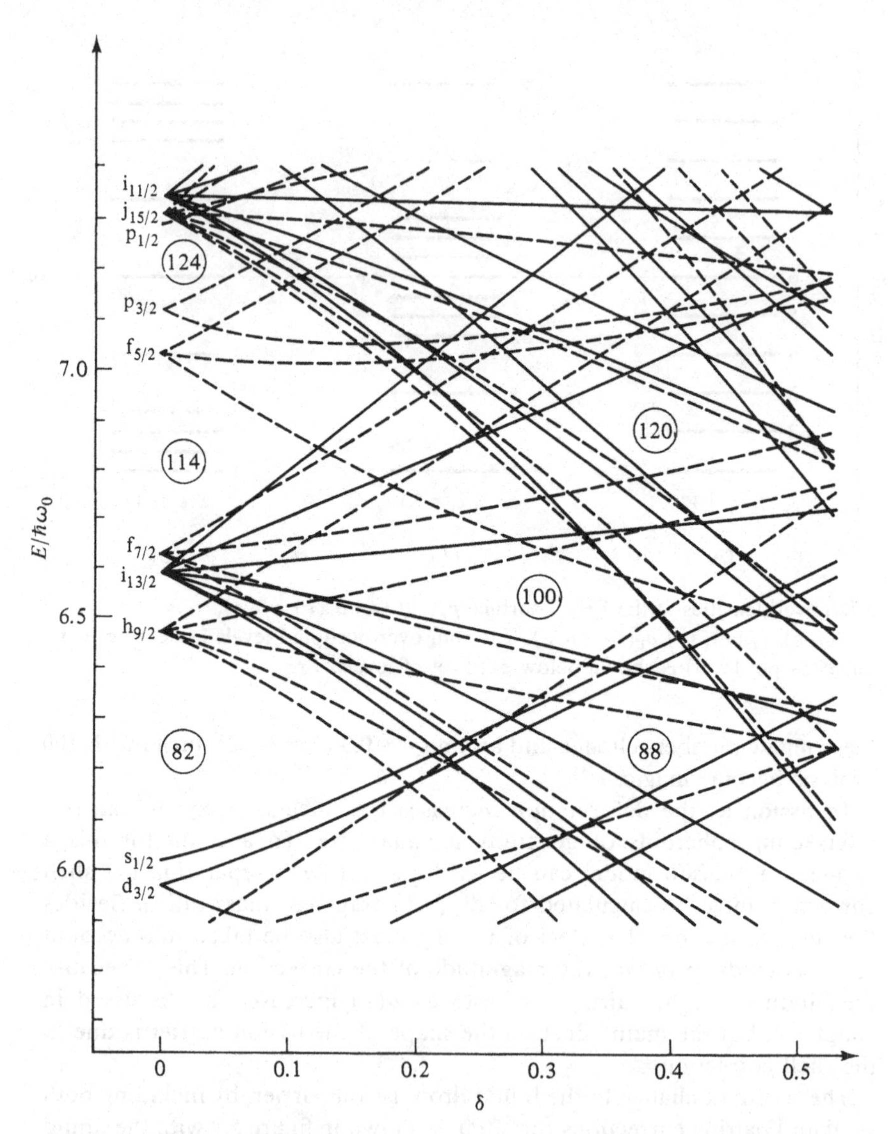

5.8. Level density variation with deformation in the Nilsson model for an odd proton above $Z = 82$. (From Bohr, A. and Mottelson, B. R., *Nuclear Structure*, Vol. 2. Benjamin (1975).)

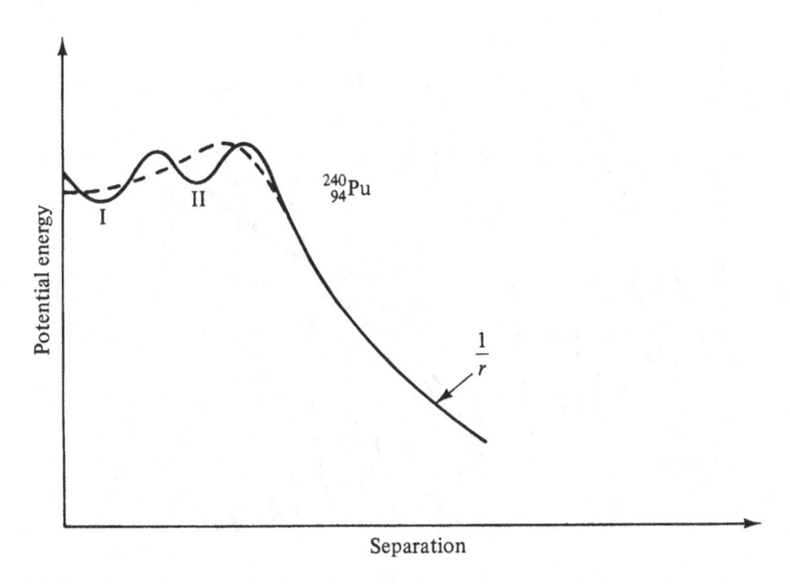

5.9. The shape of the fission barrier for ^{240}Pu predicted by the liquid drop model (dotted curve) and by the liquid drop model including both shell and pairing corrections (solid curve).

5.2.5 Fission isomeric rotational band

The occurrence of the second well, labelled II in figure 5.9, gives rise to the fission isomeric state in $^{240}_{94}Pu$. If $^{240}_{94}Pu$ is formed in the lowest state in well II it will be at 2–3 MeV excitation and a fission lifetime through the second barrier of 10^{-9} s is consistent with the reduced barrier height. Moreover, its γ-decay will be highly inhibited because its shape is so different from lower states whose deformation is much less, being similar to that of the ground state which is at the bottom of well I. This type of isomerism is called shape isomerism.

An elegant experiment which supports this interpretation of fission isomers was carried out by Specht *et al.* in 1972. They used the $^{238}_{92}U(\alpha, 2n)$ $^{240}_{94}Pu$ reaction with an incident alpha energy of 25 MeV to form members of the rotational band built on the fission isomeric state of $^{240}_{94}Pu$ and then looked for E2 transitions within this band.

These are highly converted because of the high Z of $^{240}_{94}Pu$ and the low energy of the transitions (see section 4.5, p. 135). To discriminate against the enormous electron background resulting from inner electron shell ionisation, they required coincidences between emitted electrons and fission

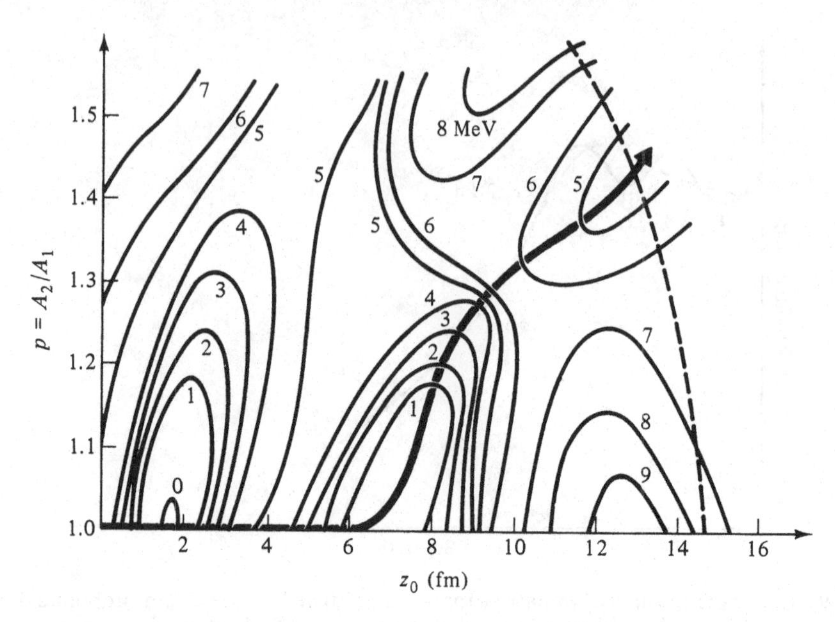

5.10. Potential energy contours and the fission path for ^{236}U. The parameter $p = A_2/A_1$ is the ratio of the fragments' masses and z_0 is their separation. (From Irvine, J. M., *Heavy Nuclei, Superheavy Nuclei and Neutron Stars.* Oxford University Press (1975).)

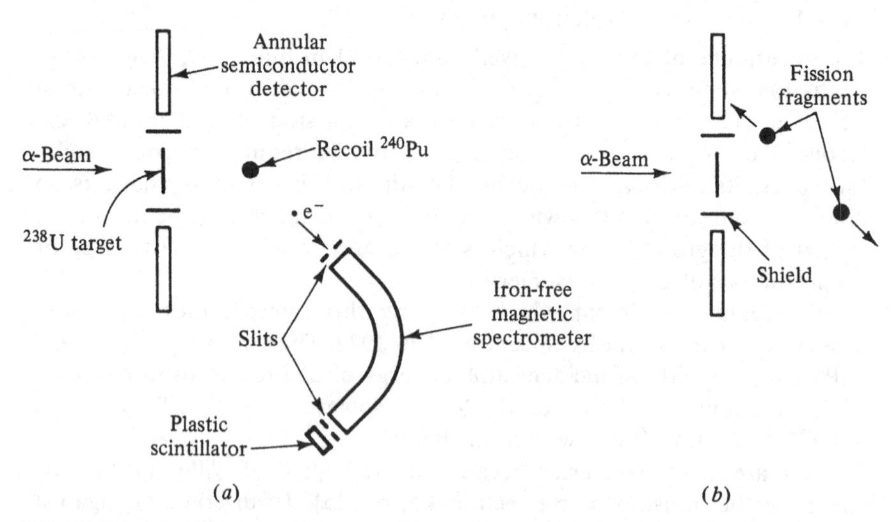

5.11. Schematic diagram of the experimental layout for detection of conversion electrons in coincidence with isomeric fission fragments.

Table 5.1

Transition	Transition energies (keV)	
	Well I	Well II
$2^+ \rightarrow 0^+$	42.82	
$4^+ \rightarrow 2^+$	95.86	46.6
$6^+ \rightarrow 4^+$	152.63	73.0
$8^+ \rightarrow 6^+$	203.3	99.4
$10^+ \rightarrow 8^+$	253.8	

fragments. However, a major problem in detecting the conversion electrons was the predominance (a factor of 10^4) of prompt over isomeric fission of ${}^{240}_{94}Pu$ formed in the ${}^{238}_{92}U(\alpha, 2n){}^{240}_{94}Pu$ reaction.

A schematic diagram (figure 5.11*a*) of their experiment shows how they overcame this problem. The target of 50 μg cm^{-2} of U evaporated on 20 μg cm^{-2} of C was placed in the hole of an annular semiconductor detector. The shield around the target stopped any prompt fission fragments from entering the detector. A ${}^{240}_{94}Pu$ nucleus formed in a low-lying excited state of the isomeric band decays by internal conversion ($t_{1/2} < 0.1$ ns) and then by fission; however, before fissioning the nucleus travels away from the target during the lifetime of the isomeric state ($t_{1/2} = 4$ ns) enabling detection of the fission fragments as illustrated in figure 5.11*b*.

The result of the coincident electron spectrum was the observation of conversion electrons corresponding to the $8^+ \rightarrow 6^+$, $6^+ \rightarrow 4^+$ and $4^+ \rightarrow 2^+$ E2 transitions within the isomeric band. (The $2^+ \rightarrow 0^+$ transition was not resolved from the chance-coincident background.) Table 5.1 gives the transition energies for both the ground state and isomeric band and the moment of inertia of the isomeric band is 2.1 times that of the ground state band. This is consistent with the nucleus being much more deformed in the isomeric state. The value is in agreement with calculation of the moment of inertia for the bottom of the second well in ${}^{240}_{94}Pu$, strongly supporting the explanation of fission isomers as being the result of shape isomerism.

5.2.6 Neutron-induced fission

Neutron capture by heavy nuclei induces fission if the compound nucleus formed in the capture is at an excitation energy greater than the fission barrier. This depends in particular on the Z of the nucleus and for a particular element on whether the target nucleus has even or odd A because of the effect of pairing. For example in the slow neutron ($E_n \approx \frac{1}{40}$ eV) capture by ^{235}U, the compound nucleus, ^{236}U, is formed at an excitation energy equal to the separation energy S_n^{236} and as $S_n^{236} > E_f$, the barrier height, prompt fission occurs. However, for capture by ^{238}U the excitation energy $S_n^{239} < E_f$ so there is no induced fission until $E_n > 1.4$ MeV, which is the difference between S_n^{239} and E_f. The difference in behaviour arises because pairing causes $S_n^{239} < S_n^{236}$. At thermal energies ($E_n \approx \frac{1}{40}$ eV) the fission cross-section σ_f^{235} is by far the largest cross-section ~600 barns arising from an excited state in ^{236}U just below threshold. By comparison the total cross-section σ_T for ^{238}U, which is mainly elastic scattering at thermal energies, is only ~10 barns.

The induced fission cross-section σ_f^{238} is not seen for neutrons below 1.4 MeV, as expected, but in contrast induced fission is seen in the sub-threshold region when the target is $^{240}_{94}$Pu (figure 5.12). The resonances observed in this region are not randomly distributed and are due to the presence of a double hump in the fission barrier of $^{241}_{94}$Pu. The ~15 eV spacing is typical of levels in well I while ~0.7 keV is typical of those in well II at the excitation energy of $^{241}_{94}$Pu (cf. figure 5.9). When these levels coincide in energy they are strongly coupled and this enhances the fission

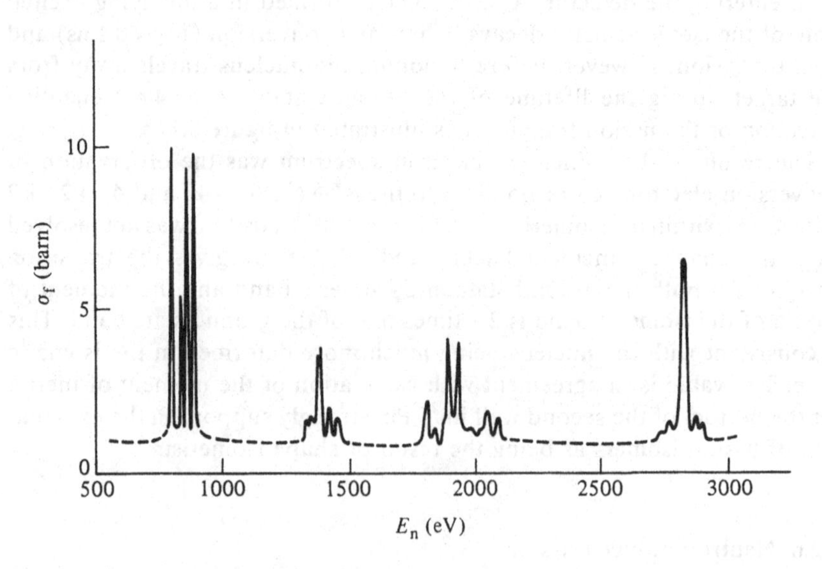

5.12. The sub-threshold neutron-induced fission cross-section for n + ^{240}Pu measured by Migneco and Theobold. (From Burcham, W. E., *Elements of Nuclear Physics*. Longman (1979).)

cross-section as the width of the barrier is much less than for other states in well I.

5.3 Nuclear power

The average energy of each neutron released in the fission of ^{235}U is ~2.3 MeV and the average number is ~2.5. As this number is greater than 1 the possibility exists of a chain reaction. In a piece of uranium containing both ^{238}U and ^{235}U, a neutron is initially most likely to scatter inelastically and through such interactions the neutron loses energy. In natural uranium, however, the neutron will most probably have been captured by ^{238}U before it has lost sufficient energy for σ_f^{235} to dominate, as the natural abundance of ^{235}U is only 0.72%, so no chain reaction occurs. If the uranium is greatly enriched in ^{235}U then a chain reaction is possible if there is a critical mass of uranium, and this forms the basis of nuclear fission weapons.

For energy production a controlled chain reaction is required and the most widely used technique is to use a moderator and control rods. The uranium fuel is formed into thin rods and these are dispersed in a medium, called a moderator, of low atomic weight and small neutron capture cross-section (e.g. H_2O, D_2O or ^{12}C). Fast neutrons emitted in the fission of a uranium nucleus have a good chance of escaping from a fuel rod and once in the moderator they elastically scatter off the low-A nuclei and thermalise. The neutrons then have an appreciable probability of inducing fission in ^{235}U in the fuel rods. This technique avoids the loss of neutrons through capture by ^{238}U and enables low-enrichment uranium to be used as a fuel.

The average time between emission and subsequent induced fission is $\sim 10^{-4}$ s and this leads to a very short time (~1 s) within which to control a chain reaction. Fortunately as well as prompt neutrons there are beta-delayed neutrons (~0.02 per fission) with a mean delay time of ~13 s, which gives a control time of ~10 mins. A reactor is designed so that it is not critical with prompt neutrons alone but can be with delayed neutrons. The reactor is controlled by rods of material with a high thermal neutron capture cross-section (e.g. Cd or B) which can be moved in or out of the moderator. The heat produced by the fission is removed by a coolant circulating through the reactor core and is used to drive a turbine and generate electricity.

An alternative technique which is under development for generating power is to use quite highly enriched fuel (~20% fissile nuclei) with control rods but no moderator. The increased enrichment enables the reactor to operate on fast neutron-induced fission. The principal attraction of this technique, which is technically more difficult, is the possibility of breeding $^{239}_{94}$Pu, which can be used as a fuel for the reactor. If, for example, a core of 20% ^{239}Pu and 80% ^{238}U is surrounded by a blanket of ^{238}U then neutrons escaping from the core are captured by the uranium nuclei and produce ^{239}Pu through the processes:

$$^{238}\mathrm{U}+\mathrm{n}\rightarrow{}^{239}\mathrm{U}\xrightarrow{\beta^-}{}^{239}\mathrm{Np}\xrightarrow{\beta^-}{}^{239}\mathrm{Pu}$$

Theoretically it is possible to design a fast-breeder reactor that produces more ^{239}Pu than is used up in the core, which would enable all of the uranium rather than just the ^{235}U fraction to be used as a fuel.

The fraction of ^{235}U 0.72% was thought to be the same everywhere on earth but in 1972 some uranium ore from Gabon was found to have only 0.4%. It was deduced that this discrepancy had been caused by a natural fission chain reaction having taken place. When the ore was deposited some 1.7×10^9 years ago the enrichment of ^{235}U would have been ~3% because of the difference in the ^{235}U and ^{238}U half-lives. At this concentration a fission chain reaction is possible if the neutrons are moderated and this can occur if the ore contains water.

5.4 Thermonuclear fusion

The release of energy by the fusion of two nuclei to form a heavier nucleus is possible for many light nuclei as the semi-empirical mass formula predicts, but does not occur at room temperatures because of the Coulomb repulsion between the nuclei. At the temperatures in the interior of the sun, though, hydrogen nuclei can come sufficiently close for fusion to occur through the weak interaction:

$$\mathrm{p}+\mathrm{p}\rightarrow{}^2_1\mathrm{H}+\mathrm{e}^++\nu+0.42\ \mathrm{MeV}$$

In this reaction one of the protons β-decays while within the range of the nuclear attractive force of the other proton and the resultant neutron and proton form a deuteron.

Consider first two protons at an energy of 1 MeV. At this energy the Coulomb barrier can be ignored, at a first approximation, and the weak interaction will be an allowed ($l=0$) Gamow–Teller transition, as the protons must be in an $^1\mathrm{S}_0$ state while the deuteron is in a $^3\mathrm{S}_1$ state. The spatial part of the matrix element (equation 3.4) will therefore be:

$$|M_{if}|^2=6\lambda^2G_\beta^2\left[\int\psi_\mathrm{d}^*(r)\psi_{2\mathrm{p}}(r)\,\mathrm{d}^3r\right]^2$$

where the normalisation volume is taken as unity. The factor 6 comes from evaluating $|\sum_{i=1}^2\langle f|\boldsymbol{\sigma}|i\rangle|^2$ summed over the two equivalent protons. The wavefunction for the two protons $\psi_{2\mathrm{p}}=(\sin kr)/kr$ over the size of the deuteron can be taken as unity. Neglecting the range of the force the deuteron wavefunction $\psi_d \approx (\beta/2\pi)^{1/2}\,(1/r)\exp(-\beta r)$ where $\beta = 0.23\ \text{fm}^{-1}$ so

$$|M_{if}|^2 = 6\lambda^2 G_\beta^2 8\pi/\beta^3$$

and the reaction cross-section $\sigma(\text{p}+\text{p}\to{}^2_1\text{H}+\text{e}^+ + \nu)$ is given by:

$$v_{\text{pp}}\sigma = \frac{2\pi}{\hbar}|M_{if}|^2 \int_0^{p_0} \frac{(4\pi)^2 p^2 (E-E_0)^2}{h^6 c^3}\,dp$$

where p_0 is the maximum positron momentum and v_{pp} is the relative velocity of the protons. Approximating $E + m_e c^2 = pc$, then for 1 MeV centre of mass energy $v_{\text{pp}} \approx 2\times10^7\ \text{m s}^{-1}$, $E_0 = 1.42$ MeV, so $E_\text{T} = E_0 + m_e c^2 = 1.93$ MeV and:

$$\sigma \approx \frac{4}{\pi^2 \hbar^7 c^6}\,\frac{\lambda^2 G_\beta^2 E_\text{T}^5}{5\beta^3 v_{\text{pp}}} \approx 10^{-23}\ \text{b}$$

To estimate the cross-section at the energies of protons in the centre of the sun (~1 keV) the Coulomb barrier penetration factor must be calculated. The cross-section σ can be expressed as:

$$\sigma = \frac{S(E)}{E}\exp(-2\gamma)$$

where $S(E) \propto E_\text{T}^5$ and $\exp(-2\gamma)$ is the Gamow factor calculated above (equations 5.1 and 5.2). This expression is derived in chapter 6, p. 173. For 1 MeV, $\sigma \approx S(E)/E$ so $S(1) \approx 10^{-23}$ MeV b. For 1 keV, $\exp(-2\gamma) \approx \exp(-22)$ so the reaction cross-section in the sun would be expected to be $\sim 7\times10^{-32}$ b and the rate $\sigma v \approx 4\times10^{-54}\ \text{m}^3\ \text{s}^{-1}$. However, this is a serious underestimate because the protons have a Maxwellian velocity distribution and the number with velocity v, $N(v)$, is given by:

$$N(v) \propto v^2 \exp(-mv^2/2kT)$$

The exceedingly strong velocity dependence of the cross-section means that the high-energy tail of the distribution contributes most. Evaluating this gives a factor of $\sim 10^4$ increase in the rate so $\sigma v \approx 10^{-50}\ \text{m}^3\ \text{s}^{-1}$. The reaction rate is given by $n^2 v\sigma V$ where n is the number of protons per m^3 and V is the volume of the sun in which protons are interacting. The density near the centre of the sun is $\sim 10^5\ \text{kg m}^{-3}$ so $n \approx 6\times10^{31}$.

The proton–proton chain of reactions, called the pp cycle, yields helium through the sequence:

$$\text{p}+\text{p}\to{}^2_1\text{H}+\text{e}^+ + \nu + 0.42\ \text{MeV}$$

$$\text{p}+{}^2_1\text{H}\to{}^3_2\text{He}+\gamma + 5.49\ \text{MeV}$$

$${}^3_2\text{He}+{}^3_2\text{He}\to{}^4_2\text{He}+\text{p}+\text{p}+12.86\ \text{MeV}$$

On average the neutrinos take away 0.26 MeV each. The positron annihilates with an electron and releases 1.02 MeV so each proton gives rise to 6.55 MeV. The observed luminosity of the sum is $3.9\times10^{26}\ \text{J s}^{-1}$ and if the pp cycle is the principal source of power this implies that there are $\sim 3.7\times10^{38}$ protons being converted a second. Equating this rate to $n^2 v\sigma V$ gives an active volume $V = 10^{25}\ \mathrm{m}^3$. This volume is $\sim 10^{-3}$ of the solar volume so the pp cycle which starts with a weak interaction can account for the observed solar luminosity.

5.4.1 Fusion reactors

The hydrogen plus hydrogen reaction is totally impractical for a fusion reactor as it is a weak process. A promising fusion reaction, on which research is currently being carried out, is:

$$^2_1\mathrm{H} + ^3_1\mathrm{H} \rightarrow ^4_2\mathrm{He} + \mathrm{n} + 17.62\ \mathrm{MeV}$$

which has an enhanced cross-section because of a resonant state in ^{5}He. A plasma of deuterium and tritium at a temperature corresponding to ~ 20 keV (~ 200 million K) is needed to produce an adequate reaction rate. Such a plasma must not be allowed to come into contact with any material as it would melt it. One way being explored is to try and contain the plasma, using a toroidal magnetic field (see figure 5.13), for sufficient time for the fusion reactions to generate an appreciable amount of energy.

In a fusion reactor a confinement time of ~ 1.5 s and a plasma density of $\sim 2 \times 10^{20}\ \mathrm{m}^{-3}$ is required. The vacuum vessel would be surrounded by a blanket to capture the neutrons produced by the fusion reactions. The

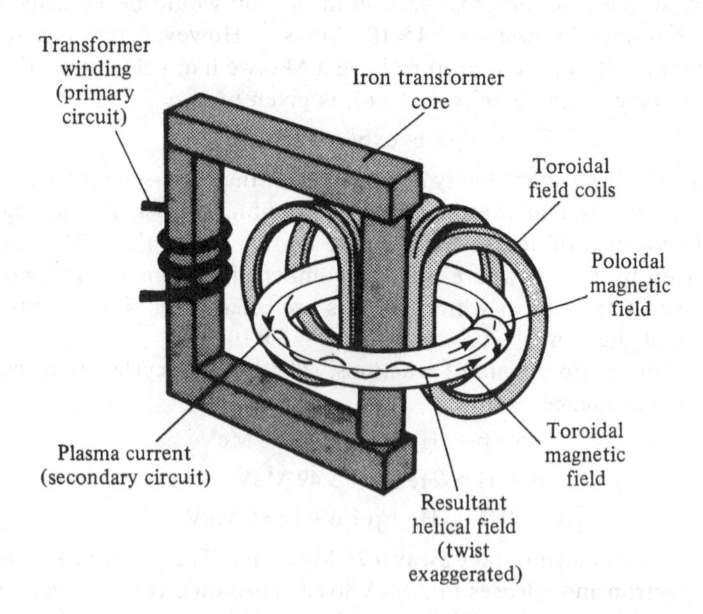

5.13. Schematic diagram of the tokamak magnetic field configuration at the Joint European Torus (JET) facility. (From a Culham laboratory leaflet.)

neutrons would interact with lithium in the blanket to produce tritium via the reactions $n+{}^7Li \rightarrow \alpha + t + n - 2.46$ MeV and $n + {}^6Li \rightarrow \alpha + t + 4.8$ MeV. The captured neutrons would heat the blanket up to ~500 °C and this heat would be used to drive turbines and generate electricity.

Another approach to controlled fusion is to use pellets of deuterium and tritium and implode them using very high powered laser beams. The implosion would raise the temperature sufficiently for the fusion reactions to occur. Currently there is still a great deal of research to be carried out on both approaches before any practical fusion reactor can be built.

5.5 Summary

In both α-decay and spontaneous fission the strong dependence on energy of the probability of tunnelling, through the barrier arising from the nuclear and Coulomb forces, is shown to explain the enormous range of lifetimes that are observed. The occurrence of fission isomers is seen to be caused by the significant effect of shell corrections to the liquid drop model of fission.

The physical principles of a nuclear fission reaction are explained and the important differences between the thermonuclear fusion of hydrogen in the sun, which is initially a weak process: $p+p \rightarrow {}^2_1H + e^+ + \nu$, and the fusion reactions in a possible fusion reactor are discussed.

5.6 Questions

1.* Show that for large Z and A the energy released when a nucleus emits an alpha particle is given by:

$$Q_\alpha = -4a_v + \tfrac{8}{3}a_s A^{-1/3} + 4a_c Z[1 - Z/3A]/A^{1/3}$$
$$-4a_a(N-Z)^2/A^2 + B({}^4\text{He})$$

where $B({}^4\text{He})$ is the alpha particle binding energy (28.3 MeV). The only naturally occurring isotopes of silver and gold are ${}^{107}_{47}$Ag and ${}^{197}_{79}$Au. Discuss the stability of these nuclei in the light of the expression for Q_α.

2. Making the assumption that the centrifugal barrier effectively increases the barrier height by $\tfrac{1}{3}[l(l+1)\hbar^2/2mR^2]$, estimate the ratio of $l=0$: $l=2$: $l=4$: $l=6$ α-decays of energy 5 MeV in the α-decay of a nucleus with $Z=90$, $A=230$.

3.* The ground state of ${}^{212}_{84}$Po decays to ${}^{208}_{82}$Pb emitting α-particles with energy about 9 MeV. There is an excited state in ${}^{212}_{84}$Po with excitation energy 2.9 MeV and $J^\pi = 16^+$ or 18^+. Suggest explanations for the following observations:

(i) The 2.9 MeV state of $^{212}_{84}$Po decays by α-decay rather than by γ-emission.
(ii) The mean lifetime of the 2.9 MeV excited state (45 s) is longer than the mean lifetime of the ground state (0.3 μs) of $^{212}_{84}$Po.

4.* Show that spherical nuclei with $Z^2/A \geq 18$ are energetically unstable against spontaneous fission into two equal fragments $(Z/2, A/2)$. Explain why this behaviour is not observed until larger values of Z^2/A are reached.
By considering the potential of two such spherical fragments when just touching and neglecting the pairing term in the semi-empirical mass formula show that nuclei with $Z^2/A > 53$ would be likely to spontaneously fission.

5.* A nuclear reactor is set up fuelled by $^{235}_{92}$U. Estimate the mass of $^{235}_{92}$U consumed in giving a power output of 100 MW for a period of 5 years. Estimate an upper limit for the production of $^{239}_{94}$Pu if the reactor core were blanketed with a layer of natural uranium.

6. A ^{40}Ca target is bombarded with a beam of ^{12}C ions for 100 ms and then the target is viewed by a silicon semiconductor detector for 100 ms with no beam on target. Protons of ~2 MeV energy are observed and their intensity is found to decrease with time with a half-life of ~50 ms. Give a possible explanation.

7.* There is a resonance in the cross-section for neutron scattering on $^{238}_{92}$U at a neutron kinetic energy of 6.7 eV; it has a width of 0.1 eV and in the resonance region the neutron predominantly suffers radiative capture. At higher energies, up to 20 eV, the scattering is predominantly elastic. Show that a neutron, with energy E, if elastically back-scattered will lose energy $\Delta E = 0.017E$. In fact, on averaging over all scattering angles the average energy loss is one half of this value. Show that after n elastic collisions, the neutron with initial kinetic energy E will have an energy of about $E_n = (1-0.0085)^n E$. Estimate the number of collisions that a 20 eV neutron will make in a large mass of $^{238}_{92}$U before it reaches the resonance energy and show that it will probably undergo radiative capture.

6 Nuclear reactions

6.1 Introduction

The study of nuclear reactions provides considerable information on the structure of nuclei as well as on the nature of their interaction. In this chapter there is first a brief discussion of some experimental details, before the general features of nuclear reactions are described. Then methods for describing reactions are developed. First some general predictions are given followed by a discussion of the simplest reactions, elastic and inelastic scattering, for both light and heavy ions. After this the theory of compound nucleus reactions is presented and the occurrence of slow neutron resonances described. Isobaric analogue resonance and isospin forbidden reactions are then discussed.

At higher incident energies direct reactions become important and the use of first-order perturbation theory, the Born approximation, for the description of pick-up and stripping reactions by both light and heavy ions is described. For heavy-ion direct reactions new features are seen reflecting the more classical behaviour of the ions: selectivity arising through kinematic matching and characteristic bell-shaped angular distributions. At energies above the Coulomb barrier, as well as compound nucleus (fusion) reactions, deep-inelastic reactions are also observed, with substantial cross-sections for very heavy ions. In these reactions there is a considerable transfer of the incident energy to internal excitation of the product nuclei. For even higher energies, a new phase of nuclear matter, a quark–gluon plasma, is predicted to be found in reactions with relativistic heavy ions, and it is in this area that nuclear and particle physics overlap.

6.1.1 Accelerators

In a typical experiment a beam of charged particles with energy ~10–100 MeV is directed by means of deflectors and quadrupole magnets through a thin foil and the scattered particles are counted by detectors. Particles with these energies are produced using accelerators and several types have been developed since the first reaction ($^7Li + p \rightarrow {}^4He + {}^4He$) was produced with accelerated ions by Cockcroft and Walton in 1932.

For low to medium energies a tandem Van de Graaf accelerator can be used. In such a machine singly-charged negative ions are injected at low energy (~50 keV) along a highly evacuated tube towards a terminal charged to a high positive voltage (~10–20 MV) where they pass through a gas (or foil), are stripped of several electrons and become positively charged, and are consequently then accelerated away from the terminal down to ground potential. For an ion stripped to a 9^+ charge state in a terminal held at 20 MV the final energy will be 200 MeV. Continuous beams of many different ions can be produced with easily varied energies.

For higher energy beams electrostatic machines are limited by electrical breakdown and alternating electric fields are used. In a linear accelerator pulses of ions are accelerated by passing them through a resonant cavity within which there is an r.f. alternating electric field along the beam direction of such frequency and phase that a pulse of ions only sees an accelerating field while within the cavity. These r.f. cavities are also used in a synchrotron accelerator. This accelerates particles round a ring around which are spaced the cavities, together with bending magnets to hold the particles in the ring.

The cyclotron accelerator, developed by Lawrence and Livingstone in the early 1930s, also uses an r.f. field. In this machine pulses of positive ions are injected at right angles to a uniform magnetic field B and describe circular orbits which pass between two D-shaped r.f. electrodes. The frequency of each orbit is Bq/m, where q and m are the charge and mass of the ion, respectively, and is independent of energy (for non-relativistic velocities). So an r.f. field of this frequency, and with the correct phase, applied to the D-shaped electrodes accelerates the particles.

6.1.2 Kinematics and conservation laws

In the first nuclear reaction observed by Rutherford in 1919 alpha particles from a radioactive source were used to bombard nitrogen, which produced oxygen 17 nuclei and protons. This is written:

$$\alpha + {}^{14}N \rightarrow {}^{17}O + p \quad \text{or} \quad {}^{14}N(\alpha, p){}^{17}O$$

In general for a reaction

$$1 + 2 \rightarrow 3 + 4$$

of which ${}^{14}N(\alpha, p){}^{17}O$ is an example, the energy released in the reaction is called the Q-value and is given by:

$$Q = (M_1 + M_2 - M_3 - M_4)c^2$$

where M_i is the mass of the ith particle. The total energy E and kinetic energy T of a particle are related by:

$$T = E - Mc^2 \tag{6.1}$$

$$\therefore \quad Q = (T_3 + T_4 - T_1 - T_2)$$

since energy is conserved. The mass excess M_x (units of energy) of a nucleus is given by:

$$M_x = (M - AM_u)c^2$$

where $M_u = 1/12$ the mass of a carbon atom ($M_u \equiv u$, the atomic mass unit (a.m.u.)) and A is the mass number of the nucleus ($A = Z + N$). So Q equals the difference in initial and final mass excesses or the difference in final and initial kinetic energies.

Besides energy, angular momentum is always conserved in reactions between nuclei and parity is to a very good approximation. Examples which show this are the states that can be formed in the capture reaction $\alpha + {}^{16}O \rightarrow {}^{20}Ne^* \rightarrow {}^{20}Ne + \gamma$. The ground state of ${}^{16}O$ and the α-particle both have $J^\pi = 0^+$ so the angular momentum of the ${}^{20}Ne^*$ state must be equal to the orbital angular momentum of the α-particle. The wavefunction describing the relative motion of the α-particle and ${}^{16}O$ nucleus is proportional to $Y_{lm}(\theta, \phi)$ so its parity is $(-1)^l$. Therefore only natural parity states should be formed, i.e. $J^\pi = 0^+, 1^-, 2^+ \ldots$.

Similar considerations apply to the decay of excited states, for example ${}^{16}O^* \rightarrow {}^{12}C(0^+) + \alpha$, which is only allowed for states in ${}^{16}O$ with natural parity. In fact a very weak α-decay from a 2^- excited state in ${}^{16}O$ has been observed, which can be attributed to the weak interaction mixing in a very small amplitude of a 2^+ state into the 2^- state. In the α-decay of ${}^{8}Be^* \rightarrow 2\alpha$ a further restriction is observed because the final state must be symmetric under particle exchange as the α-particles are bosons. This restricts the α-particle decay to states with $J^\pi = 0^+, 2^+, 4^+ \ldots$.

6.1.3 The centre of mass frame

In describing the kinematics of a nuclear reaction it is often useful to use the centre of mass frame. This is the frame in which there is no net momentum. While in the laboratory frame some of the energy of the incident particle is required to conserve momentum, in the centre of mass frame all of its kinetic energy is available for a reaction. As an example consider alpha particles of energy E_i(lab) incident on a carbon foil; here energy means kinetic energy. The radiative alpha capture reaction:

$$\alpha + {}^{12}C \rightarrow {}^{16}O^* \rightarrow {}^{16}O + \gamma \quad \text{or} \quad {}^{12}C(\alpha, \gamma){}^{16}O$$

is observed. If the Q-value of the reaction is Q_0 what is the excitation energy E_x of the initial excited state in ${}^{16}O$? The incident energy E_i(lab) is much less than $M_\alpha c^2$, so non-relativistic mechanics can be used, in which case the kinetic energy E_a available in the centre of mass is given by:

$$E_a = \frac{M_t}{M_t + M_i} \cdot E_i$$

where M_t and M_i are the masses of target and incident nuclei. Therefore

$$E_x = (\tfrac{3}{4}E_i + Q_0)$$

Similar considerations also apply when working out particle decay energies.

When describing reactions, formulae generally refer to the centre of mass frame. The momentum of either nucleus p_{cm} in the centre of mass defines $\lambda\!\!\!^{-}_{cm}$ through $p_{cm} = \hbar/\lambda\!\!\!^{-}_{cm}$. (This is correct relativistically.) Non-relativistically two nuclei whose relative velocity is v behave like one nucleus with a reduced mass $\mu = (m_i M_t)/(m_i + M_t)$ incident with velocity v on a fixed nucleus. The momentum is then given by $p_{cm} = \mu v$ and the available energy $E_a = \frac{1}{2}\mu v^2$.

6.1.4 Particle identification

In reactions many nuclei can be produced so often particle identification as well as energy measurements is required to identify a reaction.

A general technique in the identification of charged particles is to look at the energy loss per unit thickness, the stopping power dE/dx, as well as the total energy. This can be done by using a thin detector, through which all the particles of interest pass, followed by a thick detector which stops all of these particles. The first is called a ΔE and the second an E detector. An approximate expression for dE/dx for non-relativistic particles is:

$$\frac{dE}{dx} = -aMZ^2/E$$

where E is the energy, M is the mass and Z is the charge of the nucleus and a is a constant. The range $R(E_T)$ is given by this formula as:

$$R(E_T) = \tfrac{1}{2}E_T^2/(aMZ^2)$$

where E_T is the initial energy. If the thickness of the dE/dx detector is t then

$$R(E_T) - R(E) = t$$

where E is the energy deposited in the E detector. This gives a particle identification (PI) signal:

$$PI = MZ^2 = \frac{1}{2ta}(E_T^2 - E^2)$$

A more accurate empirical expression is:

$$PI \propto (E_T^{1.8} - E^{1.8})$$

and figure 6.1 shows how such a signal can be used to identify nuclei resulting from the bombardment of ^{26}Mg by 78.9 MeV ^{7}Li.

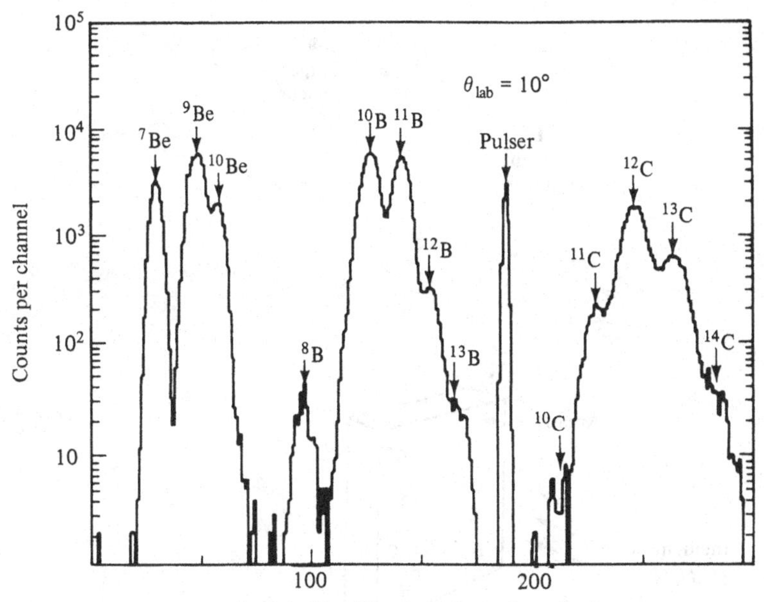

6.1. Particle identification spectrum resulting from bombardment of ^{26}Mg by 78.9 MeV ^{7}Li. The particles produced were detected at an angle $\theta_{lab} = 10°$ to the beam direction.

6.1.5 Cross-sections

The scattered particles which result from a nuclear reaction are counted by a detector, such as a silicon surface barrier detector, whose angle θ to the beam direction can be altered from outside the target chamber. The beam pipe and target chamber are kept evacuated to a pressure $\lesssim 10^{-5}$ torr so there is no appreciable loss of particles through collisions with gas molecules. Experimentally what is determined in such an experiment is the cross-section σ for a particular reaction, which is a measure of the probability that the reaction takes place and has the dimensions of area. This is a very important concept and may be visualised as the cross-sectional area (see figure 6.2*a*) within which a target and incident nucleus will interact and give rise to a particular reaction. It does not in general equal the geometrical cross-sectional area and can be much larger.

If the beam has an intensity of N_i particles per second incident over an area A of a target of thickness t which has n nuclei per unit volume (see figure 6.2*b*), then the yield Y per second for a reaction whose cross-section is σ is given by:

$$Y = N_i(nAt)\frac{\sigma}{A} = N_i nt\sigma \tag{6.2}$$

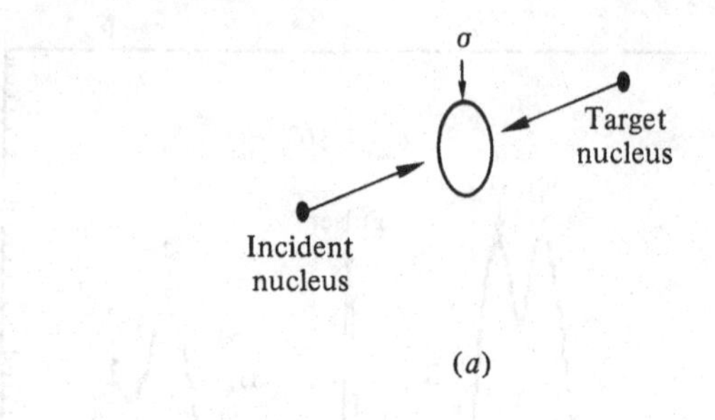

(*a*)

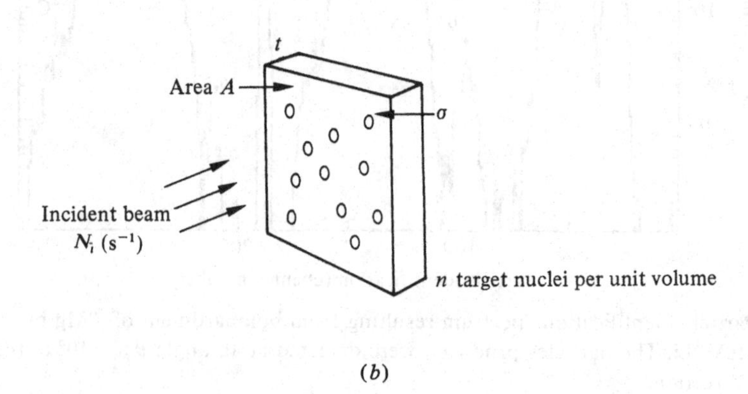

(*b*)

6.2. (*a*) Cross-section σ illustrated in the centre of mass frame. (*b*) A beam of particles incident over an area A of a target with a reaction cross-section σ.

assuming there is no appreciable loss of the beam as it goes through the target. This equation defines the cross-section σ and its units of area are barns or more typically millibarns ($10^{-3}b$).

$$1 \text{ barn} = 10^{-28} \text{ m}^2 = 10 \text{ fm} \times 10 \text{ fm}$$

In terms of the incident flux F of particles, which is defined as the number of particles per second per unit area, then:

$$Y = F(nAt)\sigma$$
$$= FN\sigma \tag{6.3}$$

where N is the number of target nuclei in the beam. These two equations (6.2 and 6.3) are important.

If there are an appreciable number of reactions then the beam will be attenuated as it passes through the target and this can be expressed by:

$$dN_i = -N_i n\sigma_t \,\mathrm{d}x$$

$$\text{so} \quad N_i(x) = N_i(0) \exp(-n\sigma_t x)$$

where σ_t is the total cross-section, i.e. the beam is exponentially attenuated with a $1/e$ attenuation length l given by:

$$l = \frac{1}{n\sigma_t} = \text{mean free path between interactions}$$

6.2 General features of nuclear reactions

Charged particles with very low incident energies only elastically scatter because of the Coulomb repulsion between the incident and target nuclei. As their energy is increased nuclear reactions are possible and there are two characteristic types of reactions observed for light ions, called compound nucleus and direct. Both are selective in the states that are excited, which can give information about the wavefunctions of both the incident and target nuclei, but are quite distinctive in other respects. For heavy ions there is a further type of reaction observed at energies above the Coulomb barrier, with a substantial cross-section for very heavy ions, which is called deep inelastic and is somewhat intermediary between compound nucleus and direct.

In the compound nucleus mechanism, first proposed by Bohr in 1936, the incident nucleus a is captured by the target nucleus A, shares its energy amongst the target nucleons and forms an excited state of the compound nucleus C^*. In this process, memory of the entrance channel $(a + A)$ is lost (except for conserved quantities such as energy and angular momentum). The excited state C^* subsequently decays to a final nucleus B and emitted particle b. The process can be represented by:

$$a + A \rightarrow C^* \rightarrow B + b$$

A classic example is a slow thermal neutron capture reaction such as:

$$\mathrm{n} + {}^{107}\mathrm{Ag} \rightarrow {}^{108}\mathrm{Ag} + \gamma \qquad E_\mathrm{n} \approx \tfrac{1}{40}\,\mathrm{eV}$$

in which the compound nucleus decays by gamma emission. This mechanism is dominant at very low energies.

As the energy of the incident nucleus is increased it is less likely to be captured, as there is more energy to be shared, and more likely to make a peripheral collision with the target nucleus during which energy (inelastic scattering) or a few nucleons (pick-up or stripping reactions) are transferred. An example is the stripping reaction:

$${}^{16}\mathrm{O}(\mathrm{d},\mathrm{p})^{17}\mathrm{O} \qquad E_\mathrm{d} = 8\ \mathrm{MeV}$$

The reaction times for direct and compound nucleus reactions are very different. A direct reaction time is of the order of the nuclear transit time ($\tau \sim 10^{-22}$ s). In a compound nucleus reaction, though, the sharing of the incident nucleus's energy with all the target nucleons makes it unlikely that an appreciable fraction of that energy is concentrated on any particular nucleon or cluster of nucleons (e.g. an α-particle), so the lifetime is typically several orders of magnitudes longer than a typical nuclear transit time.

The excitation functions (the variation of cross-section with incident energy) are also characteristically different. In a direct reaction there is a small amount of energy transfer and this would be expected to depend smoothly on the incident energy. However, in a compound nucleus reaction the energy of the incident particle must equal that required to form the compound nucleus level. This means that resonances will occur in the excitation function and an example is shown in figure 6.3. As the incident energy increases the density of levels in the compound nucleus becomes so high that individual resonances are not resolved. The width Γ of the resonances is related to the lifetime τ of the compound nucleus level by $\Gamma = \hbar/\tau$.

The incident nucleus in a direct reaction will tend to continue to travel forward and so the characteristic angular distribution of a direct reaction is forward-peaked. Its exact shape can give information in a transfer reaction

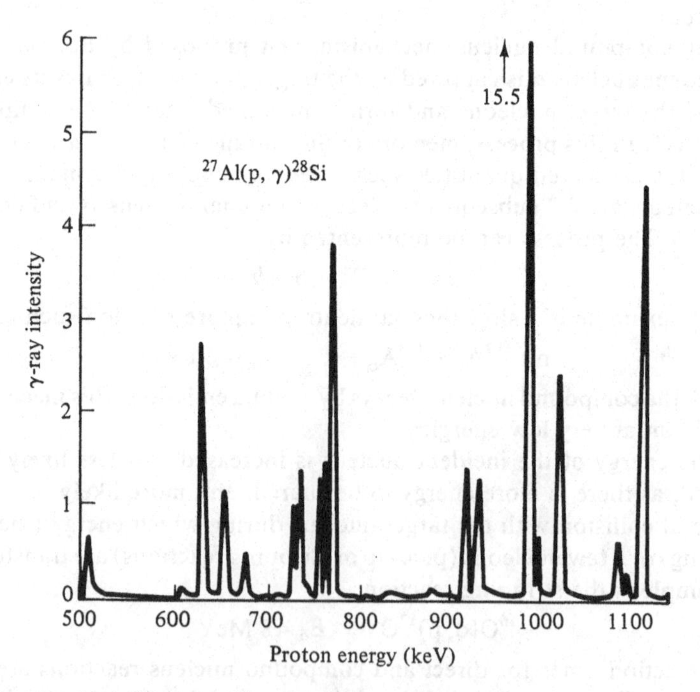

6.3. Resonances in $^{27}Al(p, \gamma)^{28}Si$ measured by Brostrom *et al.* The widths of the peaks are experimental; the natural widths are much less. (From Jones, G. A., *Properties of Nuclei.* Oxford University Press (1987).)

on the angular momentum of the transferred nucleons, though for heavy ions its shape tends to be dominated by the effect of the Coulomb interaction. In a compound nucleus reaction at low energy where an isolated compound nucleus level C^* is formed the intensity in the centre of mass of emitted particles will be equal in the direction **r** as in the direction **−r**. This follows from C^* having definite parity (see chapter 3). As there is symmetry about the beam direction this is equivalent to the angular distribution of emitted particles being symmetric about 90° to the beam direction. At higher energies where many overlapping compound nucleus levels are formed the distribution is still symmetric about 90° if a large enough number of states have been averaged over.

In a particular reaction both mechanisms will generally contribute. For example in neutron scattering off nuclei (figure 6.4) the ground and low-lying excited states will be principally excited by a direct reaction mechanism while the broad bump at higher excitation energies (lower neutron energies) mainly arises through the formation of a compound nucleus which subsequently emits a neutron. The probability that this neutron has all or none of the incident energy is low, which gives rise to the shape observed. If the compound nucleus is at a higher excitation it can be more probable to emit more than one particle (see below).

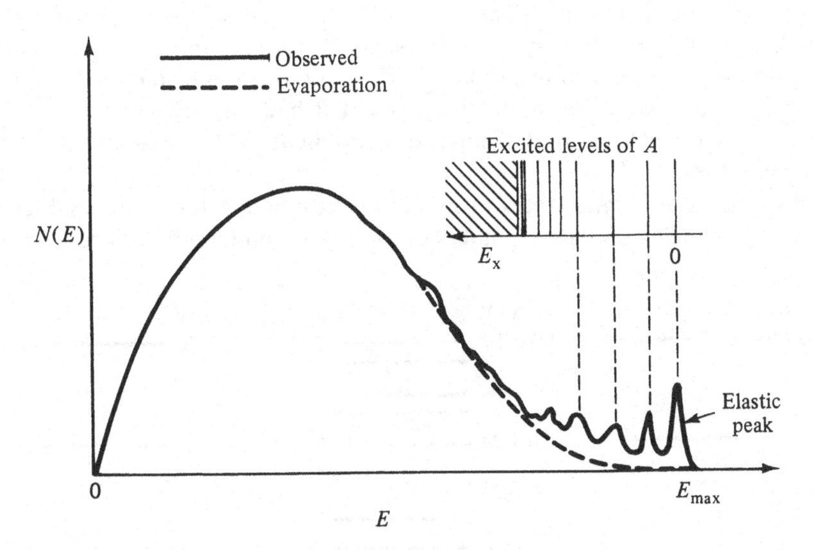

6.4. Typical spectrum of energies of inelastic neutrons emitted in a $A(\mathrm{n}, \mathrm{n}')A^*$ reaction. $N(E)$ is the number of neutrons with energy between E and $E+\mathrm{d}E$. (From Satchler, G. R., *Introduction to Nuclear Reactions.* Wiley (1980).)

For heavy ions at energies above the Coulomb barrier both compound nucleus and direct reactions are seen. In addition, for very heavy ions, there is a substantial cross-section observed for deep-inelastic reactions producing two heavy nuclei in which there has been a considerable transfer of the incident kinetic energy to internal excitation of the products. The complete fusion of two heavy nuclei is hindered by the Coulomb and centrifugal barriers acting between them.

6.2.1 Isospin in nuclear reactions

Isospin is a useful concept particularly for light-ion reactions. The nuclear force is charge independent and as discussed in chapter 1 this leads to the concept of isospin and the prediction that it will be conserved in nuclear reactions. A clear example of this is seen in the reaction $^{12}C(d, \alpha)^{10}B$. The nucleus ^{12}C is an even–even nucleus and has a $J^{\pi}=0^{+}$ and $T=0$ ground state. The deuteron and the alpha particle both have $T=0$ so only $T=0$ states in ^{10}B should be excited in the reaction. The $A=10$ level scheme is shown in figure 6.5 with the Coulomb energy and neutron–proton mass differences removed. The isobaric analogue states are indicated and in particular the 1.74 MeV excited state in ^{10}B is the $T_z=0$ member of the $T=1$ triplet of 0^{+} levels; likewise the 5.17 MeV 2^{+} level in ^{10}B has $T=1$. Both of these states should not therefore be excited in the $^{12}C(d, \alpha)^{10}B$ reaction and experimentally they are very weakly populated in comparison with $T=0$ states. This is in part caused by the $T=1$ states in ^{10}B having a small admixture of $T=0$ states of the same J^{π} through the Coulomb force. In principle, isospin impurities in the ^{12}C,d and α ground states could also contribute as could the interaction itself if it had any significant electromagnetic (or weak) as well as nuclear component as then isospin need not be conserved.

As well as analogue states in nuclei of the same A there are analogue reactions which populate members of an isospin multiplet. An example is

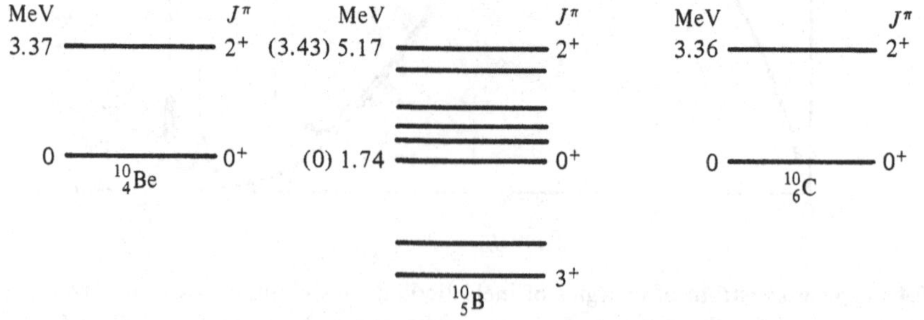

MeV J^{π} MeV J^{π} MeV J^{π}
3.37 2^{+} (3.43) 5.17 2^{+} 3.36 2^{+}
0 0^{+} (0) 1.74 0^{+} 0 0^{+}
$^{10}_{4}Be$ $^{10}_{6}C$
3^{+}
$^{10}_{5}B$

6.5. Some of the low-lying energy levels of $^{10}_{4}Be$, $^{10}_{5}B$ and $^{10}_{6}C$.

the reactions:

$$^{16}\text{O}(\text{p}, {}^3\text{H})^{14}\text{O} \quad \text{and} \quad ^{16}\text{O}(\text{p}, {}^3\text{He})^{14}\text{N}^*$$

The nucleus ^{16}O has a $T=0$ ground state and p, ^{3}He and ^{3}H all have $T=\frac{1}{2}$. Consider the (p, ^{3}H) reaction to the ^{14}O$T=1$, $J^\pi=0^+$ ground state and the analogue reaction to the ^{14}N*$T=1$, $J^\pi=0^+$ 2.31 MeV analogue state. As the structures of these states in ^{14}O and ^{14}N are identical the cross-sections to these states would be expected to have the same shape. They are not identical because the probability that the S-state pair of nucleons picked up by the proton is a pair of neutrons is twice that for a neutron–proton pair because the neutron–proton pair is equally likely to be paired up to ^{3}S$T=0$ as ^{1}S$T=1$ in the ^{16}O nucleus, unlike the S-state neutron pair which can only be coupled to ^{1}S$T=1$. This and similar predictions are borne out very well by experiment.

Barshay and Temmer pointed out that isospin conservation can be tested in reactions of the form:

$$A+B \to C+C'$$

where C and C' are members of the same isospin multiplet and where the isospin of either A or B is zero. If T is conserved and the analogue states C and C' have pure isospin then the differential cross-section will exhibit symmetry about 90° in the centre of mass system. This is because only one isospin value can contribute in the reaction as A or B has $T=0$, so the product of the space and spin wavefunctions has definite symmetry. Hence even and odd angular momenta are associated with orthogonal spin wavefunctions. The angular distribution, which is proportional to the total wavefunction squared, will therefore only contain even powers of cos θ. An example of one of these reactions is: d$+^4$He$\to ^3$H$+^3$He, for which $T=0$.

6.2.2 Detailed balance

Independent of the reaction mechanisms the cross-sections for the reactions:

$$a+A \to B+b \quad \text{and} \quad b+B \to A+a$$

which are inverse reactions, are related by time-reversal invariance. Fermi's golden rule states:

$$w_{if} = \frac{2\pi}{\hbar}|M_{if}|^2 \rho_f$$

where w_{if} is the reaction probability per unit time, M_{if} is the matrix element between initial and final states and ρ_f is the density of final states. This equation is valid for strong interactions and does not require the perturbation to be weak. To calculate the matrix element $|M_{if}|$, however, generally requires approximations and if $|M_{if}|$ is evaluated using first-order perturbation theory then the incoming and outgoing waves are assumed to be unaltered by the interaction, and this is then called the Born approximation. At the same centre of mass energy, time-reversal invariance requires:

$$|M_{if}|^2 = |M_{fi}|^2$$

The cross-sections σ_{if} and w_{if} are related by:

$$\sigma_{if} v_i = w_{if} \tag{6.4}$$

(taking unit normalisation volume) where v_i is the relative velocity in the entrance channel. Since ρ_f is given by:

$$\rho_f = \frac{4\pi}{h^3} p_f^2 \frac{dp_f}{dE} = \frac{4\pi}{h^3} p_f^2 \frac{1}{v_f}$$

then

$$\sigma_{if} v_i = \frac{2\pi}{\hbar} |M_{if}|^2 \frac{4\pi}{h^3} p_f^2 \frac{1}{v_f}$$

so for spinless particles:

$$\frac{\sigma_{if}}{p_f^2} = \frac{\sigma_{fi}}{p_i^2}$$

For particles with spin the relation, called the reciprocity theorem, becomes:

$$g_i \frac{\sigma_{if}}{p_f^2} = g_f \frac{\sigma_{fi}}{p_i^2}$$

where the statistical factors are:

$$g_f = (2I_B + 1)(2I_b + 1) \quad \text{and} \quad g_i = (2I_A + 1)(2I_a + 1)$$

The $(2I+1)$ factors are the numbers of spin substates for each particle.

6.2.3 Excitation functions

The equation for the cross-section (neglecting statistical factors) σ_{if}:

$$\sigma_{if} = \frac{2\pi}{\hbar} |M_{if}|^2 \frac{4\pi}{h^3} \frac{p_f^2}{v_i v_f} \tag{6.5}$$

can also be used to predict qualitatively the energy dependence of reactions in regions away from resonances. For neutrons, since the single-particle potential is ~50 MeV deep, then $|M_{if}|^2$ would be expected to vary only slowly with incident energy, while for charged particles at low energies the additional transmission factor for penetration through the Coulomb barrier would be expected to dominate $|M_{if}|^2$. The appropriate factor for charged particles of low energy is the ratio of the charged to neutral particle transmission factors, $(1/v)\exp(-2\gamma)$, which is the ratio for s-wave particles, where $\exp(-2\gamma)$ is the Gamow factor defined in chapter 5. These considerations predict the following energy dependences:

(i) Low-energy neutron capture (n,γ) – ρ_f constant so $\sigma \propto 1/v$.
(ii) Low-energy neutron elastic scattering – $\rho_f \propto v_f = v_i$ so σ constant.

(iii) Neutron inelastic scattering near threshold – $v_i \sim$ constant so $\sigma \propto v_f$.

(iv) Low-energy charged particle induced exothermic (Q positive) reactions – any energy dependence in the final channel is dominated at very low energies by the transmission factor in the entrance channel so $v_i \sigma \propto (1/v_i) \exp(-2\gamma_{IN})$ or $\sigma = [S(E)/E] \exp(-2\gamma_{IN})$ where $S(E)$ is expected to vary very slowly with energy. This formula is useful for calculating fusion reaction rates in stars (see chapter 5).

The predicted energy dependence of compound nucleus formation cross-sections also follows from a partial wave treatment assuming the nucleus acts like a strongly absorbing sphere (see below p. 201).

6.3 Elastic scattering

6.3.1 Rutherford scattering

The scattering of charged particles by a nucleus without loss of energy in the centre of mass frame is called elastic scattering. If the interaction is just the Coulomb interaction then it is called Rutherford scattering and the cross-section for this process can be worked out classically. For such a derivation to be justified the de Broglie wavelength of the incident particle should be small compared with the distance of closest approach. In the original Rutherford scattering experiments by Geiger and Marsden in 1913, 7.68 MeV alpha particles were scattered by gold. The wavelength $\lambda\!\!\!^{-}$ is given by:

$$\lambda\!\!\!^{-} = \frac{\hbar}{p} = \hbar c(2m_\alpha c^2 E)^{-1/2} \approx 1 \text{ fm}$$

while the distance of closest approach d_0 is given by

$$d_0 = \frac{Z_1 Z_2 e^2}{4\pi\varepsilon_0 E_\alpha} \approx 30 \text{ fm}$$

so a classical derivation would be expected to give a good approximation. However it is an accident of the $1/r$ Coulomb potential that the result is correct even for Z_1, Z_2 and incident energies where $d_0 < \lambda\!\!\!^{-}$.

Classically the elastic scattering process is as shown in figure 6.6. Consider the target initially as fixed. The Coulomb interaction with the incoming particle of mass m and charge ze only depends on r, i.e. is central, so angular momentum is conserved. Therefore

$$mvb = \text{a constant} = k$$

$$= mr^2 \frac{d\theta}{dt}$$

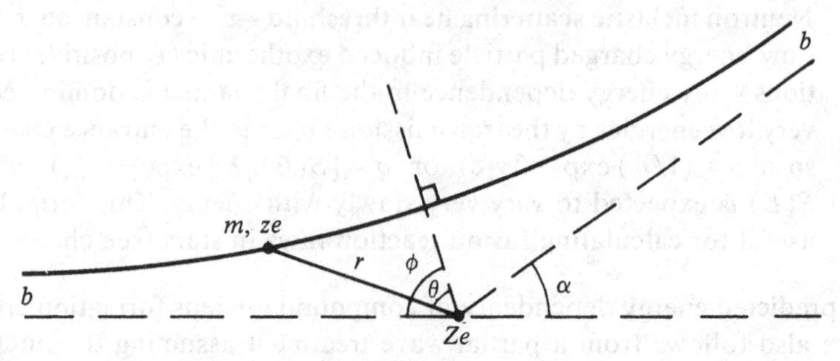

6.6. Diagram illustrating the Rutherford scattering of a charged particle.

The change in linear momentum is equal to the integral of the impulse so:

$$2mv \sin \tfrac{1}{2}\alpha = 2\int_0^\infty \frac{Zze^2}{4\pi\varepsilon_0 r^2} \cos\theta \, dt$$

$$= 2\int_0^\phi \frac{Zze^2 \cos\theta}{4\pi\varepsilon_0 k} m \, d\theta$$

$$= \frac{mZze^2}{2\pi\varepsilon_0 k} \cos \tfrac{1}{2}\alpha \qquad (\phi = 90 - \tfrac{1}{2}\alpha)$$

so

$$\cot \tfrac{1}{2}\alpha = \frac{mv^2 4\pi\varepsilon_0}{Zze^2} b = Kb$$

If the incident flux of particles is F_0 then the number dN passing between b and $b + db$ is given by:

$$dN = F_0 2\pi b \, db = \frac{\pi F_0}{K^2} \frac{\cos \tfrac{1}{2}\alpha}{\sin^3 \tfrac{1}{2}\alpha} d\alpha \tag{6.6}$$

where $K = mv^2 4\pi\varepsilon_0 / Zze^2$. The solid angle $d\Omega$ that the particles are scattered into is:

$$d\Omega = 2\pi \sin\alpha \, d\alpha = 4\pi \sin \tfrac{1}{2}\alpha \cos \tfrac{1}{2}\alpha \, d\alpha \tag{6.7}$$

Combining equations 6.6 and 6.7 gives the Rutherford scattering formula:

$$\frac{d\sigma}{d\Omega} = \frac{1}{F_0}\frac{dN}{d\Omega} = \left(\frac{Zze^2}{16\pi\varepsilon_0 E}\right)^2 \sin^{-4}(\tfrac{1}{2}\alpha)$$

where $E = \frac{1}{2}mv^2$ is the energy of the incident particle. To take account of the finite mass M of the target the mass m must be replaced by the reduced mass $\mu \equiv mM/(m+M)$ and the angle α becomes the scattering angle in the centre of mass frame.

The Rutherford formula can also be obtained using Fermi's golden rule:

$$w = \frac{2\pi}{\hbar} |M_{if}|^2 \rho_f$$

where w is the scattering probability per unit time, M_{if} is the matrix element between initial and final states of the perturbing potential V and ρ_f is the density of final states. For $V = Zze^2/4\pi\varepsilon_0 r$ and a scattering from momentum $\mathbf{p}_i = \hbar\mathbf{k}_i$ to $\mathbf{p}_f = \hbar\mathbf{k}_f$ then $\psi_i = L^{-3/2}\exp(\mathrm{i}\mathbf{k}_i\cdot\mathbf{r})$ and $\psi_f = L^{-3/2}\exp(\mathrm{i}\mathbf{k}_f\cdot\mathbf{r})$ where L^3 is the normalisation volume.

As shown in chapter 1 (equations 1.1 and 1.2) the matrix element M_{if} is given by:

$$M_{if} = \frac{Zze^2}{\varepsilon_0 q^2 L^3}$$

where $\hbar\mathbf{q} = \mathbf{p}_i - \mathbf{p}_f$ is the momentum transfer in the scattering so $q = (2mv/\hbar)\sin\frac{1}{2}\alpha$. The relations:

$$\frac{v\,\mathrm{d}\sigma}{L^3} = w \quad \text{and} \quad \rho_f = 4\pi p_f^2 \frac{\mathrm{d}p_f}{\mathrm{d}E}\frac{L^3}{h^3}\frac{\mathrm{d}\Omega}{4\pi} \tag{6.8}$$

with $p_f = mv$ and $\mathrm{d}p_f/\mathrm{d}E = 1/v$ then give Rutherford's formula.

6.3.2 Elastic scattering at energies above the Coulomb barrier

As the energy of an incident charged particle is increased the probability of elastic scattering becomes small when the Coulomb barrier is overcome as inelastic, transfer and compound nucleus processes tend to occur with the result that the nucleus behaves rather like a strong absorbing sphere. Quantum-mechanically the beam of incident particles is described by a travelling wave with $\lambda = h/p$. The effect of an absorbing sphere on such a wave will be to give rise to diffraction effects just like in the scattering of light. If Coulomb effects can be neglected, as in the scattering of α-particles at high energies, then the incident wave is a plane wave and a Fraunhofer-like diffraction pattern is seen. Assuming strong absorption the nucleus acts like a black disc of radius R and the differential cross-section is equal to the corresponding Fraunhofer diffraction pattern, i.e.:

$$\frac{\mathrm{d}\sigma}{\mathrm{d}\Omega} = k^2R^4\left[\frac{J_1(qR)}{qR}\right]^2 \tag{6.9}$$

where $qR \approx kR\theta$ for small θ, $k = h/\lambda$ and J_1 is the Bessel function of the first order. For large qR:

$$[J_1(qR)]^2 \approx (2/\pi qR)\sin^2(qR - \pi/4)$$

which has successive zeros separated by $\delta\theta \approx \pi/kR$. An example is seen in figure 6.7, which shows the angular distribution of 52 MeV deuterons elastically scattered off ^{54}Fe. For this reaction $k = 2.15\ \mathrm{fm}^{-1}$, $\delta\theta \approx 0.23$ rad, predicting $R \approx 6.3$ fm. In general the estimate for R based on the sizes of the interacting nuclei, $R = r_0(A_1^{1/3} + A_2^{1/3})$ with $r_0 = 1.2$ fm, slightly underestimates the strong interaction radius, and this reflects the range of the nuclear interaction.

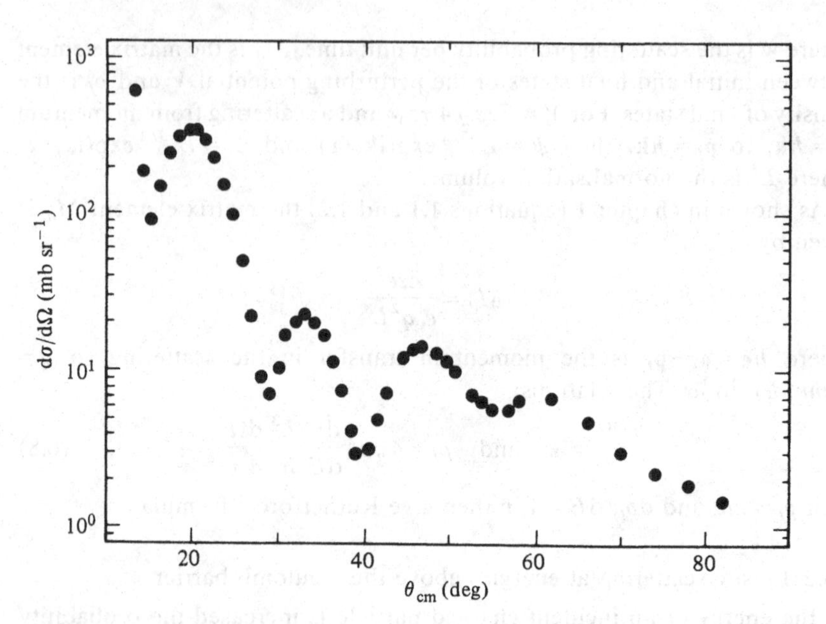

6.7. The angular distribution of 52 MeV deuterons elastically scattered off ^{54}Fe. The angle θ_{cm} is in the centre of mass frame. (From Hintenberger, F. *et al.*, *Nucl. Phys.* **A115** (1968) 570.)

When the Coulomb interaction cannot be ignored, as in the scattering of heavy ions such as ^{16}O off ^{208}Pb, the incident particles on either side of the nucleus are deflected and the resulting two wavefronts do not interfere. The effect is to give an angular distribution very like the Fresnel diffraction pattern from the edge of a circular disc. A comparison of the optical and observed distribution for ^{16}O scattering off ^{208}Pb is shown in figure 6.8.

6.3.3 Optical model for elastic scattering

The quantum-mechanical description can be improved considerably by describing the interaction not just by a strong absorption but by a potential V, which includes both the nuclear and Coulomb potentials, and the absorption by an imaginary potential $\mathrm{i}W$. A complex potential means that the wavevector K for the particle within the nuclear potential is complex:

$$K = K_r + \mathrm{i}K_m$$

and the wavefunction becomes:

$$\psi = \exp(\mathrm{i}Kz) = \exp(\mathrm{i}K_r z) \cdot \exp(-K_m z)$$

so

$$|\psi|^2 = \exp(-2K_m z)$$

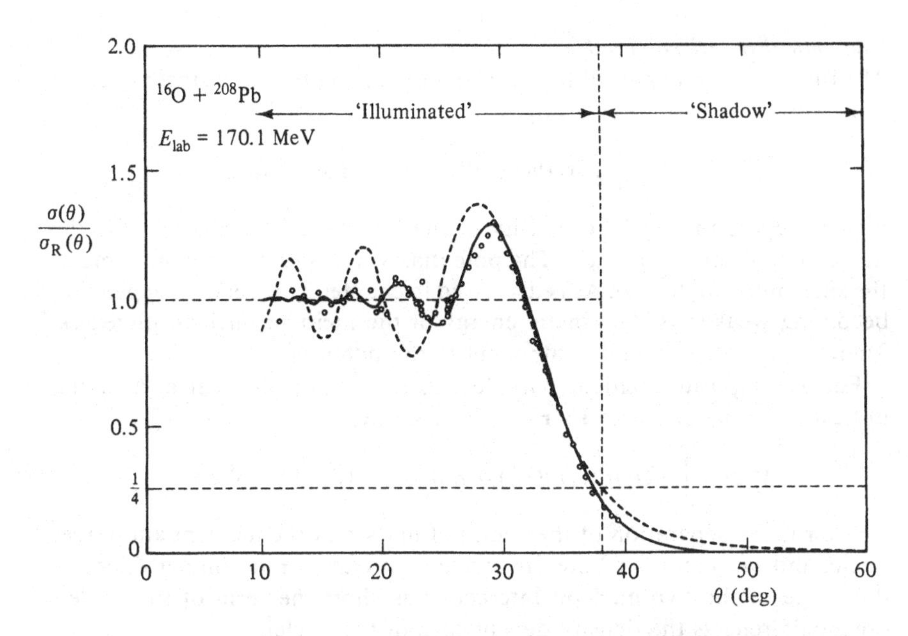

6.8. Fresnel diffraction pattern in heavy-ion scattering. Plotted is the ratio of observed to Rutherford cross-section for 170.1 MeV ^{16}O elastically scattered off ^{208}Pb. The dotted curve is the Fresnel diffraction pattern shape predicted by the strong-absorption model. (From Frahn, W. E., *Ann. Phys.* **72** (1972) 524.)

The incident wave is therefore attenuated with a mean free path λ_m given by

$$\lambda_m = \frac{1}{2K_m} \tag{6.10}$$

The wavevector K satisfies:

$$\frac{\hbar^2 K^2}{2m} = U + E$$

where $U = V + \mathrm{i}W$. Solving for K_r and K_m yields:

$$K_r^2 - K_m^2 = \frac{2m}{\hbar^2}(V+E) \quad \text{and} \quad K_r K_m = \frac{m}{\hbar^2} W$$

For W and $E \ll V$ then:

$$K_m \approx \frac{W}{\hbar}\sqrt{\frac{m}{2V}} \tag{6.11}$$

Using a complex potential enables the scattering of both strongly and weakly absorbed particles such as low-energy neutrons to be described.

The optical model potential

The form of the real part of the optical model potential for a single nucleon at $\mathbf{r}_s$ with $\mathbf{r}_{st} = \mathbf{r}_s - \mathbf{r}_t$ is given by:

$$V(\mathbf{r}_s) = \int \rho(\mathbf{r}_t) v(\mathbf{r}_{st})\, \mathrm{d}^3 r_t \qquad \text{(single folding)}$$

where $\rho(\mathbf{r}_t)$ is the nucleon density of the target nucleus and $v(\mathbf{r}_{st})$ is the nucleon-nucleon interaction. The potential $V(\mathbf{r}_s)$ is essentially the same as the shell model potential. As is discussed in chapter 7, it is energy dependent becoming weaker as the kinetic energy of the incident nucleon increases. There is also a spin-orbit component to the potential.

For a composite incident particle, e.g. a ^{16}O ion, the real part of the optical potential at $\mathbf{r}$ with $\mathbf{r} = \mathbf{r}_s - \mathbf{r}_i$ is given by:

$$V(\mathbf{r}) = \int \rho(\mathbf{r}_i)\rho(\mathbf{r}_t) v(\mathbf{r}_{st})\, \mathrm{d}^3 r_i\, \mathrm{d}^3 r_t \qquad \text{(double folding)}$$

where $\mathbf{r}$ is the separation of the centre of masses of the incident and target nuclei and $\rho(\mathbf{r}_i)$ and $\rho(\mathbf{r}_t)$ are the incident and target nucleon densities. As the range of the two-nucleon interaction is short the form of the folded potentials reflects the density dependence of the nuclei.

The imaginary part of the optical potential reflects the loss of incident particle flux through nuclear reactions. For nucleon scattering at low energies reactions tend to occur at the nuclear surface (see chapter 7, p. 237) but at higher energies ($E > 50$ MeV) interactions tend to occur throughout the nucleus. For heavy ions, absorption is strong and reactions tend to occur at the nuclear surface.

A very common analytic form used to parametrise the optical potential is:

$$U(r) = -Vf(r, R, a) - \mathrm{i}\, Wf(r, R', a') - \mathrm{i}\, W_{\mathrm{D}} g(r, R', a')$$

where f is the Woods–Saxon form factor

$$f(r, R, a) = (\mathrm{e}^x + 1)^{-1}, \qquad x = (r - R)/a$$

and g is its derivative

$$g(r, R', a') = -4a \frac{\mathrm{d}}{\mathrm{d}r} f(r, R', a')$$

Typical shapes of f and g are shown in figure 6.9.

The analysis of elastic scattering

The Schrödinger equation for the scattering of a particle by a fixed potential $U(r)$ is:

$$\left[-\frac{\hbar^2}{2m}\nabla^2 + U(r)\right]\psi(r) = E\psi(r).$$

The solution $\psi(r)$ is called the distorted wave and describes the elastic

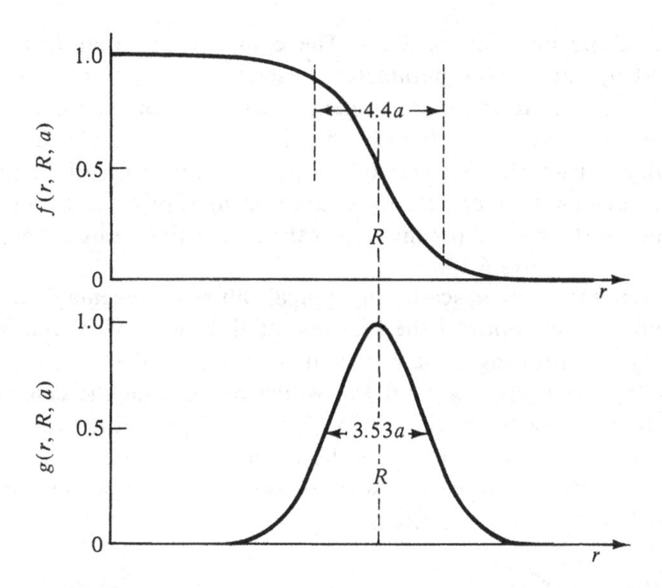

6.9. The Woods–Saxon and Woods–Saxon derivative shapes used for optical model potentials with $a/R \approx 1/9$. (From Satchler, G. R., *Introduction to Nuclear Reactions*. Wiley (1980).)

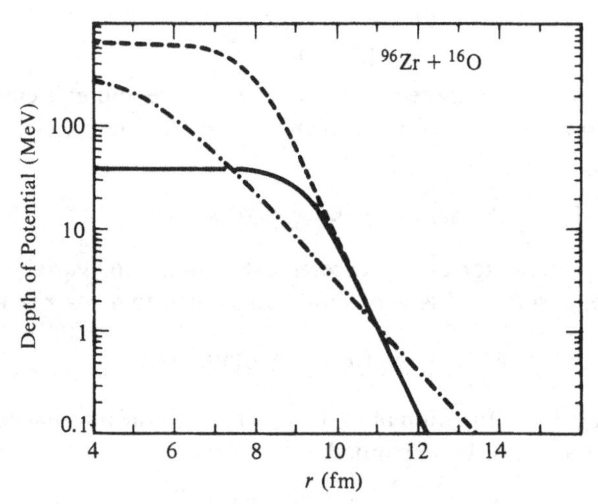

6.10. The real-part of the nuclear contribution to three optical model potentials which fit $^{16}O + ^{96}Zr$ elastic scattering at $E_{lab}(^{16}O) = 60$ MeV. (From Becchetti, F. D. *et al.*, *Nucl. Phys.* **A203** (1973) 1.)

scattering of the incident particles. The complex potential $U(r)$ can be determined by varying the parameters describing $U(r)$ to obtain the best fit to the elastic scattering data. In general there is not a unique potential but a family of potentials which fit the elastic scattering. For heavy ions, in particular, for which there is strong absorption the shape of the potential within the nucleus is not critical and the scattering is principally determined by the value of the optical potential near the interaction radius. An example of this is seen in figure 6.10.

For low-energy nucleon scattering, typical values for the imaginary potential W and for the depth of the real potential V are 3 MeV and 50 MeV, respectively. Substituting these values in equations 6.10 and 6.11 above for the mean free path gives $\lambda_m \approx 10$ fm, which means that the chance of an interaction is relatively small. The excitation energy in the target nucleus is the incident nucleon energy plus the nucleon separation energy, which is typically ~8 MeV. This weak absorption is in agreement with the independent-particle model of the nucleus.

6.4 Inelastic scattering

In a peripheral collision there is the possibility of energy transfer, which will give rise to inelastic scattering. This direct surface interaction is likely to excite collective vibrational and rotational states. The differential cross-section for such a reaction can be calculated using equations 6.5 and 6.8. For the reaction $a+A \rightarrow A^*+a$ then:

$$\frac{d\sigma}{d\Omega}=\frac{\mu^2}{(2\pi\hbar^2)^2}\frac{k_f}{k_i}|M_{if}|^2$$

where k_f and k_i are the wavevectors in the entrance and exit channels and μ is the reduced mass. Assuming plane waves the matrix element M_{if} is given by:

$$M_{if}=\int \psi_{A^*}^* \psi_a^* \exp(-i\mathbf{k}_f \cdot \mathbf{r}_f)\, V \psi_A \psi_a \exp(i\mathbf{k}_i \cdot \mathbf{r}_i)\, d\tau \qquad (6.12)$$

where V is the interaction and τ represents the integration variables. Making the approximation that V is very short range such that $\mathbf{r}_i=\mathbf{r}_f=\mathbf{r}$ then:

$$M_{if}=\int \exp(i\mathbf{q}\cdot\mathbf{r})\langle A^*a|V|Aa\rangle\, d^3r$$

The interaction V is a function of $\mathbf{r}=(r, \theta, \phi)$ and of the internal coordinates τ_i of A and a and can be expanded as:

$$V=\sum_{lm} V_{lm}(r, \tau_i)\, Y_{lm}(\theta, \phi) \qquad (6.13)$$

The V_{lm} cause transitions involving a transfer of angular momentum l and a parity change $(-1)^l$. For inelastic scattering off spin zero nuclei then only one l is involved (e.g. $0^+ \to 2^+$) and M_{if} becomes:

$$M^l_{if} = \int \exp(i\mathbf{q}\cdot\mathbf{r})\langle A^*a|V_{lm}|Aa\rangle Y_{lm}(\theta,\phi)\,d^3r$$

$$= \int \exp(i\mathbf{q}\cdot\mathbf{r})f(r)Y_{lm}(\theta,\phi)\,d^3r$$

Now $\exp(i\mathbf{q}\cdot\mathbf{r})$ can be expanded in terms of spherical Bessel functions and spherical harmonics:

$$\exp(i\mathbf{q}\cdot\mathbf{r}) = \sum_l i^l[4\pi(2l+1)]^{1/2} j_l(qr)Y_{l0}(\theta)$$

where $j_l(qR) \equiv (\pi/2qr)^{1/2}J_{l+1/2}(qr)$. Since the spherical harmonics are orthogonal then M^l_{if} becomes:

$$M^l_{if} = \int i^l[4\pi(2l+1)]^{1/2} j_l(qr)f(r)r^2\,dr$$

If the nucleus acts like a black non-circular disc of radius $\approx R$, then V_{lm} arises from the surface rim and M^l_{if} involves regular Bessel functions, e.g.

$$|M^2_{if}|^2 \propto \tfrac{1}{4}[J_0(qR)]^2 + \tfrac{3}{4}[J_2(qR)]^2$$

For large values of the argument:

$$[J_l(qR)]^2 \approx (2/\pi qR)\sin^2(qR - \tfrac{1}{2}l\pi + \pi/4)$$

so inelastic scattering from a 0^+ to an even J^+ state is predicted to oscillate out of phase with the elastic scattering, which is proportional to $[J_1(qR)]^2$ (eqn. 6.9), while to an odd J^- state it should be in phase. An example of this, which is called the Blair phase rule, is shown in figure 6.11. The theoretical curve for the inelastic scattering is obtained by using, instead of plane waves in the expression for M_{if} above (eqn 6.12), distorted waves which describe the elastic scattering of $a+A$ by the curve shown. This is called the distorted wave Born approximation (DWBA).

6.4.1 The collective model and inelastic scattering

For a deformed nucleus, the perturbation which causes inelastic scattering is the difference between the spherical part of the optical potential U_0, which gives rise to elastic scattering, and the total potential U, i.e.:

$$V = U - U_0$$

The total potential will follow the shape of the nucleus and will therefore depend on $|\mathbf{r}-\mathbf{R}|$ where $\mathbf{R}$ is the radius of the nucleus in the direction $\mathbf{r} = (r, \theta, \phi)$. If the radius R is parametrised by

$$R(\theta', \phi') = R_0\left[1 + \sum_{lm'} \alpha^*_{lm'} Y_{lm'}(\theta', \phi')\right]$$

where the angles (θ', ϕ') are with respect to the symmetry axis of the nucleus, which is at an angle Ω to the incident particle (the beam) direction, then

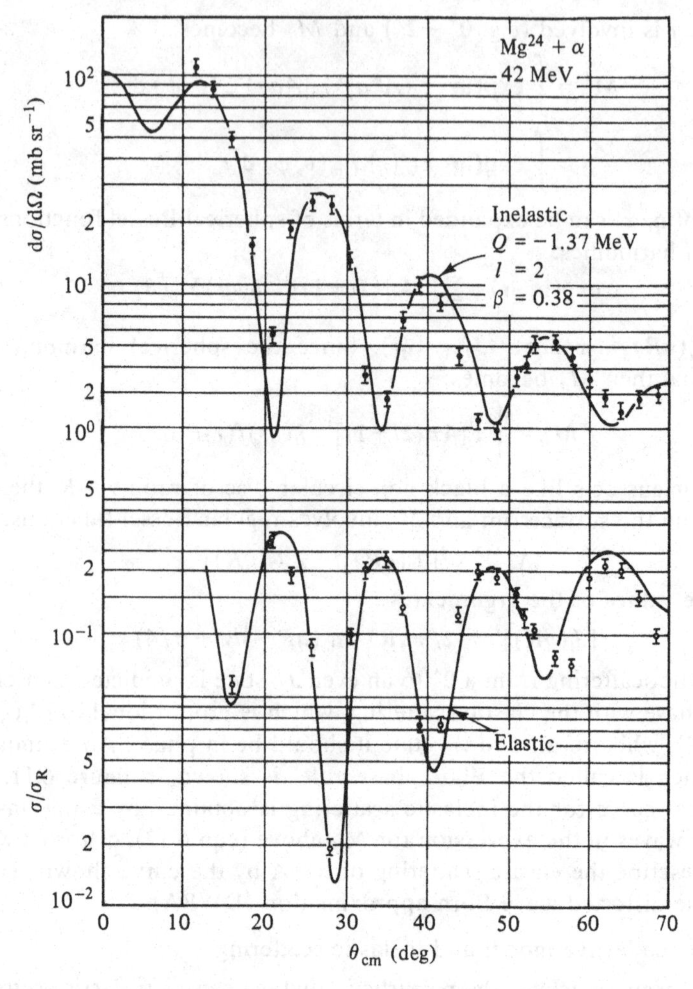

6.11. An example of the Blair phase rule. Shown are the elastic and inelastic scattering to the 1.37 MeV 2^+ state of 42 MeV α-particles off ^{24}Mg. The solid curves are the optical model and DWBA fits to the data. (From Rost, E., *Phys. Rev.* **128** (1962) 2708.)

the radius in the direction **r** is given by:

$$R(\theta, \phi) = R_0\left[1 + \sum_{lm}\sum_{m'} \alpha^*_{lm'} Y_{lm}(\theta, \phi) D^l_{mm'}(\Omega)\right]$$

where the relation (appendix page 265)

$$Y_{lm'}(\theta', \phi') = \sum_m Y_{lm}(\theta, \phi) D^l_{mm'}(\Omega)$$

has been used. The function $D^l_{mm'}(\Omega)$ is called the rotation matrix. The potential U can be expanded in a Taylor series:

$$U = U_0 + \left(\frac{\partial U}{\partial R}\right)_{R_0}(R - R_0) + \cdots$$

$$\therefore \quad V = \sum_{lm} R_0 \left(\frac{\partial U}{\partial R}\right)_{R_0} \sum_{m'} \alpha^*_{lm'} D^l_{mm'}(\Omega)\, Y_{lm}(\theta, \phi)$$

Hence in this model the $V_{lm}(r, \tau_i)$ of equation 6.13 are given by:

$$V_{lm}(r, \tau_i) = R_0 \left(\frac{\partial U}{\partial R}\right)_{R_0} \sum_{m'} \alpha^*_{lm'} D^l_{mm'}(\Omega)$$

Assuming a Woods–Saxon form for U then $(\partial U/\partial R)_{R_0}$ and hence V_{lm} is peaked at the surface (see figure 6.9) as assumed above.

For an axially-symmetric even–even nucleus then only $\alpha^*_{l0} = \alpha_l$ ($\alpha_2 \equiv \beta$ see p. 52) are not zero. Also the eigenfunctions of the rotational levels are $D^J_{M0}(\Omega)$ so in the matrix element of V only the term with $l = J$ will contribute as all others will be orthogonal. Therefore:

$$\langle A^* a|V|Aa\rangle \propto R_0 \left(\frac{\partial U}{\partial R}\right)_{R_0} \alpha_J$$

so the magnitude of the cross-section is proportional to the deformation parameter squared, $|\alpha_J|^2$. Fitting to experimental data gives values of α_J in good agreement with those obtained from electromagnetic transition rates, which in the collective model also depend on $|\alpha_J|^2$.

6.4.2 Coulomb excitation

An important example of an inelastic scattering reaction is that caused by the Coulomb interaction. For low enough incident energies the excitation is only via the long-range Coulomb interaction and there is no nuclear interaction. The cross-section for such a reaction can be calculated approximately by assuming the incident particle is on a classical Rutherford trajectory and integrating the probability of a transition along the trajectory. The differential cross-section can then be written:

$$\frac{\mathrm{d}\sigma}{\mathrm{d}\Omega} = \frac{\mathrm{d}\sigma}{\mathrm{d}\Omega}(\text{Rutherford}) \cdot P(\text{trajectory})$$

where P(trajectory) is the integrated transition probability. If the perturbing interaction between the incident charged particle q and the nucleus is V then the probability amplitude for a transition from the ground state to an excited state f is given by:

$$a_{if} = \frac{1}{\mathrm{i}\hbar} \int V_{if} \exp(\mathrm{i}\omega t)\, \mathrm{d}t$$

where $\hbar\omega = E_f - E_i$, $P(\text{trajectory}) = |a_{if}|^2$ and

$$V_{if} = \int \psi_f^* V \psi_i \, d\tau$$

As a specific simple example consider the excitation of the first excited 2^+ rotational state from the 0^+ ground state of a deformed nucleus by a charged particle which undergoes an 180° back-scattering. The perturbation V in this case arises through the interaction of the electric field of the incident particle with the quadrupole moment of the nucleus (see p. 37). The matrix element is:

$$V_{if} = \frac{1}{2} \frac{qeQ_{if}}{4\pi\varepsilon_0 r^3}$$

with the quadrupole transition matrix element:

$$Q_{if} = \frac{1}{e} \sum_i \int \psi_f^* q_i (3z_i^2 - r_i^2) \psi_i \, d\tau$$

where q_i is the charge of the ith nucleon. The probability amplitude is then:

$$a_{if} = (qeQ_{if}/8\pi\varepsilon_0 i\hbar) \int \frac{1}{r^3} \exp(i\omega t) \, dt$$

For 180° back-scattering the relation between the separation r and velocity v of the incident particle is given in terms of the distance of closest approach s and initial velocity v_0 by:

$$v = \frac{dr}{dt} = \pm v_0 \left(1 - \frac{s}{r}\right)^{1/2}$$

since by conservation of energy:

$$\tfrac{1}{2}mv^2 + \frac{qZe}{4\pi\varepsilon_0 r} = \tfrac{1}{2}mv_0^2 \quad \text{and} \quad s = (qZe/2\pi\varepsilon_0 m v_0^2)$$

At a separation of $2s$ the velocity is $v_0/\sqrt{2}$ so the time taken for the incident particle to go from $2s \to s$ is of the order of $2\sqrt{2}s/v_0$. Because of the $1/r^3$ factor in the integral in the expression for a_{if} the duration of the interaction is therefore of the order of $10s/v_0$. If:

$$10 \frac{\omega s}{v_0} \ll 1$$

then $\exp(i\omega t) \approx 1$ during the interaction hence:

$$\begin{aligned} a_{if} &\approx (qeQ_{if}/8\pi\varepsilon_0 i\hbar) \int \frac{dt}{r^3} \\ &= \frac{2A}{v_0} \int_s^\infty \frac{dr}{r^3(1 - s/r)^{1/2}} \\ &= \frac{8A}{3s^2 v_0} \end{aligned}$$

Therefore the differential cross-section is given by:

$$\frac{d\sigma}{d\Omega}=\frac{d\sigma}{d\Omega}(180^\circ \text{ Rutherford})\cdot\frac{64A^2}{9s^4v_0^2}$$

where

$$\frac{d\sigma}{d\Omega}(180^\circ \text{ Rutherford})=\frac{s^2}{16}$$

Substituting yields

$$\frac{d\sigma}{d\Omega}(180^\circ)=\frac{m^2v_0^2}{36Z^2\hbar^2}|Q_{if}|^2$$

$$=0.133\frac{mE}{Z^2}|Q_{if}|^2 \text{ barn sterad}^{-1}$$

where m is in a.m.u., E the incident particle energy is in MeV and Q_{if} is in barns. For a $0^+\rightarrow 2^+$ transition in a rotational band the intrinsic wavefunctions in the collective model are the same for both states and $Q_{if}=Q_0$, the quadrupole moment of the ground state. The cross-section increases both with the mass and energy of the incident particle so heavy ions are more effective than light ions in Coulomb excitation. The cross-section, however, is independent of the charge of the incident particle.

6.5 Compound nucleus reactions

At low energies and small impact parameters a nuclear reaction is most likely to occur by the formation of a compound nucleus. Such reactions are characterised by resonances occurring when the energy of the incident particle is such that a long-lived compound nucleus state is formed. An example is shown in figure 6.12. The shape reflects the probability that the incident particle energy matches the energy of the compound nucleus state, which does not have a unique sharp value because it is unstable. The time dependence of such a state corresponds to the level having a complex energy eigenvalue, $E_r+i\Gamma/2$, for then:

$$\psi(t)=\exp\left[\frac{i}{\hbar}\left(E_r+\frac{i\Gamma}{2}\right)t\right]$$

$$\text{so}\quad |\psi(t)|^2=\exp\left(-\frac{\Gamma t}{\hbar}\right)$$

The width of the level is therefore Γ and its lifetime $\tau=\hbar/\Gamma$.

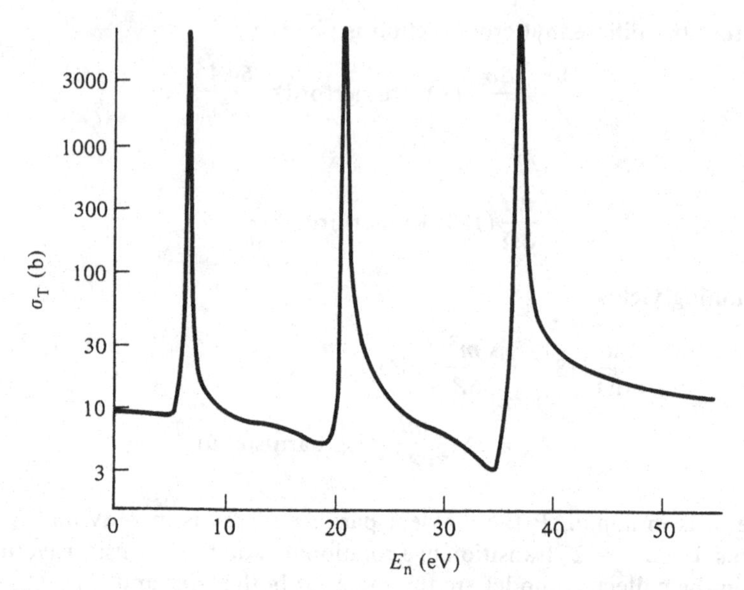

6.12. Resonances in the total absorption cross-section for low-energy neutrons incident on ^{238}U. The peaks correspond to the formation of levels in ^{239}U. (From Jones, G. A., *Properties of Nuclei*, Oxford University Press (1987).)

The wavefunction $\psi(t)$, which is created in the reaction at $t=0$, can be represented as a superposition of states of definite energy:

$$\psi(t)=\frac{1}{\sqrt{2\pi}}\int_{-\infty}^{\infty} a(E)\exp\left(\frac{\mathrm{i}Et}{\hbar}\right)\mathrm{d}t$$

The amplitude $a(E)$ is therefore the Fourier transform of $\psi(t)$:

$$a(E)=\frac{1}{\sqrt{2\pi}}\int_{0}^{\infty} \psi(t)\exp\left(\frac{-\mathrm{i}Et}{\hbar}\right)\mathrm{d}t$$

The limits are 0 to infinity because $\psi(t)=0$ for $t<0$. Substituting for $\psi(t)$ gives:

$$|a(E)|^2\propto\frac{1}{(E_r-E)^2+\Gamma^2/4}$$

The probability distribution for the compound nucleus level having an energy E is called a Breit–Wigner shape and reproduces the energy dependence of resonant reaction cross-sections (not elastic). For elastic scattering cross-sections there is interference between direct and compound elastic scattering (apparent between the resonances in figure 6.12).

To calculate the magnitude of a resonant reaction cross-section, time-independent perturbation theory can be used. Unit normalisation volume will be assumed and the spin factors g taken as 1. The cross-section σ_{if} is:

$$\sigma_{if} = \frac{2\pi}{\hbar} |M_{if}|^2 \frac{4\pi}{h^3} \frac{p_f^2}{v_i v_f} \tag{6.5}$$

The decay rate of the compound nucleus level r to the final state is:

$$\frac{1}{\tau_f} = \frac{\Gamma_f}{\hbar} = \frac{2\pi}{\hbar} |M_{rf}|^2 \frac{4\pi}{h^3} \frac{p_f^2}{v_f} \tag{6.14}$$

By second-order time-independent perturbation theory:

$$M_{if} = \sum_c \frac{M_{ic} M_{cf}}{E - E_c}$$

where c are the intermediary states. If only one state is significant with $E_c = E_r + i\Gamma/2$ then:

$$|M_{if}|^2 = \frac{|M_{ir}|^2 |M_{rf}|^2}{(E_r - E)^2 + \Gamma^2/4} \tag{6.15}$$

Now the level width Γ_i is given by:

$$\frac{\Gamma_i}{\hbar} = \frac{2\pi}{\hbar} |M_{ir}|^2 \frac{4\pi p_i^2}{h^3 v_i} \tag{6.16}$$

Combining equations 6.5, 6.14, 6.15 and 6.16 yields:

$$\sigma_{if} = \frac{\pi \lambda\!\!\!^{-2} \Gamma_i \Gamma_f}{(E_r - E)^2 + \Gamma^2/4}$$

This is the Brei–Wigner formula for incident, target and compound nucleus all with $J = 0$. With spin σ_{if} is multiplied by $(2J_c + 1)/(2J_i + 1)(2J_t + 1)$.

In this derivation the connection between the incoming and outgoing particles and the compound nucleus is through the matrix elements M_{ri} and M_{fr}. A closer link to an independent-particle description of the compound nucleus is provided by describing the reaction using partial waves.

6.5.1 A partial wave description of a reaction

A spinless uncharged particle incident on a target can be described by a wavefunction $\exp(ikz)$ which can be expanded as a sum of eigenfunctions of orbital angular momentum:

$$\exp(ikz) = \sum_{l=0}^{\infty} i^l (2l+1) j_l(kr) P_l(\cos\theta)$$

where $P_l(\cos\theta)$ are the Legendre polynomials and $j_l(kr)$ are the spherical Bessel functions. For large kr:

$$\begin{aligned} j_l(kr) &\Rightarrow (kr)^{-1} \sin(kr - \tfrac{1}{2} l\pi) \\ &= \frac{i^{1-l}}{2kr} [(-1)^l \exp(-ikr) - \exp(ikr)] \end{aligned}$$

which is the sum of ingoing and outgoing spherical waves. If the reaction can be described by a central potential $V(r)$ then $[H, V]=0$ and angular momentum is conserved. This means that in an interaction the only effect of $V(r)$ can be to alter the phase and magnitude of the outgoing spherical wave by a factor η_l. The wavefunction describing the elastically scattered particles after the reaction will therefore be at large distances:

$$\psi_{sc} = \psi_f - \psi_i$$

$$\psi_{sc} = \psi_f - \exp(ikz)$$

$$\psi_{sc} = \sum_{l=0} \frac{i(2l+1)}{2kr}[1-\eta_l]\exp(ikr)P_l(\cos\theta)$$

where ψ_i is the initial wavefunction, and ψ_f is the final wavefunction which describes the particle after the interaction. The flux of scattered particles is $|\psi_{sc}|^2 v$ so for a given l the total number of scattered particles per second N_l is:

$$N_l = \frac{(2l+1)^2}{4k^2r^2}|1-\eta_l|^2 v \int P_l^2(\cos\theta) r^2 \sin\theta \, d\theta \, d\phi$$

$$= \pi\lambda\!\!\!^-{}^2(2l+1)|1-\eta_l|^2 v$$

As the incident flux is v the partial scattering cross-section $\sigma_{s,l}$ is therefore given by:

$$\sigma_{s,l} = (2l+1)\pi\lambda\!\!\!^-{}^2|1-\eta_l|^2$$

If $|\eta_l| < 1$ then there is absorption of the incoming particles arising through reactions and:

$$\sigma_{r,l} = (2l+1)\pi\lambda\!\!\!^-{}^2(1-|\eta_l|^2)$$

There can therefore be scattering with no reaction but not a reaction without some scattering. If there is only elastic scattering then $|\eta_l|$ is 1 and η_l can be written in terms of a phase shift δ_l:

$$\eta_l = \exp(i2\delta_l)$$

so

$$\sigma_{s,l} = (2l+1)4\pi\lambda\!\!\!^-{}^2 \sin^2\delta_l \qquad (6.17)$$

For $l=0$ the wavefunction ψ_0, which is valid for all r, is given by:

$$\psi_0 = \frac{i}{2kr}[\exp(-ikr) - \eta_0 \exp(ikr)]$$

$$= \frac{\sin(kr+\delta_0)}{kr}\exp(i\delta_0)$$

$$= \frac{u_0(r)}{kr}\exp(i\delta_0)$$

which defines the radial function $u_0(r)$. The effect of a simple square well on $u_0(r)$ is shown in figure 6.13.

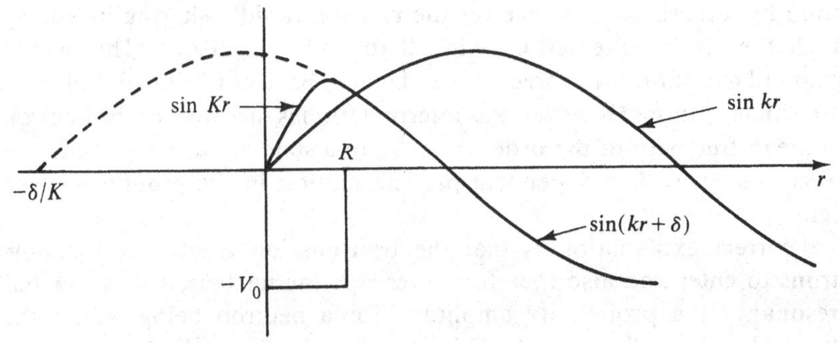

6.13. Illustration of the phase shift δ introduced by a simple square well of depth V_0 to the radial function $u_0(r)$. K is the wavenumber of the particle within the well.

6.5.2 Limiting cross-sections

The maximum reaction cross-section for a given partial wave is:

$$\sigma_{r,l}(\max) = (2l+1)\pi\lambda\!\!\!^{-2}$$

The classical impact parameter b_l corresponding to an angular momentum l is given by $b_l = l\lambda\!\!\!^{-}$ so this maximum cross-section corresponds to the area of the incoming plane wave with impact parameters lying between b_l and b_{l+1}. Its value, even for $l=0$, which is $\pi\lambda\!\!\!^{-2}$, can be very much larger than the size (cross-sectional area) of a nucleus, which is a reflection of the fact that the particles are described by wavefunctions and not by classical mechanics. A similar phenomenon is found in the absorption of light by atoms where the resonant absorption cross-section can be much greater than the size of an atom.

If the nucleus is strongly absorbing such that all particles with impact parameters $b_l \leq R$ are absorbed then this will correspond to $\eta_l = 0$ for $l \leq kR$ and $\eta_l = 1$ for $l > kR$. The total reaction cross-section σ_r will then be given by:

$$\sigma_r = \pi\lambda\!\!\!^{-2} \sum_{l=0}^{kR} (2l+1) = \pi\lambda\!\!\!^{-2}(kR+1)^2 = \pi(R+\lambda\!\!\!^{-})^2$$

The elastic scattering cross-section σ_{el} in this case equals σ_r. For $\lambda\!\!\!^{-} \ll R$ the reaction cross-section equals the geometric cross-section.

6.5.3 Slow neutron-induced resonances

Very low-energy (near thermal) neutron capture resonances were seen in 1935 by Fermi and others which corresponded to narrow and hence relatively long-lived states in the compound nucleus at an excitation energy of ~8 MeV. This was interpreted by Bohr in 1936 in his compound nucleus reaction hypothesis as evidence for the neutron rapidly sharing its energy through the strong interaction with all the other nucleons. This would correspond to a short mean free path and would be in conflict with independent-particle motion. However, this interpretation is incorrect as the neutron has a mean free path of the order of 10 fm in a state at ~8 MeV excitation quite consistent with independent-particle motion in the ground state of nuclei.

The correct explanation is that the transmission coefficient for slow neutrons to enter and also therefore to escape the nucleus is very low but on resonance the probability amplitude for a neutron being within the nucleus is large (see figure 6.14). This phenomenon is very like what occurs in the transmission of light by a Fabry–Perot etalon. Off a resonant frequency the reflection is high and the amplitude within the etalon very low. When the incident frequency is on resonance a pulse of light will be delayed by the etalon as the intensity within the etalon builds up and will be transmitted with none reflected. For neutrons on resonance there is a time delay likewise. The high amplitude for the neutron to be within the nucleus and the many reflections at the nuclear surface makes the path length many times the mean free path so the neutron energy is shared with the other nucleons. The effect is the same as from Bohr's hypothesis but is compatible with independent-particle motion.

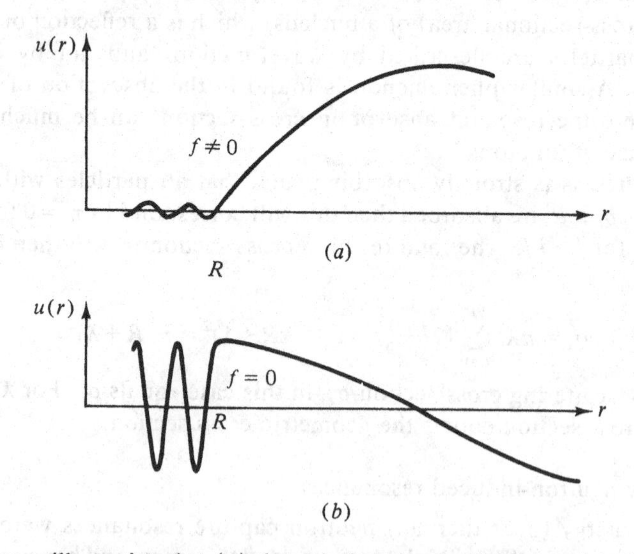

6.14. Diagram illustrating that (*a*) the wavefunction inside the nucleus is small compared to that outside unless (*b*) the logarithmic derivative at the surface is zero.

To obtain a connection between the resonant cross-section and the wavefunction of the resonant state the wavefunctions describing the neutron inside and outside the nucleus must be matched at the nuclear surface.

Boundary conditions at the nuclear surface

Very low-energy neutrons in the region outside the nucleus will be described by the $l=0$ wavefunction:

$$\psi = \frac{\mathrm{i}u}{2kr}$$

with

$$u = \exp(-\mathrm{i}kr) - \eta \exp(+\mathrm{i}kr)$$

The wavefunction for the neutron within the nucleus is unknown but its magnitude and derivative at the nuclear surface must equal that of the wavefunction for the neutron outside the nucleus. This can be achieved by matching the logarithmic derivatives f of the wavefunctions defined by:

$$f = \frac{R}{u}\left(\frac{\partial u}{\partial r}\right)_R$$

This requires:

$$\eta = \frac{f + \mathrm{i}kR}{f - \mathrm{i}kR} \exp(-2\mathrm{i}kR) \tag{6.18}$$

If f is real then $|\eta| = 1$ and there are no reactions. Let $f = x - \mathrm{i}y$, then $|\eta| \leq 1$ and:

$$\sigma_{\mathrm{el}} = \pi \lambda\!\!\!^{-2}|1-\eta|^2 = \pi \lambda\!\!\!^{-2}\left|[\exp(2\mathrm{i}kR) - 1] - \frac{2\mathrm{i}kR}{x - \mathrm{i}(y + kR)}\right|^2$$

$$\sigma_{\mathrm{reac}} = \pi \lambda\!\!\!^{-2}[1 - |\eta|^2] = \pi \lambda\!\!\!^{-2}\left[\frac{4kRy}{x^2 + (y + kR)^2}\right]$$

The reaction cross-section σ_{reac} has a maximum when the real part (x) of the logarithmic derivative is zero. This corresponds to $(\partial u/\partial r)_R = 0$ and the internal wavefunction having a large amplitude on resonance, as shown in figure 6.14. Near a resonance the dependence of x on energy can be approximated by:

$$x = (E - E_{\mathrm{r}})\left(\frac{\partial x}{\partial E}\right)_{E_{\mathrm{r}}}$$

Defining:

$$\Gamma_{\mathrm{n}} = -2kR \Big/ \left(\frac{\partial x}{\partial E}\right)_{E_{\mathrm{r}}} \tag{6.19}$$

$$\Gamma = -2(kR + y) \Big/ \left(\frac{\partial x}{\partial E}\right)_{E_{\mathrm{r}}}$$

then:

$$\sigma_{\text{elas}} = \pi \lambda\!\!\!^{-2} \left| [\exp(2ikR) - 1] + \frac{i\Gamma_n}{(E - E_r) + \frac{1}{2}i\Gamma} \right|^2$$

$$\sigma_{\text{reac}} = \frac{\pi \lambda\!\!\!^{-2} \Gamma_n (\Gamma - \Gamma_n)}{(E - E_r)^2 + \Gamma^2/4}$$

which identifies Γ_n and Γ as the neutron and total widths respectively. In the expression for σ_{elas} the second term is much larger on resonance than the first if k is very small as then $kR \approx 0$ and the first term is approximately zero. Off resonance the second term becomes insignificant and:

$$\sigma_{\text{elas}} \approx 4\pi R^2$$

showing that the first term corresponds to scattering from a hard-sphere of radius R. This would produce a phase shift $\delta = -kR$, which substituted in the expression (equation 6.17) above gives a cross-section of $4\pi R^2$. On resonance there is interference between the two terms and figure 6.15 shows an example of this characteristic interference.

The phase shift δ on resonance

If only elastic scattering takes place, then the cross-section σ can be described in terms of a phase shift δ which for $l = 0$ is:

$$\begin{aligned} \sigma &= 4\pi \lambda\!\!\!^{-2} \sin^2 \delta \\ &= 4\pi \lambda\!\!\!^{-2} / (1 + \cot^2 \delta) \end{aligned} \tag{6.20}$$

Neglecting hard-sphere scattering, i.e. neglecting the radius of the nucleus, then a resonance will correspond to cot δ going through zero as the incident energy goes through E_r, the resonance energy. If the resonance is sharp then:

$$\begin{aligned} \cot \delta &= \cot \delta_r + \left(\frac{\partial}{\partial E} \cot \delta \right)_{E_r} (E - E_r) \\ &= -(E - E_r) \left(\frac{\partial \delta}{\partial E} \right)_{E_r} \\ &\equiv -(E - E_r) \frac{2}{\Gamma_n} \end{aligned}$$

which defines

$$\Gamma_n = 2 \Big/ \left(\frac{\partial \delta}{\partial E} \right)_{E_r} \tag{6.21}$$

Substituting in equation 6.20 gives the Breit–Wigner formula:

$$\sigma = \frac{\pi \lambda\!\!\!^{-2} \Gamma_n^2}{(E - E_r)^2 + \Gamma_n/4}$$

From $\Gamma \tau = \hbar$ the delay τ is related to the energy dependence of the phase

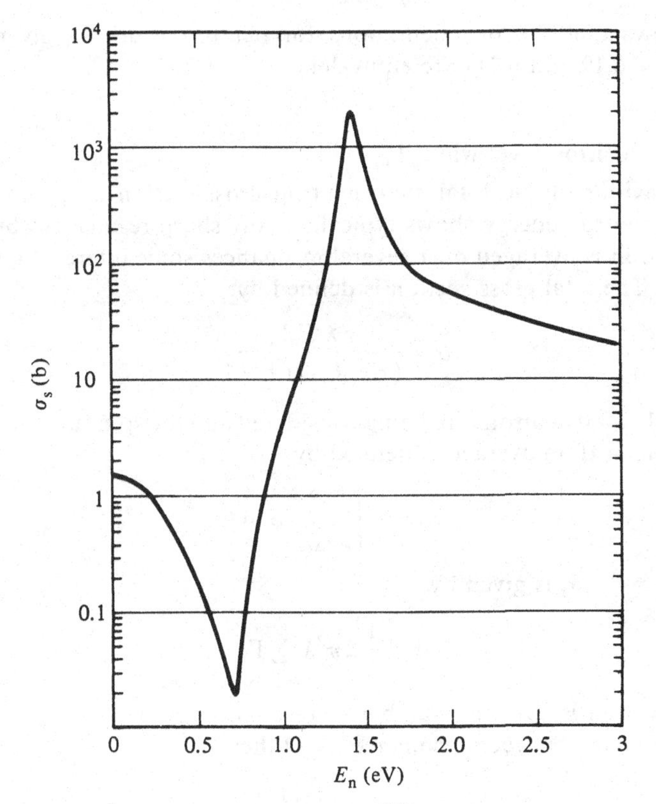

6.15. The neutron scattering cross-section for ^{115}In showing the characteristic interference between the hard-sphere and resonant contributions to the cross-section. (From Segrè, E., *Nuclei and Particles.* Benjamin (1977).)

shift by:

$$\frac{\hbar}{\Gamma_n} = \tau = \frac{\hbar}{2}\left(\frac{\partial \delta}{\partial E}\right)_{E_r}$$

For $l = 0$ scattering the wavefunction u is:

$$u = \sin(kr + \delta)$$

and so

$$f = x = kR \cot(kR + \delta)$$

Since the energy dependence of x near a resonance is dominated by that of the phase δ:

$$\left(\frac{\partial x}{\partial E}\right)_{E_r} \approx -[kR \operatorname{cosec}^2(kR + \delta)]\left(\frac{\partial \delta}{\partial E}\right)_{E_r}$$

$$= -kR\left(\frac{\partial \delta}{\partial E}\right)_{E_r}$$

This shows that the two definitions for neutron width Γ_n given above (equations 6.19 and 6.21) are equivalent.

6.5.4 The neutron level width Γ_n

The behaviour of the total slow neutron cross-section as a function of incident neutron energy shows typically many sharp resonances but if the cross-section is averaged over several resonances some interesting features are seen. The total cross-section is defined by:

$$\sigma_t = \frac{\pi \lambda\!\!\!^{-2} \Gamma_n \Gamma_t}{(E - E_r)^2 + \Gamma_t^2/4}$$

where only $l = 0$ neutrons are being considered and the spin statistical factor is taken as 1. If an average is defined by:

$$\bar{\sigma}_t = \frac{1}{\Delta} \int_{E-\Delta/2}^{E+\Delta/2} \sigma_t \, dE$$

then if $\Delta \gg \Gamma_n$, $\bar{\sigma}_t$ is given by:

$$\bar{\sigma}_t = \frac{1}{\Delta} 2\pi^2 \lambda\!\!\!^{-2} \sum_i \Gamma_n^i$$

where the sum is over the number of resonances in the internal Δ. If the average spacing between resonances is D then:

$$\bar{\sigma}_t = 2\pi^2 \lambda\!\!\!^{-2} \frac{\bar{\Gamma}_n}{D}$$

where $\bar{\Gamma}_n / D$ is called the neutron strength function and is a measure of how the neutron cross-section depends on energy.

To understand the observed energy variation it is useful to consider the expected behaviour for the idealised case of a neutron incident on a square well of depth V and radius R equal to that of a nucleus, i.e. single-particle motion in a potential. Resonances would be expected whenever the internal wavelength K just matched the size of the well such that $KR = (n + \frac{1}{2})\pi$, as then the logarithmic derivative f is zero at R. On resonance the internal wavefunction is large and matches the external wavefunction at $r = R$. The internal wavefunction is $\sin Kr$ so:

$$f = x = KR \cot KR$$

$$\hbar^2 K^2 / 2m = V + E$$

$$\therefore \quad \frac{\partial x}{\partial E} = -\frac{mR^2}{\hbar^2}$$

Therefore the width of such a single-particle resonance is given by:

$$\Gamma_{sp} = -2kR\bigg/\left(\frac{\partial x}{\partial E}\right) = \frac{2k\hbar^2}{mR} = \frac{4k}{K}\cdot\frac{\hbar K}{2mR}\hbar$$

The factor $T_0 = 4k/K$ is the transmission factor for a neutron crossing the potential discontinuity at $r = R$ and $\hbar K/2mR = v/2R$ is the frequency with which a neutron would strike the nuclear surface (v is the velocity of the neutron within the nucleus). The product gives the rate of decay $= 1/\tau_{sp} = \Gamma_{sp}/\hbar$. The width Γ_{sp} is energy dependent and for $E_n = 1$ eV and $R = 5$ fm, $\Gamma_{sp} = 3.6$ keV. Observed neutron widths at this energy are generally much smaller and a quantity γ^2 is introduced, defined by:

$$\Gamma_n = 2kR\gamma^2$$

which is energy independent and is a measure of how much a level is like a single-particle level. For a single-particle state $\gamma_{sp}^2 = \hbar^2/mR^2$.

If a neutron interacted strongly within a nucleus then the single-particle strength would be evenly split amongst all the levels of the compound nucleus and no concentration would be seen for values of $KR = (n+\frac{1}{2})\pi$. This would imply Γ_{sp} divided equally among the levels lying between single-particle levels. The number of such levels is (D_{sp}/D) where D_{sp} is the spacing between single-particle levels and D is the average level spacing. The spacing D_{sp} is given by:

$$D_{sp} = \frac{\hbar^2}{2m}(K_1^2 - K_2^2) \approx \frac{\hbar^2 \pi K}{mR}$$

$$\text{so} \quad \bar{\Gamma} = \Gamma_{sp}/(D_{sp}/D)$$

$$\therefore \quad \frac{\bar{\Gamma}}{D} = \left(\frac{\Gamma}{D}\right)_{sp} \approx \frac{2k}{\pi K}$$

For $E_n = 1$ eV and a well depth of 40 MeV, $\bar{\Gamma}/D = 10^{-4}$.

Experimentally it is difficult to measure the energy variation of $\bar{\Gamma}/D$ within a nucleus as it is just for slow neutrons that only the $l = 0$ partial wave contributes. However, it is easy to vary the target nucleus and hence vary R. The results are plotted in figure 6.16 and show that there is a concentration of strength at $A \approx 50$ and at $A \approx 160$ which correspond to where the $3s_{1/2}$ and $4s_{1/2}$ single-particle levels are just bound (i.e. $KR = (2+\frac{1}{2})\pi$ and $KR = (3+\frac{1}{2})\pi$). The splitting of the bump near $A \approx 160$ is caused by the nuclei in this region being deformed, which has roughly the effect of giving rise to two effective values of R in $KR = (n+\frac{1}{2})\pi$, corresponding to the two principal radii of the nuclei, and hence two bumps rather than one. The conclusion is that the neutron's motion is still approximately independent-particle at ~8 MeV excitation and its energy is not very rapidly shared with all the other nucleons.

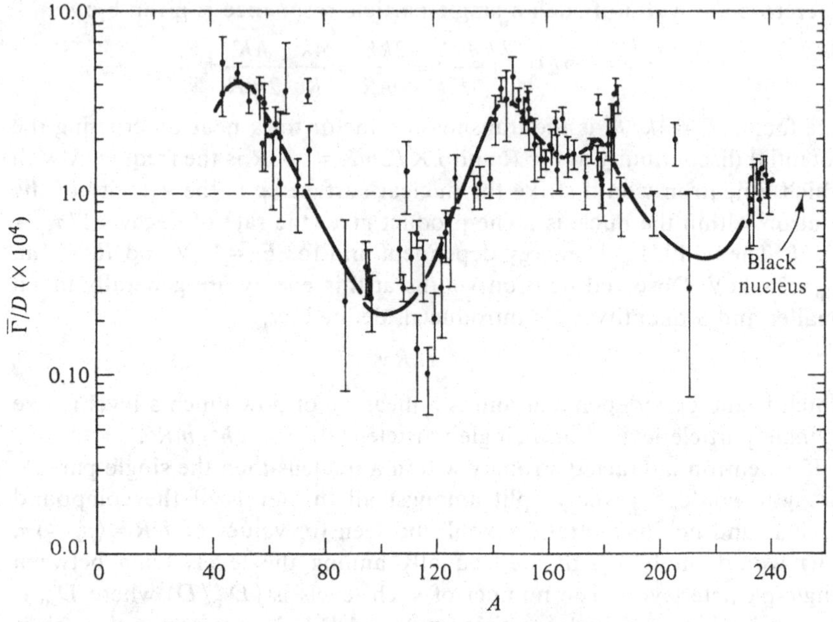

6.16. The $l=0$ neutron strength function $\bar{\Gamma}/D$ for various nuclei. (From Cohen, B. L., *Concepts of Nuclear Physics*. McGraw-Hill (1971).)

A comparison of Γ_n and Γ_γ

The neutron width $\Gamma_n = 2kR\gamma^2$ is proportional to the velocity of the incident neutron. The gamma width Γ_γ depends on the particular compound nucleus level but, as this is at ~8 MeV excitation and there are many levels to which it may decay by E1, M2 or E2 transitions, it would not be expected to vary very strongly from level to level, nor from nucleus to nucleus. The ratio of neutron scattering to neutron capture, which is given by Γ_n/Γ_γ, is therefore expected to be roughly proportional to the neutron velocity. Empirically this is about 1 when E_n is of the order of 10 eV. This means that thermal resonances are mostly absorption resonances while for $E_n \gg 10$ eV they are mostly scattering resonances.

6.5.5 Isobaric analogue resonances

The $T=2$ isobaric analogue state of the ^{28}Mg 0^+ $T=2$ ground state is at 15.28 MeV excitation in ^{28}Si. At this excitation the level density of states with lower isospin ($T=1$ or 0) is high and the purity of the $T=2$ state would not be expected to be so good as for states in lighter nuclei at lower excitations. This is because the Coulomb force will be greater and the mixing

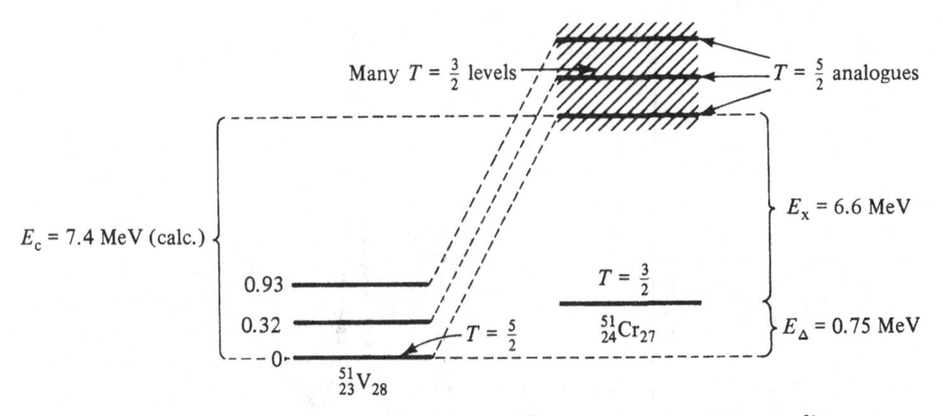

6.17. An illustration of the low-lying levels of $^{51}_{23}V$ and their analogues in $^{51}_{24}Cr$ showing their relative energies. (From Wilkinson, D. H., *Nuclear Physics Lecture Notes.* Oxford Nuclear Physics Laboratory.)

larger because by second-order perturbation theory it depends inversely on the average level spacing.

So it was thought for heavier nuclei ($Z \gtrsim 20$) that isospin would not be a good quantum number. In 1961, however, a sharp spike was seen in the neutron energy spectrum of the $^{51}_{23}V(p, n)^{51}_{24}Cr$ reaction at $E_p = 20$ MeV. This spike corresponds to a direct exchange of the proton and neutron in the target such as is seen in the head-on collision of two equal mass particles: the struck particle continues with the incident energy and the incident particle stops. In such a collision the proton will be in the same state as the ejected neutron, i.e. the analogue state to the ^{51}V ground state will be formed. The levels involved are shown in figure 6.17.

Experimentally for a large number of nuclei it is found that the calculated excitation energy E_c of the analogue state, which equals the Coulomb energy difference less the neutron-H atom mass difference, is equal to the measured excitation energy E_x plus the ground state mass difference E_Δ, showing that analogue states do exist in heavy nuclei. This is mainly for heavy nuclei because the degree to which the protons contribute to isospin mixing is reduced by a factor $1/(N-Z-2)$ compared to simple estimates.

These analogue states have also been studied as resonances in reactions such as $^{89}Y(p, n)^{89}Zr$ where the yield of neutrons is measured as a function of E_p (see figure 6.18). The two resonances seen in the neutron yield correspond to the formation and decay of the lowest two $T = 6$ levels in $^{90}_{40}Zr$, which are the analogues of the ground and first excited states in $^{90}_{39}Y$. The separation in energy of ~200 keV is close to the separation 202.4 keV of the ground and first excited states in ^{90}Y. The low-lying levels in ^{90}Zn have $T = 5$ ($T_z = 5$) and in $^{89}_{40}Zn$ $T = 4\frac{1}{2}$. The neutron decay from the $T = 6$

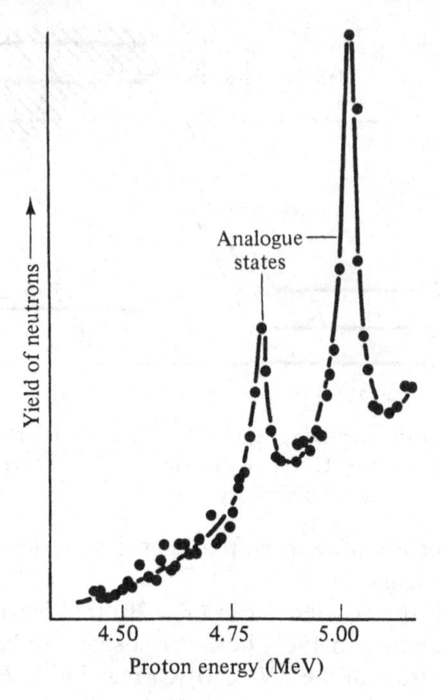

6.18. Excitation function for the reaction $^{89}Y(p, n)^{89}Zr$. (From Jones, G. A., *Properties of Nuclei.* Oxford University Press (1987).)

levels in $^{90}_{40}Zr$ is therefore forbidden and arises through mixing with $T = 5$ levels of the same J^{π}; it is through the $T = 5$ admixtures that the decay occurs.

The situation is somewhat similar to that for the neutron strength function; there the single-particle strength is split amongst many levels over a range of the order of MeV but still concentrated at the excitation energies corresponding to single-particle states in a potential. In the present case the $T = 6$ level is split by the Coulomb interaction amongst many $T = 5$ levels (of the same J^{π}) but only over a range of the order of a few tens of keV, which reflects the strength of the Coulomb interaction.

These $T = 6$ levels in ^{90}Zr are the analogues of the low-lying states of ^{90}Yr, which would be expected to have a structure somewhat like $^{89}Y(gs) + n$. Therefore the $T = 6$ levels in ^{90}Zr will be somewhat like $^{89}Y(gs) + p$, i.e. the proton spectroscopic factor would be expected to be reasonably large. Their proton decay widths however will be reduced because of the Coulomb barrier so the isospin allowed proton width is quite narrow and hence the resonances are quite sharp.

6.5.6 Isospin-forbidden reactions

A classic but rare example of where the Coulomb interaction causes a large mixing of different isospin states is found in ^{8}Be. The relevant part of the level scheme is shown in figure 6.19. In the ^{10}B(d, α)^{8}Be reaction both levels at 16.93 and 16.63 MeV are excited about equally but if one were the analogue of the ^{8}Li and ^{8}B ground states then it would not be excited. They also both breakup to $\alpha+\alpha$, which is forbidden for a $T=1$ state. The explanation is that the Coulomb force has mixed two very close 2^+ levels with $T=0$ and $T=1$ to give two about equally mixed isospin states.

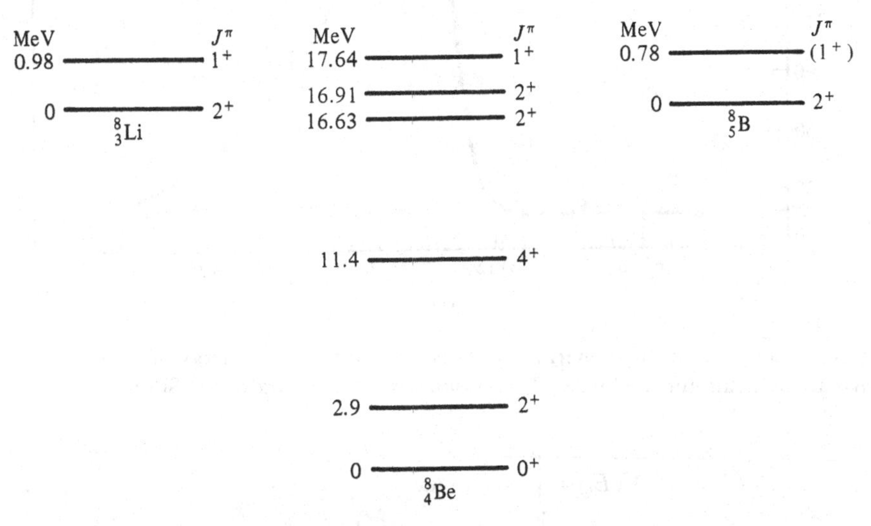

6.19. Partial energy level diagram of ^{8}Li, ^{8}Be and ^{8}B, showing the low-lying $T=1$ analogue states. The strength of the lowest $T=1$ analogue level in ^{8}Be is split between the 16.63 and 16.91 2^+ levels.

Isobaric analogue states may also be studied using an isospin-forbidden reaction. An example is the ^{24}Mg(α, γ)^{28}Si* reaction to the lowest $T=2$ state in ^{28}Si, which is at 15.28 MeV excitation. Figure 6.20 shows the yield curve of high-energy γ-rays as the energy of the α-beam was swept through the $T=2$ resonance. The target was a thin film of ^{24}Mg (supported on high-purity gold) whose thickness was equivalent to ~5 keV energy loss for the incident α-particles. The γ-ray yield $Y(E_i)$ at a particular incident α-energy is given by integrating the Breit–Wigner formula:

$$\sigma(E)=\frac{\pi\lambda\!\!\!^{-2}\Gamma_\alpha\Gamma_\gamma}{(E-E_r)^2+\Gamma^2/4}$$

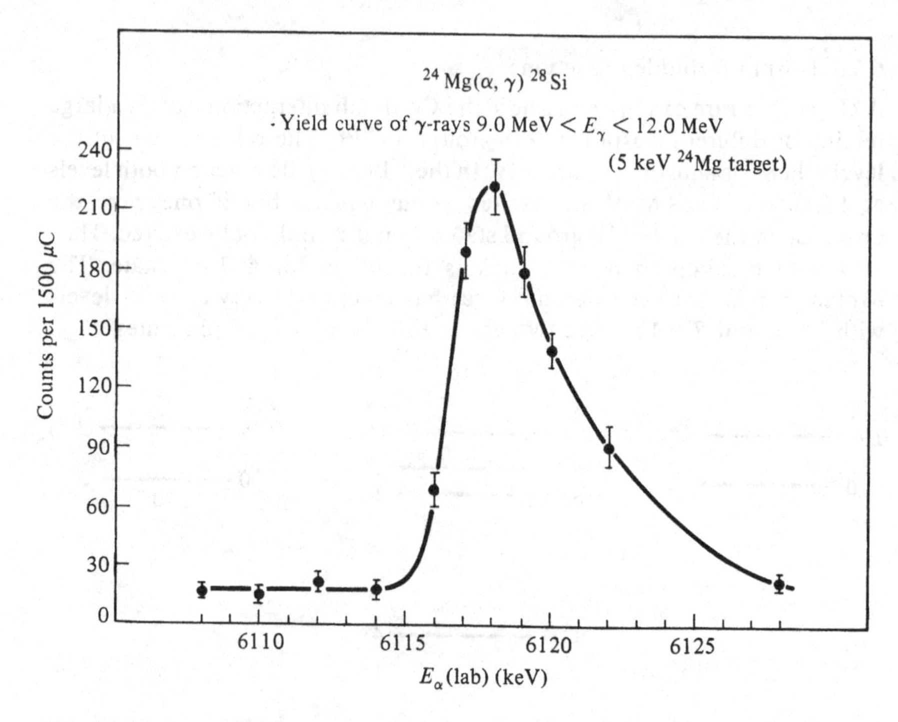

6.20. Yield curve of high-energy γ-rays as a function of the energy of the α-particle beam for the lowest $T=2$ resonance in the $^{24}Mg(\alpha, \gamma)^{28}Si$ reaction.

so

$$Y(E_i) = \int n\sigma(E)\,\mathrm{d}x$$

$$= [n/(\mathrm{d}E/\mathrm{d}x)] \int_{E_f}^{E_i} \sigma(E)\,\mathrm{d}E$$

where n is the number of ^{24}Mg nuclei per unit volume and $\mathrm{d}E/\mathrm{d}x$ is the stopping power for the α-particle at the incident α-particle energy. If $(E_i - E_r) \gg \Gamma$ and $(E_r - E_f) \gg \Gamma$ then:

$$Y(E_i) = 2\pi^2 \lambda\!\!\!^{-2} \frac{\Gamma_\alpha \Gamma_\gamma}{\Gamma} [n/(\mathrm{d}E/\mathrm{d}x)]$$

From the leading edge of the yield curve $\Gamma \leq 1$ keV so the peak yield is proportional to $\Gamma_\alpha \Gamma_\gamma / \Gamma$. If $\Gamma_\alpha/\Gamma \approx 1$ then the yield of γ-rays is proportional to Γ_γ. The gamma-decay of the $T=2$ 0^+ state is by $\Delta T = 1$ M1 transitions to 1^+ $T=1$ states near 11 MeV excitation, so Γ_γ is of the order of 1 eV. These states near 11 MeV decay to the ground and first excited state also by $\Delta T = 1$ M1 transitions with the emission of 9 MeV and 11 MeV γ-rays, which can be more easily detected than the primary 4 MeV decay γ-rays because of the ~6 MeV background γ-rays arising from neutron capture γ-rays and the contaminant $^{13}C(\alpha, n)^{16}O^*$ reaction. An allowed α-width at an excitation energy of 15 MeV would be of the order of 1 MeV so the $T=0$ isospin impurity in the $T=2$ state need only be very small for $\Gamma_\alpha \gtrsim \Gamma_\gamma$. As a resonance is seen in the γ-ray yield, Γ_α must be greater than or comparable to all of the other decay widths for then Γ_α/Γ will not be very different from 1. For the 0^+ $T=2$ state in ^{28}Si it is found experimentally that $\Gamma \approx \Gamma_\alpha$ and since $\Gamma \leq 1$ keV from the γ-ray excitation curve (see figure 6.20) the admixture of $T=0$, $J^\pi = 0^+$ states is less than an order of 0.1%.

6.5.7 Compound nucleus reaction cross-sections

A useful approximation when describing compound nucleus reactions is to assume that the nucleus is strongly absorbing. For an s-wave reaction this assumption is equivalent to taking the internal wavefunction as:

$$u = \exp(-\mathrm{i}Kr) \qquad r < R$$

corresponding to an ingoing wave only. The logarithmic derivative at the nuclear surface $f = -\mathrm{i}KR$ so the matching condition (equation 6.18) becomes:

$$\eta = \frac{-\mathrm{i}KR + \mathrm{i}kR}{-\mathrm{i}KR - \mathrm{i}kR}\exp(-2\mathrm{i}kR)$$

$$= \frac{K-k}{K+k}\exp(-2\mathrm{i}kR)$$

so

$$\sigma_{\text{abs}} = \pi\lambda\!\!\!^{-2}(1-|\eta|^2)$$

$$= \pi\lambda\!\!\!^{-2}\frac{4Kk}{(K+k)^2}$$

$$= \pi\lambda\!\!\!^{-2}T_0$$

where T_0 is the transmission coefficient for an s-wave particle crossing the potential discontinuity at $r=R$. This expression can be generalised to other partial waves and to charged particles and can be combined with the independence hypothesis, that the way a compound nucleus decays is independent of the way it was formed, to give the following expression for the cross-section $\sigma_{\alpha\beta}$ for the reaction $A(a, b)B$:

$$\sigma_{\alpha\beta} = \pi\lambda\!\!\!^{-2}T_\alpha T_\beta \Big/ \sum_\gamma T_\gamma \tag{6.22}$$

This treatment of compound nucleus reactions is called the Hauser–Feshbach theory and can be used to predict relative cross-sections such as $(\alpha, 2p)/(\alpha, 2n)$.

The independence hypothesis for compound nucleus reactions in the continuum was checked in a classic experiment by Goshal in 1950. The compound nucleus ^{64}Zn at the same excitation energy was formed in two different ways:

$$\alpha + {}^{60}\text{Ni} \rightarrow {}^{64}\text{Zn}^* \quad \text{and} \quad \text{p} + {}^{63}\text{Cu} \rightarrow {}^{64}\text{Zn}^*$$

The independence hypothesis (equation 6.22 above) predicts:

$$\sigma_{\alpha\beta}/\sigma_{\alpha\gamma} = T_\beta/T_\gamma = \sigma_{\delta\beta}/\sigma_{\delta\gamma}$$

i.e. the cross-section ratios for two emission channels, such as n and pn, are independent of the formation channel (α or p). The excitation curves plotted in figure 6.21 show that such cross-section ratios are to a good approximation the same.

This figure also demonstrates that the most probable number of particles emitted by a compound nucleus increases with increasing excitation energy.

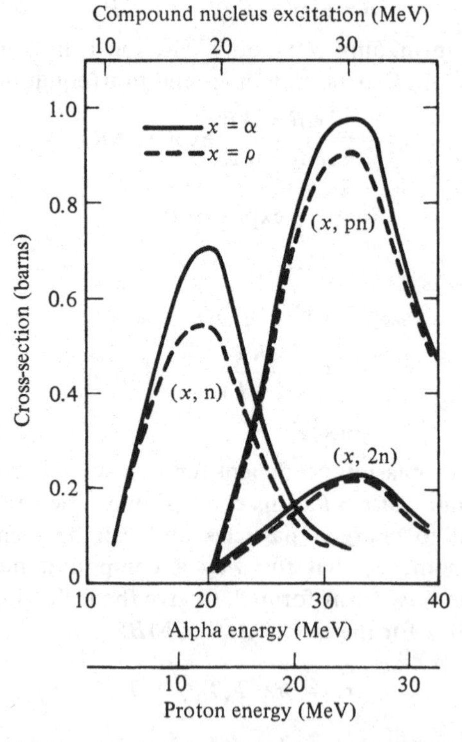

6.21. Excitation curves for the formation of the compound nucleus ^{64}Zn by proton bombardment of ^{63}Cu and by α-bombardment of ^{60}Ni. The similarity of cross-section ratios provides evidence for the independence hypothesis for compound nucleus reactions in the continuum. (From Satchler, G. R., *Introduction to Nuclear Reactions.* Wiley (1980).)

This is because the probability $p(E)$ that one neutron is emitted with energy E is proportional to the phase space available, which depends on $E^{1/2}$, and to the density of final states ρ, which increases with increasing excitation energy in the residual nucleus. Using a Fermi gas model the level density ρ is given by:

$$\rho(\varepsilon) \propto \exp(b\varepsilon^{1/2})$$

where ε is the excitation energy and b is a constant. The relation between E and ε is $\varepsilon = E_0 - E$ where E_0 is the maximum neutron energy so

$$p(E) \propto E^{1/2} \exp(b(E_0 - E)^{1/2})$$

The maximum value of $p(E)$ is when E is of the order of a few MeV so if the residual nucleus is left with sufficient excitation energy another particle will be emitted.

6.6 Direct reactions

The cross-section for a direct pick-up or stripping reaction can be calculated using the Born approximation in an analogous way to the cross-section for inelastic scattering discussed above. For the reaction $a + A \to B + b$:

$$\frac{d\sigma}{d\Omega} = \frac{\mu_i \mu_f}{(2\pi\hbar^2)^2} \frac{k_f}{k_i} |M_{if}|^2$$

where k_i and k_f are the wave vectors in the entrance and exit channels and μ_i and μ_f are the reduced masses. Assuming plane waves and making the approximation that V is very short range such that $\mathbf{r}_i = \mathbf{r}_f = \mathbf{r}$ then:

$$M_{if} = \int \exp(i\mathbf{q} \cdot \mathbf{r}) \langle Bb | V | Aa \rangle d^3 r$$

Expanding V as

$$V = \sum_{lm} V_{lm}(r, \tau_i) Y_{lm}(\theta, \phi)$$

then when only one V_{lm} is involved, i.e. there is only one l value transferred:

$$M^l_{if} = \int i^l [4\pi(2l+1)]^{1/2} j_l(qr) f(r) r^2 \, dr \qquad (6.23)$$

where now

$$f(r) = \langle Bb | V_{lm} | Aa \rangle$$

If V_{lm} is assumed to peak at the surface then $f(r) = 0$ unless $r \approx R$ so:

$$\frac{d\sigma}{d\Omega} \propto |j_l(qR)|^2$$

The angular dependence is contained in q since:

$$q^2 = k_i^2 + k_f^2 - 2k_i k_f \cos\theta$$

where θ is the angle between k_i and k_f and is the scattering angle. The first and largest peak is at an angle for which qR is a little greater than l and subsequent peaks occur as qR increases by π. An example is shown in figure 6.22, in which the differential cross-section for the reaction ^{56}Fe(d, p)^{57}Fe is plotted. The approximate relation $qR = l$ for the first peak is what a simple semi-classical argument would predict: as $\hbar q$ is the transferred momentum then the maximum angular momentum transferred is $\hbar qR$ so for $l\hbar$ transfer $l \approx qR$.

For the pick-up reaction ^{41}Ca(p, d)^{40}Ca the matrix element $\langle Bb|V|Aa\rangle$ can be simplified. The low-lying states of ^{41}Ca are expected to be very like ^{40}Ca+n and a spectroscopic factor S_l can be defined by:

$$\langle\psi(^{40}\text{Ca}^*)|\psi(^{41}\text{Ca})\rangle = S_l|\phi_n^l\rangle$$

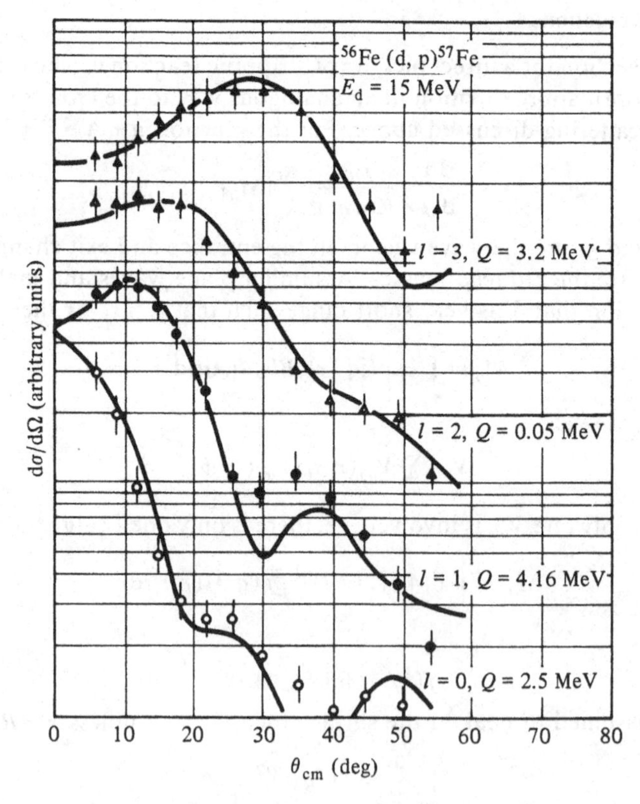

6.22. The differential cross-section $d\sigma/d\Omega$ for the ^{56}Fe(d, p)^{57}Fe reaction with 15 MeV incident deuterons. (From Satchler, G. R., *Introduction to Nuclear Reactions.* Macmillan (1980).)

where ϕ_n^l is the wavefunction of the neutron in an l state about ^{40}Ca and $S_l \le 1$ is the extent to which the state in ^{41}Ca is a single-particle state. In this reaction $\psi_A = \psi(^{41}\text{Ca})$, $\psi_B = \psi(^{40}\text{Ca}^*)$, $\psi_b = \psi_d$ and $\psi_a = \psi_p$. Ignoring the spins of the deuteron and proton then ψ_p is unity and:

$$\langle Bb|V|Aa\rangle = S_l\langle\psi_d|V_{pn}|\phi_n^l\rangle$$

where V_{pn} is the interaction between the incident p and the neutron in the target. If this is taken to have zero range then:

$$\langle\psi_d|V_{pn} = V_0\delta(\mathbf{r}_p - \mathbf{r}_n)$$

and

$$M_{if} = V_0 S_l \int \exp(i\mathbf{q}\cdot\mathbf{r}_n)\phi_n^l(\mathbf{r}_n)\,d^3r_n$$

Therefore the reaction cross-section is proportional to the probability (square of the probability amplitude M_{if}) that the bound state neutron has a momentum equal to q, which is the momentum transferred to the incoming proton when it picks up the neutron, and to the square of the spectroscopic factor S_l.

The bound state wavefunction can be written:

$$\phi_n^l(\mathbf{r}_n) = r^{-1}u_l(r_n)Y_{lm}(\theta, \phi)$$

so in equation 6.23 above $f(r) = r^{-1}u_l(r_n)$. If the reaction is assumed localised to the surface, i.e. $\delta(r_n - R) = 0$ then:

$$\frac{d\sigma}{d\Omega} \propto S_l^2[j_l(qR)]^2$$

This example shows that the magnitude of the cross-section is a measure of the spectroscopic factor S_l and hence is a test of the nuclear wavefunction describing the state. As in the case above of inelastic scattering, much better agreement is obtained with experiment if distorted waves are used instead of plane waves.

6.6.1 Direct reactions with heavy ions

For energies near the Coulomb barrier direct pickup and transfer reactions occur with heavy ions. However, because of the strong Coulomb interaction the angular distributions are less characteristic of the transferred angular momentum of the exchanged particle. For angles much less than the grazing angle θ_g the probability of transfer is small while for angles much greater than θ_g inelastic processes predominate and a bell-shaped angular distribution is seen. An example is shown in figure 6.23.

To extract spectroscopic factors using heavy-ion direct reactions it is important not to make the zero-range approximation but to consider the finite range of the nuclear interaction, which is equivalent to not neglecting

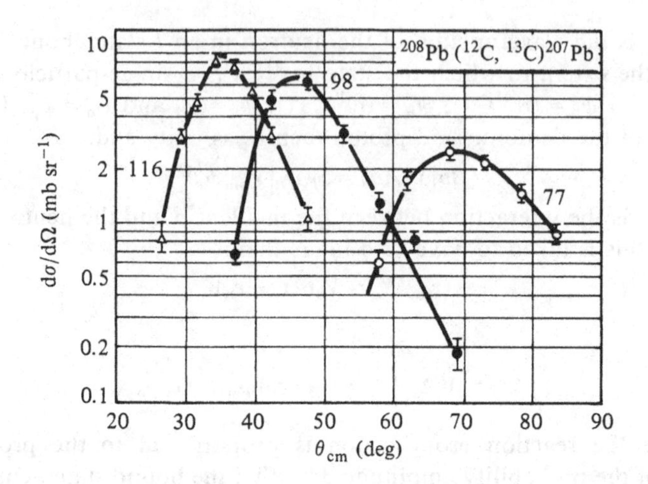

6.23. Characteristic bell-shaped angular distribution seen in heavy-ion transfer reactions near the Coulomb barrier. Shown is the differential cross-section for the $^{208}Pb(^{12}C, ^{13}C)^{207}Pb$ reaction at incident energies of 77, 98 and 116 MeV populating the ground states of ^{207}Pb and ^{13}C. (From Larsen, J. S. *et al.*, *Phys. Lett.* **42B** (1972) 205.)

recoil in the theoretical description. While the angular distributions are not very distinctive, selective excitation of states can occur in heavy-ion reactions because of kinematic considerations.

6.6.2 Kinematic factors in heavy-ion reactions

Kinematic factors can have a significant effect on what states are populated in a heavy-ion reaction and examples are shown in figures 6.24 and 6.25. In figure 6.24*a* it can be seen that the states preferentially excited in the $^{54}Fe(^{16}O, ^{12}C)^{58}Ni$ reaction at $E(^{16}O) = 46$ MeV are near 7 MeV excitation while at $E(^{16}O) = 60$ MeV (figure 6.24*b*) they are near 13 MeV excitation. This corresponds to a preferred Q-value of -8 MeV at $E(^{16}O) = 46$ MeV and -14 MeV at $E(^{16}O) = 60$ MeV. In the $^{54}Fe(^{14}N, ^{13}C)^{55}Co$ reaction shown in figure 6.25*a* the Q-value to the ground state is $Q_0 = -2.5$ MeV and as states near 5 MeV are preferentially populated $Q_{opt} \approx -7.5$ MeV. For the $^{54}Fe(^{16}O, ^{15}N)^{55}Co$ reaction (see figure 6.25*b*) $Q_0 = -7.1$ MeV so its Q_{opt} is also ≈ -7.5 MeV since states near the ground state are mainly populated.

This behaviour can be understood qualitatively using a simple semi-classical model. Consider a stripping reaction at forward angles as depicted in figure 6.26. For high enough incident energies the Coulomb energy differences between the incident and exit channels will be relatively small. The cross-section would be expected to be large when the incoming and

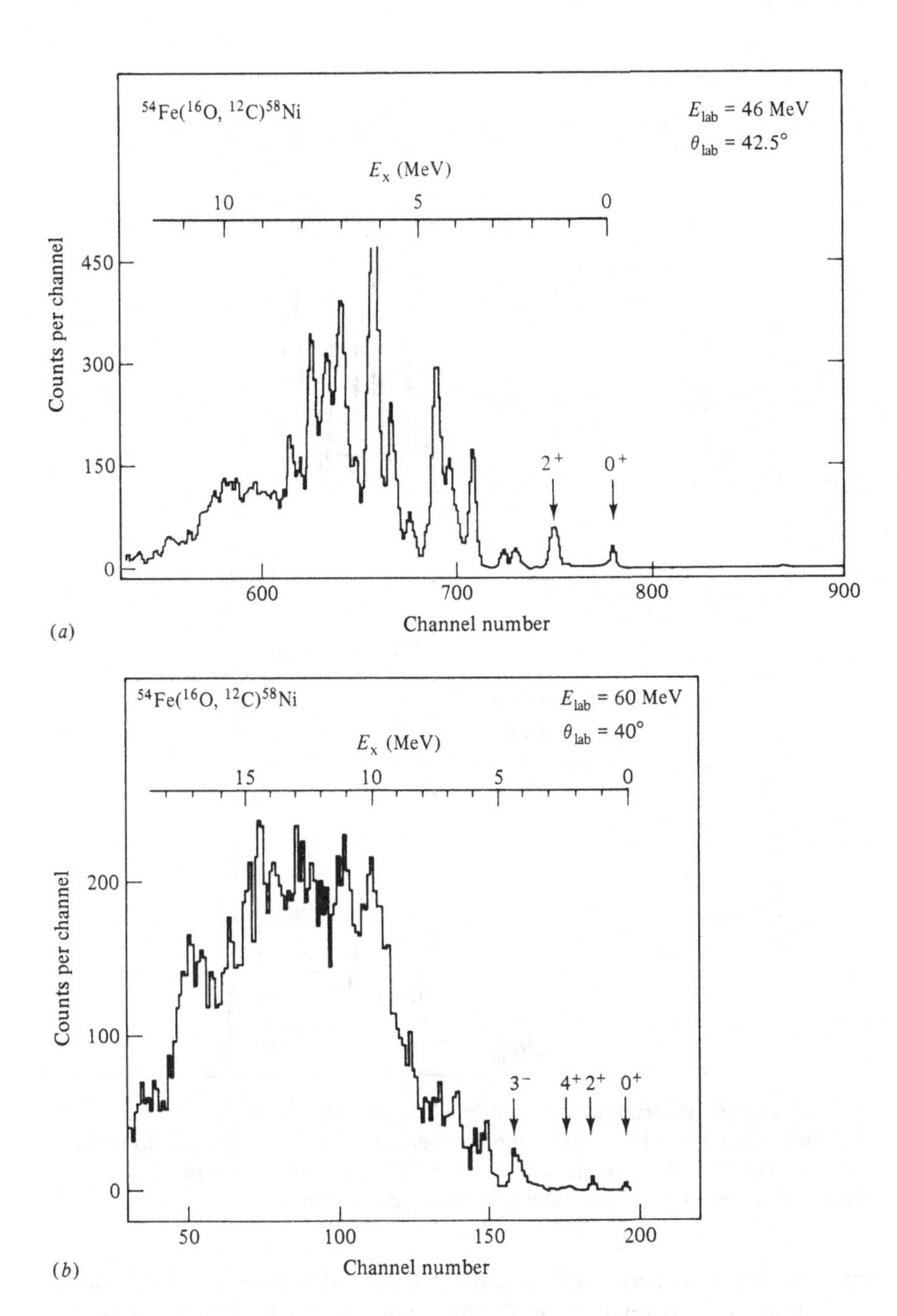

6.24. (*a*) Energy spectrum of ^{12}C ions from the reaction $^{54}Fe(^{16}O, ^{12}C)^{58}Ni$ at 46 MeV bombarding energy and $\theta_{lab} = 42.50$. (From Bouche, P. *et al.*, *Phys. Rev.* **C6** (1972) 577.) (*b*) Same as (*a*) but at a bombarding energy of 60 MeV and $\theta_{lab} = 40°$. (From Christensen, P. R. *et al.*, *Nucl. Phys.* **A207** (1973) 33.)

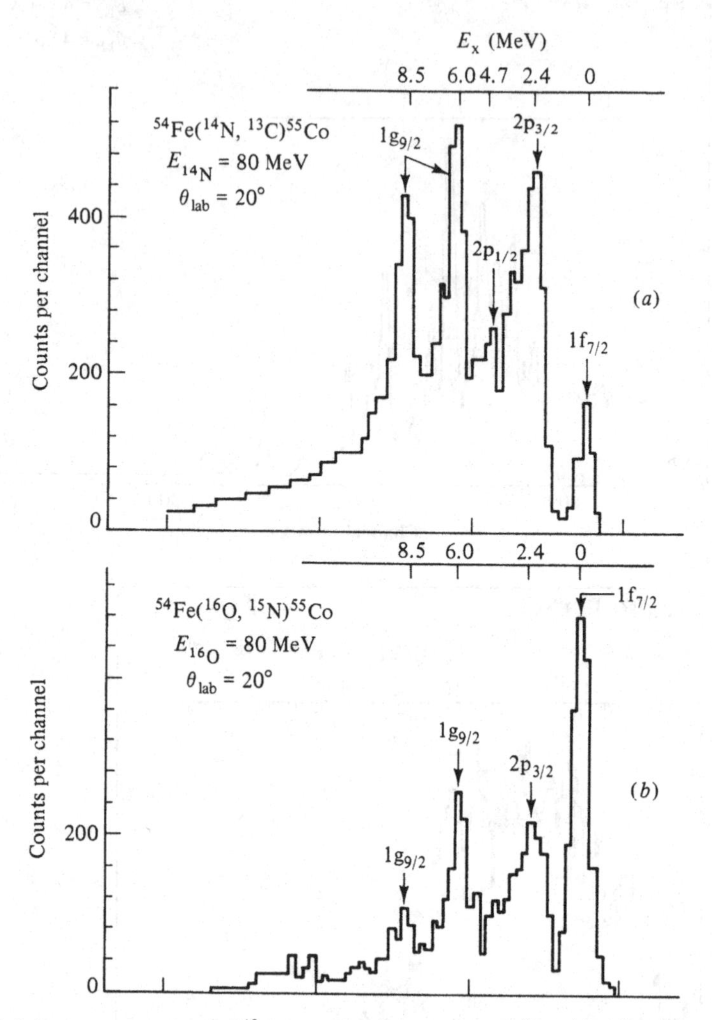

6.25. (*a*) Energy spectrum of ^{13}C ions from the reaction $^{54}Fe(^{14}N, ^{13}C)^{55}Co$. (*b*) Energy spectrum of ^{15}N ions from the reaction $^{54}Fe(^{16}O, ^{15}N)^{55}Co$. Both (*a*) and (*b*) taken at 80 MeV bombarding energy and $\theta_{lab} = 20°$. (From Bass, R., *Nuclear Reactions with Heavy Ions.* Springer-Verlag (1980).)

outgoing trajectories are well matched, which will happen if the velocities of the incident and final nuclei are the same i.e. $v_f = v_i$. When this is the case the final kinetic energy T_f of m_f and $(m_x + m_t)$ is given by:

$$T_f = T_i - \tfrac{1}{2}\left(\frac{m_x m_t}{m_x + m_t}\right) v_i^2$$

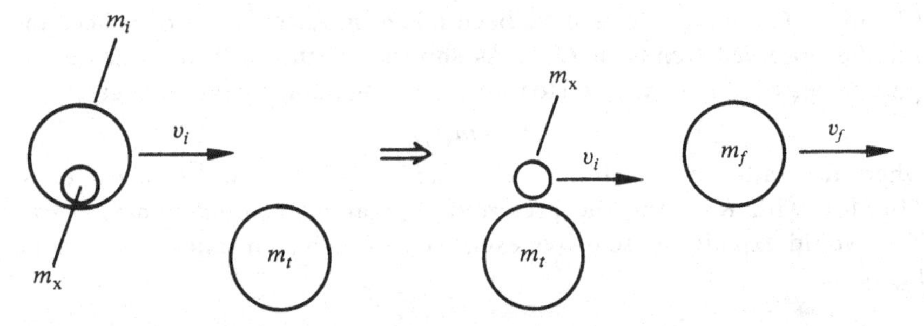

6.26. Diagram illustrating a heavy-ion stripping reaction at 0°. The velocity v_i is the relative velocity of incident and target nuclei; $m_f = m_i - m_x$.

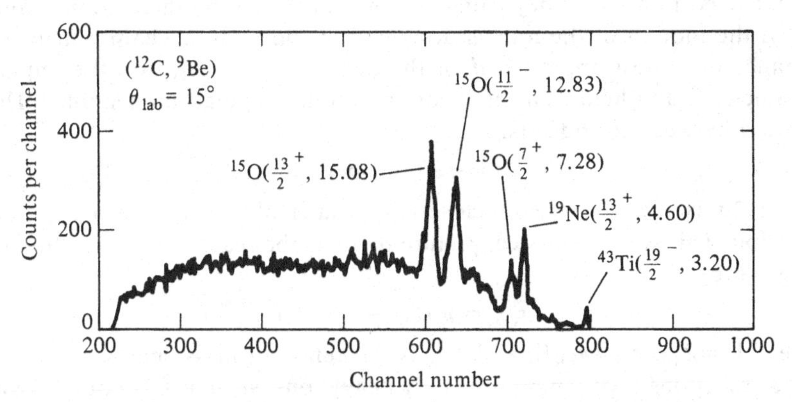

6.27. Pulse height spectrum for the three-nucleon transfer reaction (^{12}C, ^{9}Be) with a mixed target containing ^{12}C, ^{16}O and ^{40}Ca, at a bombarding energy of 114 MeV and $\theta_{lab} = 15°$. (From Anyas-Weiss, N. *et al.*, *Phys. Rep.* **12** (1974) 201.)

where m_x, m_t and v_i are as indicated in figure 6.26. The Q-value for a reaction is the difference between final and initial kinetic energies (equation 6.1) so the optimum Q-value, Q_{opt}, is given by:

$$Q_{opt} = T_f - T_i = -T_i\left[\frac{m_x m_t}{m_i(m_x + m_t)}\right]$$

For the ^{54}Fe(^{16}O, ^{12}C)^{58}Ni reaction $Q_{opt} = -0.233\,T_i$ predicting $Q_{opt} = -10.7$ MeV at $T_i = 46$ MeV and $Q_{opt} = -14.0$ MeV at $T_i = 60$ MeV; cf. -8 MeV and -14 MeV experimentally. For the ^{54}Fe(^{14}N, ^{13}C)^{55}Co reaction at $T_i = 80$ MeV $Q_{opt} = -5.6$ MeV and for the ^{54}Fe(^{16}O, ^{15}N)^{55}Co reaction $Q_{opt} = -4.9$ MeV; cf. $Q_{opt} \approx -7.5$ MeV experimentally. So although this model is very approximate, as the reactions have been assumed to occur at

0° and no Coulomb effects have been taken into account, it does account for the observed trends in Q_{opt}. As shown in figure 6.26 the transferred particle m_x would be expected to transfer an angular momentum $\hbar l$ given by:

$$\hbar l = m_x v_i R$$

where R is radius of transfer. For the reaction $^{12}C(^{12}C, ^{9}Be)^{15}O$ at $E(^{12}C) =$ 114 MeV with $R = 3$ fm, the preferred angular momentum transfer $l \approx 6$. This would explain the selective excitation of high spin states as shown in figure 6.27.

6.7 Deep-inelastic reactions

In a reaction at energies above the Coulomb barrier two heavy nuclei would be expected to react if they came closer than the strong interaction radius R_s. If the motion of the ions is described by classical mechanics then the distance of closest approach d of the heavy ions depends on the impact parameter b and hence on the incoming orbital angular momentum l. The relation between d and b is:

$$d = a + (a^2 + b^2)^{1/2}$$

where $2a$ is the distance of closest approach of the ions in a head-on collision. This simple classical picture predicts the reaction cross-section to be given by:

$$\sigma_{reac} = \pi R_s^2(1 - V/E_{cm})$$

where $V = Z_1 Z_2 e^2/4\pi\varepsilon_0 R_s$ and E_{cm} is the centre of mass energy.

For reactions involving moderately heavy ions, such as ^{20}Ne on ^{107}Ag at $E_{lab} = 165$ MeV, the cross-section for complete fusion (i.e. to form a compound nucleus) is a significant fraction of σ_{reac}. However for very heavy ions, such as ^{136}Xe on ^{209}Bi at $E_{lab} = 1130$ MeV, the probability of complete fusion is found to be a small fraction of σ_{reac} and there is a substantial cross-section for reactions producing two heavy nuclei in which there has been a considerable transfer of the incident kinetic energy to internal excitation of the products. In these dissipative collisions, called deep-inelastic reactions, the incoming heavy ion reacts strongly with the target nucleus for a short time during which kinetic energy is dissipated among internal degrees of freedom and some nucleons are exchanged before the resulting two nuclei separate. Because of their high excitation energy and large Z these nuclei may decay by fission as well as by particle and γ-ray emission.

If there is complete dissipation the two final nuclei have a relative kinetic energy equal to their Coulomb energy when they were interacting. This is seen in the reaction of ^{136}Xe on ^{209}Bi at $E_{lab}(^{136}Xe) = 1130$ MeV illustrated in figure 6.28, which also shows that there is a correlation between the amount of dissipation and the mass exchange.

In terms of the simple classical model described above, the total reaction cross-section σ_b for impact parameters up to b is given by $\sigma_b = \pi b^2$. Since $l = kb$ the differential cross-section $d\sigma/dl = 2\pi l/k^2$. A schematic diagram showing how $d\sigma/dl$ depends on l for a typical reaction involving quite heavy ions is shown in figure 6.29. The hatched areas are proportional to the cross-section for the different processes. For $l \approx l_{\text{grazing}}$ direct reactions are seen while for low l (i.e. near head-on) compound nucleus reactions

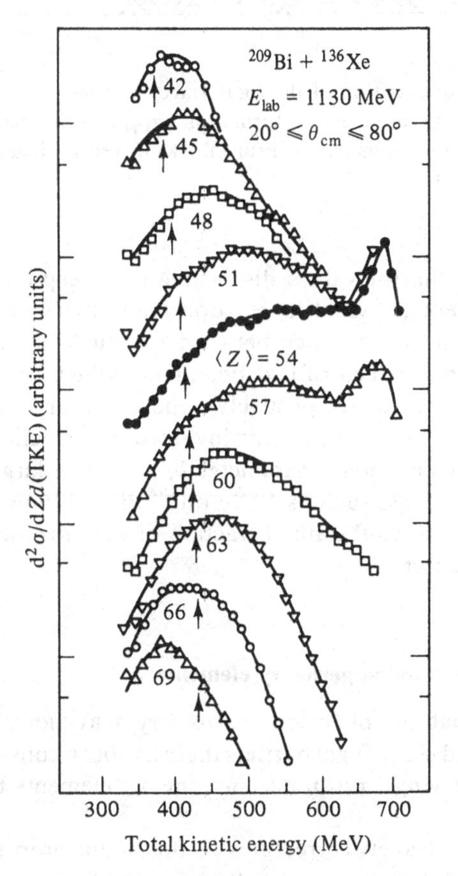

6.28. Plot of $d^2\sigma/dZd(\text{TKE})$ as a function of the total kinetic energy (TKE) for different elements produced in the $^{209}Bi + ^{136}Xe$ reaction. Each bin contains three elements. The ordinate is a logarithmic scale. The arrows indicate Coulomb energies for touching spheres. (From Schröder, W. V. and Huizenga, J. R., *Ann. Rev. Nucl. Sci.* **27** (1977) 465.)

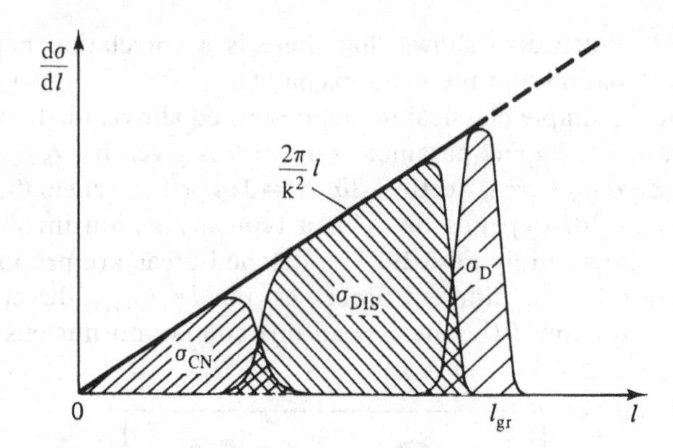

6.29. Schematic decomposition of the total reaction cross-section into the cross-sections of compound nucleus formation (σ_{CN}), of dissipative collisions (σ_{DIS}) and of direct reactions (σ_D). (From Bock, R. (ed.), *Heavy-Ion Collisions*. North-Holland (1980).)

occur. For intermediary l values dissipative (i.e. deep-inelastic) reactions are seen. The extent to which a reaction leads to complete fusion or a deep-inelastic event is a balance between the nuclear attraction and the repulsive Coulomb and centrifugal potentials, which are proportional to Z_1Z_2/r and to $l(l+1)/r^2$, respectively, where r is the separation of the heavy ions. As the masses of the ions involved in a reaction increase, both Z_1Z_2 and, for a given impact parameter b, l become larger. The result is that for very heavy ions, such as ^{136}Xe on ^{209}Bi, only low l reactions lead to fusion and there is a substantial fraction of the cross-section leading to deep-inelastic reactions.

6.8 Fusion reactions and superheavy elements

Although the probability of fusion of two very heavy ions is small because of the Coulomb and centrifugal barriers there has been considerable research directed towards trying to produce superheavy elements by the fusion of two such ions.

The calculation of shell corrections to the liquid drop model using the Strutinski method (chapter 5) not only explained fission isomers as arising because of a second minimum in the potential energy versus deformation curve but also predicted an island of stability for nuclei with $Z \approx 114$ and $N \approx 184$. For these numbers of protons and neutrons the shell effects are predicted to give nuclei with lifetimes possibly greater than the age of the earth. The great difficulty experimentally is that when an attempt is made to fuse two very heavy nuclei to form such an element, the compound nucleus is formed at such high excitation that it fissions rather than loses energy by a cascade of evaporated particles.

However, several very heavy nuclei have been produced using very selective isotope separators, such as the SHIP instrument at GSI, Darmstadt. Their decay lifetimes bear out the predictions made from calculations of the shell effects. For example, for the nucleus $^{260}106$ ($Z = 106$, $A = 260$) the spontaneous fission lifetime is increased by 15 orders of magnitude over a liquid drop model estimate. It decays approximately equally by fission and by the emission of an α-particle of 9.76 MeV into $^{256}104$, which then disintegrates by fission. The half-life of $^{260}106$ is 3.6 ms.

The heaviest element yet produced is $^{266}109$ by the reaction ^{209}Bi(^{58}Fe, n)$^{266}109$ at an incident energy of 5.15 MeV/a.m.u. The production cross-section is in the range 3–50 pb. This is called a cold-fusion reaction because the elements ^{58}Fe and ^{209}Bi both have closed shells and hence high binding energy, which results in the compound nucleus being formed at a relatively low excitation energy. The element $^{266}109$ decays by α-decay with an energy of 11.10 MeV and has a lifetime of $\sim$5 ms.

As far as superheavy elements are concerned, it appears that it is the difficulty of formation rather than their lifetime not being very long that is the probable reason why they have not been observed.

6.9 Summary

At energies well below the Coulomb barrier both light and heavy ions undergo Rutherford scattering. As the incident energy is increased quantum features are seen. In elastic and inelastic scattering diffraction and interference phenomena occur which are well described using the Born approximation. For light and heavy ions at energies above the Coulomb barrier and for low-energy neutrons in particular compound nucleus reactions are significant. The occurrence of resonances in the cross-sections for these reactions corresponds to the formation of long-lived states in the compound nucleus.

At higher incident energies direct reactions for both light and heavy ions become more important. The angular distributions are characteristic of the angular momentum transfer in light-ion induced reactions but are generally more featureless for heavy-ion induced reactions. The distributions are well described using the distorted wave Born approximation (DWBA). For heavy-ion reactions at energies above the Coulomb barrier deep-inelastic reactions are observed, particularly for reactions involving very heavy ions.

In these reactions there is a considerable transfer of the incident kinetic energy to internal excitation of the product nuclei.

6.10 Questions

1.* When the $^{12}C(\alpha, \gamma)$ cross-section is measured, a peak is found at an α-particle energy of 7045 keV. Explain this observation and find the energy of the γ-ray emitted. Discuss whether it would be possible to confirm your explanation by measurement of the $^{15}N(p, \gamma)$ cross-section as a function of proton energy. (The mass excesses of the proton, α-particle, ^{15}N and ^{16}O are 7289, 2425, 100 and −4737 keV, respectively.)

2.* The cross-section for the reaction $^{10}_{5}B(n, \alpha)^7Li$ is 630 b for incident neutrons of 1 eV. This reaction is often used as a means of detecting neutrons by filling an ionisation chamber with $^{10}BF_3$ gas. Such a chamber, with the gas at STP, of rectangular cross-section is exposed to a broad beam of neutrons of energy 1 eV; if the chamber thickness (in the beam direction) is 0.1 m, the cross-sectional area is 2×10^{-2} m^2 and the counting rate is 10^3 s^{-1}, what is the incident neutron flux?

3.* Classify the following reactions as direct or compound nucleus: (*a*) $^{19}F(p, \gamma)^{20}Ne$ $E_p \sim 2$ MeV, (*b*) $^{40}Ca(p, d)^{39}Ca$ $E_p \sim 40$ MeV (*c*) $^{48}Ca(^{16}O, ^{15}N)^{49}Sc$ $E_{^{16}O} \sim 50$ MeV, (*d*) $^{197}Au(^{13}C, 5n)^{205}At$ $E_{^{13}C} \sim$ 85 MeV.

4.* The following γ-ray and α-particles are seen when resonances in the $^{19}_{9}F + p$ reaction are studied (α-particle energies in the centre of mass):
   (i) At E_p(lab) = 668 keV, γ-rays with 6.13, 6.92 and 7.12 MeV and α-particles with 1.30, 1.47 and 2.10 MeV.
   (ii) At E_p(lab) = 843 keV, no γ-rays and 7.14 MeV α-particles.
   (iii) At E_p(lab) = 874 keV, γ-rays with 6.13, 6.92 and 7.12 MeV and α-particles with 1.46, 1.62 and 2.25 MeV.

   Draw a diagram to show the levels in ^{20}Ne and ^{16}O involved in these reactions.

5. ^8Be particles produced in a reaction with tens of MeV energy decay into two α-particles with approximately equal energy as the Q-value for $^8Be_{gs} \rightarrow \alpha + \alpha$ is only 92 keV. If both α-particles are detected by a $\Delta E - E$ detector telescope, show that the PI signal would be close to that of a ^7Li particle.

6.* List the isospin of the $^{14}_{7}N$, $^{15}_{6}C$ and $^{20}_{10}Ne$ states which can be reached in the following reactions: (i) $^{10}_{5}B(^7_3Li, ^3_1H)^{14}_{7}N$, (ii) $^{16}_{8}O(d, \alpha)^{14}_{7}N$, (iii) $^9Be(^7Li, p)^{15}C$, (iv) $^{14}C(d, p)^{15}C$, (v) $^{16}_{8}O(\alpha, \gamma)^{20}_{10}Ne$, (vi) $^{18}_{8}O(^3He, n)^{20}_{10}Ne$.

7.* For the reaction $p+p \to d+\pi^+$, $\sigma(\alpha \to \beta)$ is $0.18 \pm 0.06 \times 10^{-31}$ m^2 for incident protons of kinetic energy 340 MeV in the laboratory frame, whilst for the inverse reaction $\sigma(\beta \to \alpha)$ is $3.1 \pm 0.3 \times 10^{-31}$ m^2 for incident pions with kinetic energy 29 MeV in the laboratory frame. Use these data to calculate the spin of π^+, taking $J_p = \frac{1}{2}$ and $J_d = 1$ as known.

8.* The ground state of ^{8}Be is unstable and its only decay mode is into two α-particles with a Q-value of 92 keV. Calculate the laboratory energy of an α-particle that would excite this resonance in scattering by a helium target and calculate the cross-section on resonance. When the laboratory energy is increased to 6.06 MeV a second resonance, also decaying into two α-particles, is observed with a cross-section on resonance of 2.16 b and a width of 1 MeV. What is the J^π of this state?

9.* In slow neutron capture show that at a resonance peak (neglecting any background), $\pi\sigma^2 = \lambda^2 g \sigma_s$ where σ and σ_s are the total cross-section and the elastic scattering cross-section for neutrons and g is the statistical factor in the Breit–Wigner formula.

10.* Explain why at $E_p = 15$ MeV in the reaction $^{65}_{29}$Cu(p, n)$^{65}_{30}$Zn, the energy spectrum of the neutrons shows a continuous distribution on which is superimposed one sharp peak corresponding to a reaction Q-value of -9.4 MeV and to ^{65}Zn being left at an excitation energy of 7.3 MeV.

11. Calculate whether the approximate relation $qR = l$ for the first peak is satisfied for the data on the ^{56}Fe(d, p)^{57}Fe reaction shown in figure 6.22.

12. Derive the expression for Q_{opt} given on p. 209.

13.* In Rutherford scattering show that the scattering angle θ in the centre of mass frame and the distance of closest approach d are related by: $b = a \cot(\frac{1}{2}\theta)$ and $d = a + (a^2 + b^2)^{1/2}$, where b is the impact parameter and $2a$ is the distance of closest approach in a head-on collision. Deduce the expression given on p. 210 for the reaction cross-section $\sigma_{reac} = \sigma_0(1 - V/E_{cm})$ where $\sigma_0 = \pi R_s^2$ and $V = Z_1 Z_2 e^2 / 4\pi\varepsilon_0 R_s$.
The fusion cross-section for $^{32}_{16}$S on $^{24}_{12}$Mg as a function of the laboratory energy of ^{32}S is as follows: 70 MeV 150 mb; 80 MeV 414 mb; 90 MeV 670 mb; 110 MeV 940 mb. Use these data to test the hypothesis that $\sigma_{reac} = \sigma_{fus}$ at these energies.

14. What is the minimum energy proton beam required to produce anti-protons via the reaction $p+p = p+p+p+\bar{p}$ (*a*) with a hydrogen target, (*b*) with an iron target, (*c*) with a uranium target.

7 The nuclear force

The fundamental strong interaction is that between quarks arising from the exchange of spin 1 particles called gluons. Nucleons are made up of three quarks: a neutron of two down quarks each of charge $-\frac{1}{3}$ and an up quark of charge $+\frac{2}{3}$; a proton of one down and two up quarks. The gluon exchange force has the property of increasing with increasing quark separation with the result that quarks are confined. The nuclear force between nucleons arises from the exchange of mesons, which are made up of a quark–antiquark pair. It therefore bears somewhat of the same relationship to the strong interaction as the Van der Waals force between molecules does to the electrostatic interaction. However, this does not mean that a good description is not possible of the nuclear force using a meson exchange picture but only that its connection with the fundamental quark-gluon interaction is very complicated. The simplest system to study the nuclear force is that of two nucleons and that is first discussed before the properties of the nuclear force within nuclei are considered.

7.1 Low-energy nucleon–nucleon scattering

7.1.1 Partial wave treatment of nucleon–nucleon scattering

The interaction between two nucleons can be described using partial wave theory. For low-energy interactions the wavenumber k of the reduced mass is such that $kR \ll 1$ when R is the range of the interaction (~ 1.4 fm). This means that the cross-section σ is given by:

$$\sigma = \frac{4\pi}{k^2} \sin^2 \delta_0 \tag{7.1}$$

where δ_0 is the $l = 0$ phase shift. An approximate expression for δ_0, valid at low energies, may be derived for scattering by a potential. The time-independent Schrödinger equation assuming a central potential is:

$$-\frac{\hbar^2}{2m}\nabla^2\psi + V(r)\psi = E\psi$$

where m is the reduced mass. Defining $\psi = [u(r)/r]Y_{lm}(\theta, \phi)$, where $Y_{lm}(\theta, \phi)$ is the spherical harmonic function, leads to the function $u(r)$ satisfying:

$$-\frac{\hbar^2}{2m}\frac{\mathrm{d}^2u}{\mathrm{d}r^2} + V(r)u = Eu$$

For a simple square well of radius R and depth V_0 the solutions $u_1(r)$ within the well and $u_2(r)$ outside are:

$$u_1(r) = A\sin Kr$$
$$u_2(r) = B\sin(kr + \delta_0)$$

where $K^2 = 2m(V_0 + E)/\hbar^2$, A and B are normalisation constants and δ_0 is the $l = 0$ phase shift. If $E = 0$ then $u_2(r)$ satisfies:

$$-\frac{\hbar^2}{2m}\frac{\mathrm{d}^2u}{\mathrm{d}r^2} = 0$$
$$\therefore \quad u_2(r) = C + Dr$$

where C and D are constants. If the well is deep enough to form a bound state then $KR > \pi/2$ as $u_1(r) = A\sin Kr$ and $u_2(r) = Be^{-\beta r}$ with $\beta^2 = 2mE_\mathrm{B}/\hbar^2$, so matching the logarithmic derivatives (equivalent to matching u and $\mathrm{d}u/\mathrm{d}r$) at R gives $\tan KR = -K/\beta$. Some of these solutions are illustrated in figure 7.1.

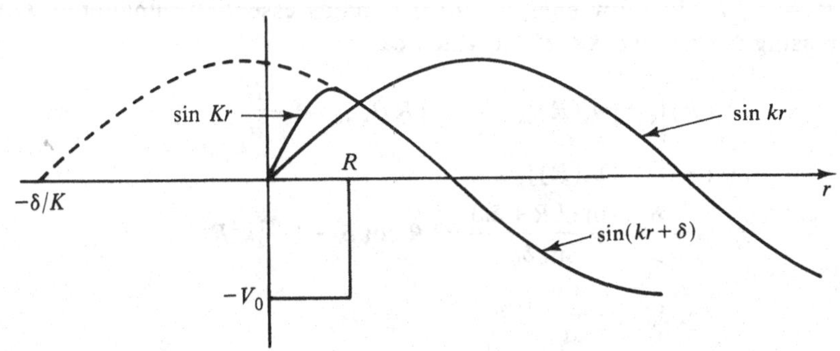

7.1. Diagram illustrating the positive s-wave phase shift introduced by an attractive square-well potential of depth V_0.

7.1.2 Scattering length and effective range

It is useful to define a length (see figure 7.2) called the scattering length which is the value of the intercept of $u_2(r)$ with the r-axis for $k = 0$.

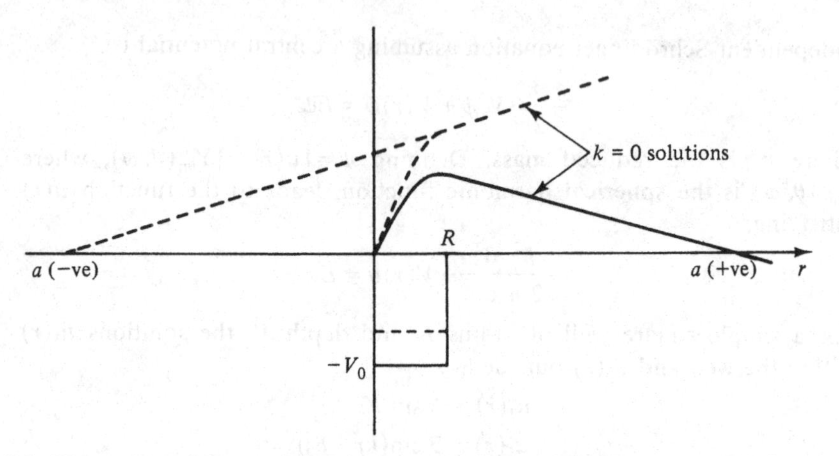

7.2. Illustration of the scattering length a. If the potential is deep enough (solid line) to form a bound state a is positive, if not (dashed line) then a is negative.

Normalising $u_2(r)=1$ for $r=0$ gives the solutions:

$$u_2(r)=1-\frac{r}{a} \qquad k=0$$

$$u_2(r)=\frac{\sin(kr+\delta_0)}{\sin\delta_0} \qquad k\neq 0$$

The wavefunction $u_1(r)$ within the well satisfies the boundary condition $u_1(R)=u_2(R)$. For low energies $u_1(r)$ remains essentially unaltered with increasing k provided $k \ll K$, in which case:

$$[u_1(R)]_k=[u_1(R)]_{k=0}=[u_2(R)]_{k=0}=1-\frac{R}{a}$$

$$[u_1(R)]_k=[u_2(R)]_k$$

$$\therefore \quad 1-\frac{R}{a}=\frac{\sin(kR+\delta_0)}{\sin\delta_0}=kR\cot\delta_0+1-\tfrac{1}{2}k^2R^2\cdots$$

$$\therefore \quad \cot\delta_0=-\frac{1}{ka}+\tfrac{1}{2}kR\cdots$$

Substituting this into equation 7.1 predicts the scattering cross-section σ_0 to be:

$$\sigma_0=4\pi a^2[(1-\tfrac{1}{2}aRk^2)^2+a^2k^2]^{-1} \tag{7.2}$$

where R in this equation is called the effective range of the nuclear potential.

For a bound state of energy $-E_B$ then $u_2(r)=e^{-\beta r}$ with $\beta^2=2mE_B/\hbar^2$.

Equating $u_2(R)$ to $1-R/a$ in this case gives:

$$1-\frac{R}{a}=1-\beta R+\tfrac{1}{2}\beta^2R^2$$

$$\therefore \quad \frac{1}{a}=\beta-\tfrac{1}{2}\beta^2R$$

and so relates the binding energy E_B to the scattering length and effective range. For the deuteron, in which the neutron and proton are in a triplet state, the value of $\beta=0.2316\ \text{fm}^{-1}$ and $a_t=5.425$ fm (see below) so $R_t=1.76$ fm.

7.1.3 Low-energy neutron–proton scattering

The neutron and proton both have an intrinsic spin of a half which can combine to give $S=1$ (triplet) or $S=0$ (singlet) states. The cross-section for $l=0$ scattering for unpolarised neutrons and protons is therefore:

$$\sigma_0=\tfrac{3}{4}\sigma_0^t+\tfrac{1}{4}\sigma_0^s \tag{7.3}$$

where t stands for triplet and s for singlet scattering. (The ratio of 3:1 is the ratio of magnetic substates.) Equation 7.2 together with equation 7.3 is plotted in figure 7.3 together with the experimental data. The agreement is very good. The values of a_s, a_t and R_t are determined from very low-energy

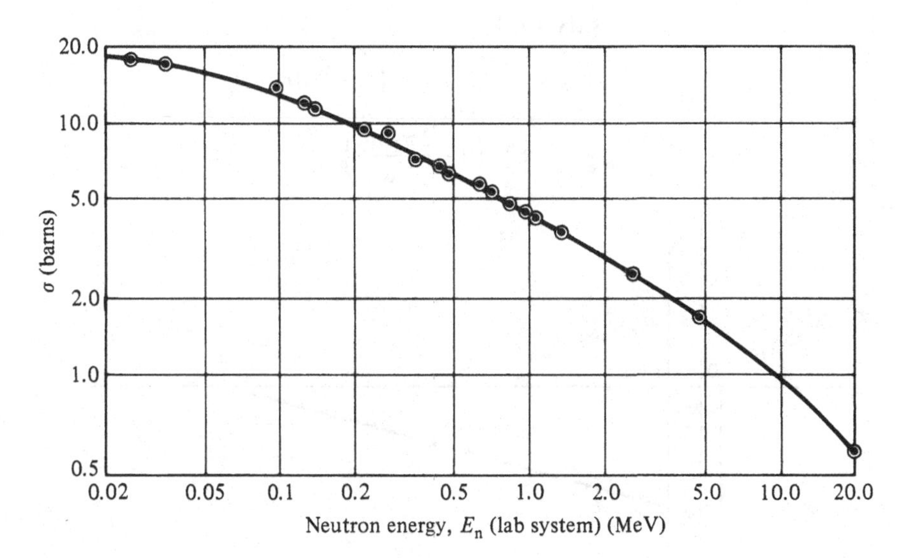

7.3. Total cross-section for p–n scattering. The theoretical curve (solid curve) is based on $a_t=5.38$ fm, $a_s=-23.7$ fm, $R_t=1.70$ fm and $R_s=2.40$ fm. (From Enge, H. A., *Introduction to Nuclear Physics*. Addison-Wesley (1966).)

scattering (see below) and deuteron data and only R_s is free. (More recent data give $R_s = 2.70$ fm.)

This assumes that the scattering is just between one neutron and one proton. For sufficiently low energies an incident neutron can interact with several protons or other nuclei at once, i.e. the scattering can be coherent. For very low energies $\cot \delta_0 = -1/ka$ and $\sigma_0 = 4\pi a^2$. From partial wave theory the amplitude of the scattered wave is $(e^{i\delta_0}/k) \sin \delta_0 \to -a$ as $k \to 0$ because $\delta_0 \to \pi$ as $k \to 0$ therefore the total wavefunction is:

$$\psi = e^{ikz} - a\, e^{ikr}/r$$

where e^{ikz} represents the incident neutron's wavefunction.

7.1.4 Refractive index for slow neutrons

If a thin slab of material containing n nuclei m^{-3} is in a beam of neutrons (figure 7.4) then the resulting wavefunction will be given by:

$$\psi = e^{ikz} - nta \int_0^\infty \frac{e^{ikr}}{r} 2\pi s \, ds$$

where a is the scattering length for each nucleus, s the radius of an annulus and t is the thickness of the slab with $t \ll 1/k$ so all nuclei act coherently. From figure 7.4:

$$z^2 + s^2 = r^2$$

$$\therefore \quad s\,ds = r\,dr$$

$$\therefore \quad \psi = e^{ikz} - 2\pi nta \int_z^\infty e^{ikr}\,dr$$

$$= e^{ikz}\left(1 - \frac{2\pi inta}{k}\right)$$

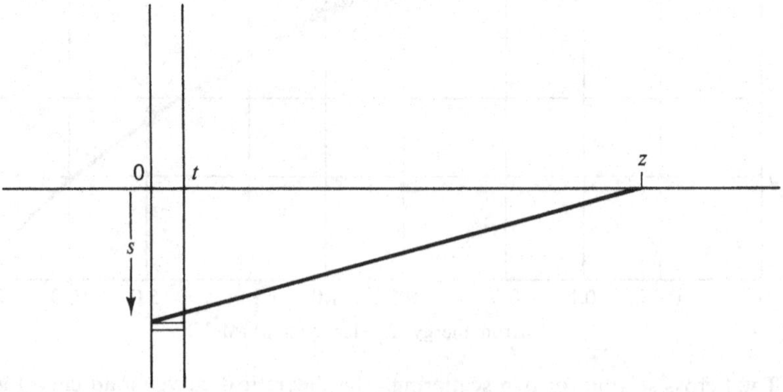

7.4. The scattering of slow neutrons by a thin slab of material containing n nuclei/m^3 and of thickness t.

This is just like in the scattering of light and is equivalent to a refractive μ as this would change the incident wave e^{ikz} to:

$$\exp[ik(z-t)+i\mu kt] = e^{ikz}\, e^{ik(\mu-1)t} = e^{ikz}[1+ik(\mu-1)t]$$

since $kt \ll 1$ so that:

$$(\mu-1) = -2\pi n\frac{a}{k^2}$$

Therefore if a is positive $\mu<1$ and neutrons can be totally reflected if $\sin i>\mu$ where i is the angle of incidence. This phenomenon enables the scattering length a to be measured.

Reflection takes place when the grazing angle ϕ_g ($\phi_g = 90-i$) is less than:

$$\phi_g = 2(\pi na)^{1/2}/k$$

$$\therefore \quad k\phi_g = mv_\perp/\hbar = 2(\pi na)^{1/2}$$

where $v_\perp$ is the component of the neutron's velocity perpendicular to the surface of the material. If a beam of slow neutrons is collimated to travel horizontally initially and then falls under gravity onto a surface a distance d below, all the neutrons will be reflected if:

$$d<2\pi\hbar^2 na/m^2 g$$

This enables the scattering length a to be measured. For slow neutrons, when the scattering is coherent, a is the sum of the scattering lengths of the different nuclei in the molecules of the 'mirror'. The scattering length a_H for unpolarised neutrons coherently scattering off protons bound in a heavy molecule is:

$$a_H = 2(\tfrac{1}{4}a_s+\tfrac{3}{4}a_t)$$

where the factor 2 reflects the difference in reduced mass and the factors $\frac{1}{4}$ and $\frac{3}{4}$ the different probabilities for singlet and triplet states. The result of experiments using hydrocarbon mirrors is:

$$a_H = -3.719 \pm 0.002 \text{ fm}$$

The low-energy (but sufficiently high to be incoherent) scattering cross-section σ_H is, from above (equations 7.2 and 7.3):

$$\sigma_H = 4\pi(\tfrac{1}{4}a_s^2+\tfrac{3}{4}a_t^2)$$

$$= 20.442 \pm 0.023 \text{ barns}$$

and combining these results gives:

$$a_s = -23.714 \pm 0.013 \text{ fm}$$

$$a_t = 5.425 \pm 0.004 \text{ fm}$$

This shows that the neutron–proton force is spin dependent with the triplet state bound (the deuteron) and the single state just unbound. This difference shows up very markedly in low-energy neutron scattering off ortho-(spin 1) and para-(spin 0) hydrogen.

7.1.5 Slow neutron scattering off ortho- and para-hydrogen

Ortho-hydrogen is when the two proton spins couple up to spin 1, para-hydrogen when they couple to spin 0. For the scattering off one proton the scattering length a_H may be written as:

$$a_H = a_s\pi_s + a_t\pi_t$$

where

$$\pi_s = \tfrac{1}{4}(1 - \boldsymbol{\sigma}_n \cdot \boldsymbol{\sigma}_p)$$
$$\pi_t = \tfrac{1}{4}(3 + \boldsymbol{\sigma}_n \cdot \boldsymbol{\sigma}_p)$$

and $\boldsymbol{\sigma}$ is the Pauli spin operator $\boldsymbol{\sigma} = 2\mathbf{s}$ where $\mathbf{s}$ is the spin 1/2 operator, taking $\hbar = 1$. The operators π_s and π_t are projection operators as the following identity shows:

$$4\mathbf{S}^2 = (\boldsymbol{\sigma}_n + \boldsymbol{\sigma}_p)^2 = \boldsymbol{\sigma}_n^2 + \boldsymbol{\sigma}_p^2 + 2\boldsymbol{\sigma}_n \cdot \boldsymbol{\sigma}_p$$
$$\therefore \quad 4S(S+1) = 6 + 2\boldsymbol{\sigma}_n \cdot \boldsymbol{\sigma}_p$$

So $\pi_s = 1$, $\pi_t = 0$ for $S = 0$ and $\pi_s = 0$, $\pi_t = 1$ for $S = 1$. For the coherent scattering off two protons:

$$a = a_{H_1} + a_{H_2} = a_s(\pi_{s_1} + \pi_{s_1}) + a_t(\pi_{t_1} + \pi_{t_2})$$
$$= 2(\tfrac{1}{4}a_s + \tfrac{3}{4}a_t) + 2(\tfrac{1}{4}a_t - \tfrac{1}{4}a_s)\boldsymbol{\sigma}_n \cdot \mathbf{S}_H$$

where $\mathbf{S}_H = \frac{1}{2}(\boldsymbol{\sigma}_{p_1} + \boldsymbol{\sigma}_{p_2})$. This expression assumes that the protons are free but they are bound together which increases the reduced mass by a factor $\frac{4}{3}$. The cross-sections for scattering are therefore:

$$\sigma_{para} = \tfrac{16}{9} \cdot 4\pi(\tfrac{3}{2}a_t + \tfrac{1}{2}a_s)^2$$
$$\sigma_{ortho} = \tfrac{16}{9} \cdot 4\pi[(\tfrac{3}{2}a_t + \tfrac{1}{2}a_s)^2 + \tfrac{1}{2}(a_t - a_s)^2]$$

Substituting values for a_s and a_t gives $\sigma_{ortho} \approx 30\sigma_{para}$. Experimentally σ_{para} can be measured by cooling H_2 molecules to ~20 K at which point the hydrogen is essentially all para-hydrogen because the para-form is slightly more bound than the ortho-form.

7.1.6 Low-energy proton–proton and neutron–neutron scattering

Because the nucleons are fermions, only n–n and p–p scattering in the ^{1}S, ^{3}P, ^{1}D . . . states is possible. For the p–p scattering the effect of the Coulomb force must be subtracted and the result is for a_s^{pp} to be altered from −7.817 fm to:

$$a_s^{pp} = -17.0 \pm 0.2 \text{ fm}$$

This is quite different from $a_s^{np} = -23.714 \pm 0.013$ fm but for $a/R \gg 1$ a is very sensitive to V_0, the depth of the well. The most significant contribution to this difference is the difference in the masses of the exchanged pions: for p–p only $\pi^\pm$ exchange is possible while for n–p both π^0 and $\pi^\pm$ can be exchanged. The heavier mass of $\pi^{\pm}$ relative to π^{0} makes the range less, the depth effectively smaller and hence a larger. The magnitude of the effect is close to the observed difference.

For n–n scattering a_s^{nn} has to be measured indirectly and can be inferred from the γ-ray spectrum produced in π^- capture by deuterium:

$$\pi^- + d \rightarrow n + n + \gamma$$

If n+n were a bound system there would be a unique γ-ray energy; if there were no n–n interaction then the energy would be shared amongst all three particles (cf. β-decay). The result is illustrated in figure 7.5 and a_s^{nn} is estimated as:

$$a_s^{nn} = -16.4 \pm 1.9 \text{ fm}$$

consistent with $a_s^{pp} = a_s^{nn}$, i.e. charge symmetry.

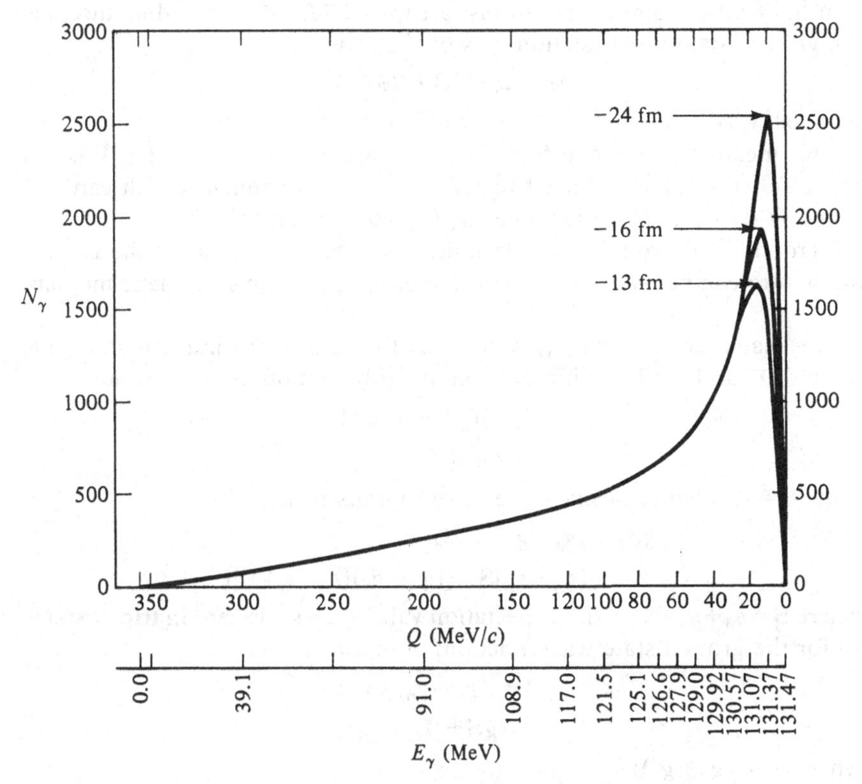

7.5. Illustration of the effect of the n–n interaction on the spectrum of γ-rays emitted in the $\pi^- + d \rightarrow n + n + \gamma$ reaction (note the non-linear energy scale). Q is the relative neutron–neutron momentum in their centre of mass system. (From Haddock, R. P. *et al.*, *Phys. Rev. Lett.* **14** (1965) 318.)

The conclusion from low-energy nucleon–nucleon scattering is that the nuclear force is both charge symmetric and charge independent to a very good approximation.

7.2 The deuteron

The deuteron has a total angular momentum $J=1$ and the triplet ground state is bound (a_t positive) by 2.2245 MeV. Its first excited state is a singlet state which is unbound (a_s negative) by 70 keV. In 1939 Kellogg *et al.* discovered from measuring the hyperfine spectrum of atomic deuteron that the deuteron had a positive quadrupole moment of +2.82 mb, which corresponds to a prolate (cigar-shaped) deformation. The ground state therefore is not a pure S($L=0$) state, as this would be spherically symmetric, but must include other L components. Parity and angular momentum conservation limit other components to just a triplet D($L=2$) state admixture, i.e. the ground state wavefunction ψ is of the form:

$$\psi = \alpha\psi(^3\mathrm{S}_1) + \beta\psi(^3\mathrm{D}_1)$$

with $|\alpha|^2 + |\beta|^2 = 1$.

This means that the n–p force is not a purely central force, i.e. V is not just a function of r, but must include a tensor component which can mix S and D states. If $V = V(r)$ then $[V, L^2] = 0$ and $[H, L^2] = 0$, in which case the ground state would have a definite L value. An estimate of the D state admixture can be made from the value of the deuteron's magnetic moment μ_D.

The magnetic moment $\boldsymbol{\mu}_\mathrm{D}$ is given by the sum of the intrinsic moments $\boldsymbol{\mu}_\mathrm{p}$ and $\boldsymbol{\mu}_\mathrm{n}$ and the contribution from the orbital motion of the proton, i.e.:

$$\begin{aligned}\boldsymbol{\mu}_\mathrm{D} &= \boldsymbol{\mu}_\mathrm{p} + \boldsymbol{\mu}_\mathrm{n} + \mu_\mathrm{N}\mathbf{l}_\mathrm{p} \\ &= g\mu_\mathrm{N}\mathbf{J}\end{aligned}$$

Now $\mathbf{l}_\mathrm{p} = \frac{1}{2}\mathbf{L}$ (centre of mass effect) so in units of μ_N:

$$\begin{aligned}g\mathbf{J} &= g_\mathrm{p}\mathbf{s}_\mathrm{p} + g_\mathrm{n}\mathbf{s}_\mathrm{n} + \tfrac{1}{2}\mathbf{L} \\ &\equiv \tfrac{1}{2}(g_\mathrm{p}+g_\mathrm{n})\mathbf{S} + \tfrac{1}{2}(g_\mathrm{p}-g_\mathrm{n})(\mathbf{s}_\mathrm{p}-\mathbf{s}_\mathrm{n}) + \tfrac{1}{2}\mathbf{L}\end{aligned}$$

where $\mathbf{S} = \mathbf{s}_\mathrm{p} + \mathbf{s}_\mathrm{n}$. Now the expectation value $\langle \mathbf{s}_\mathrm{p} - \mathbf{s}_\mathrm{n}\rangle$ is zero in triplet states so for the ground state wavefunction:

$$\begin{aligned}g\mathbf{J} &= \tfrac{1}{2}(g_\mathrm{p}+g_\mathrm{n})\mathbf{S} + \tfrac{1}{2}\mathbf{L} \\ &= \bar{g}\mathbf{S} + \tfrac{1}{2}\mathbf{L}\end{aligned}$$

where $\bar{g} = \frac{1}{2}(g_\mathrm{p}+g_\mathrm{n})$:

$$\begin{aligned}\therefore\quad \langle\psi|g\mathbf{J}^2|\psi\rangle &= \langle\psi|\bar{g}\mathbf{S}\cdot\mathbf{J} + \tfrac{1}{2}\mathbf{L}\cdot\mathbf{J}|\psi\rangle \\ &= \langle\psi|\left(\frac{\bar{g}}{2}+\frac{1}{4}\right)\mathbf{J}^2 - \left(\frac{\bar{g}}{2}-\frac{1}{4}\right)(\mathbf{L}^2-\mathbf{S}^2)|\psi\rangle\end{aligned}$$

Substituting $|\psi\rangle = \alpha|\psi(^3S_1)\rangle + \beta|\psi(^3D_1)\rangle$ gives:

$$g = \tfrac{1}{2}(g_p + g_n) + \tfrac{3}{4}(1 - g_p - g_n)|\beta|^2$$

Now $\mu_D = 0.8574\mu_N$, $\mu_n = -1.9131\mu_N$ and $\mu_p = +2.793\mu_N$, which yields $|\beta|^2 = 0.039 = 4\%$. This value assumes that the binding of the neutron and proton is not affecting the values of their magnetic moments. The radius of the deuteron, ~2.1 fm, compared with that of the nucleons, ~0.8 fm, makes the effects of meson exchange quite small but a more reasonable estimate for $|\beta|^2$ is $|\beta|^2 \approx 2$–8%.

An example of a tensor force is that between two magnetic dipoles and the nuclear 'tensor' potential V_T is of the same form:

$$V_T = f(r)[3(\boldsymbol{\sigma}_1 \cdot \mathbf{r})(\boldsymbol{\sigma}_2 \cdot \mathbf{r})/r^2 - \boldsymbol{\sigma}_1 \cdot \boldsymbol{\sigma}_2]$$

The potential therefore depends not only on the separation r but also on the orientation of the intrinsic spins to $\mathbf{r}$. Such a potential can be understood as arising from the exchange of mesons. The pions have negative intrinsic parity, therefore when a nucleon emits a pion it must be in a p-state, which is not spherically symmetric, in order to conserve parity. Furthermore angular momentum must be conserved so the orientation of the p-state relative to the nucleon spin is not arbitrary and this gives rise to a $(\boldsymbol{\sigma}_1 \cdot \mathbf{r}) \times (\boldsymbol{\sigma}_2 \cdot \mathbf{r})$ dependence in the potential.

The small admixture of the D state in the ground state of the deuteron does not mean that the strength of the tensor force is weak. The radial function $u(r)$ for an l state satisfies the equation:

$$-\frac{\hbar^2}{2m}\frac{d^2u}{dr^2} + \left[V(r) + \frac{l(l+1)\hbar^2}{2mr^2}\right]u = Eu$$

The term $l(l+1)\hbar^2/2mr^2$ is called the centrifugal potential. It is a repulsive potential and has the effect of reducing the admixture of the D state by the tensor force. The tensor interaction between nucleons is an important part of the nuclear force, in particular in producing saturation (see below).

7.3 High-energy nucleon–nucleon scattering

The information from low-energy scattering is on the strength and range of the potential but is not about its shape, as expected since $\lambda \gg R$. For higher energies a phase shift analysis can be made and the results for some of the lower l states are shown in figure 7.6. These are decreasing with increasing energy and become negative for the S states near 300 MeV. A repulsive potential produces a negative phase shift. However, a short-range repulsive core to an attractive potential will not produce any significant phase shift at low energies, but at high energies ($E \gg V_0$) the effect of the attractive well will become negligible and the phase shift will become

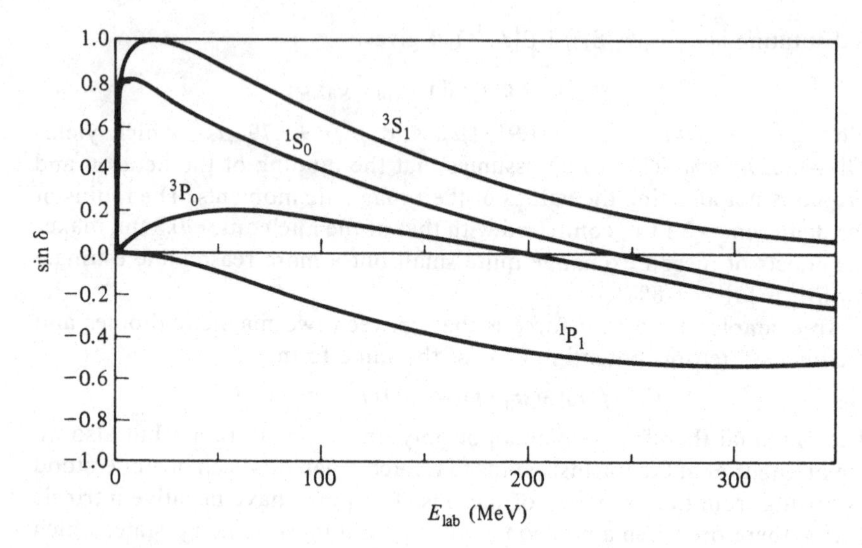

7.6. A sketch of the energy dependence of the $^1S_0$, $^3S_1$, $^1P_1$ and $^3P_0$ phase shifts. (From Lock, W. O. and Measday, D. F., *Intermediate Energy Nuclear Physics.* Methuen (1970).)

negative. Analysis of the data implies a repulsive core of ~0.5 fm with a height of at least several hundred MeV. In terms of meson exchange theory the exchange of a vector meson (such as a ρ $J^\pi = 1^-$) can give rise to a repulsive potential and its range will be given by $\hbar/m_x c$ where m_x is the mass of the exchange particle, as discussed in chapter 3, p. 79.

Besides providing evidence for a repulsive core the high-energy proton-proton scattering experiments indicate that there is a spin–orbit component to the nucleon–nucleon potential. This would be expected from the existence of an effective spin–orbit interaction in the shell model, and is also predicted by meson exchange theory.

7.4 Exchange forces

The direct experimental evidence for charged $\pi^\pm$ meson exchange being part of the n–p force is from the n–p differential cross-section at high energies, which shows a pronounced backward as well as forward peak (see figure 7.7). At such energies the Born approximation can be used to estimate the cross-section as:

$$\frac{d\sigma}{d\Omega} = |f(q)|^2$$

where $f(q) = -(m/2\pi\hbar^2) \int V(r)\, e^{i\mathbf{q}\cdot\mathbf{r}/\hbar}\, d^3r$ and $\mathbf{q}$ is the momentum transfer,

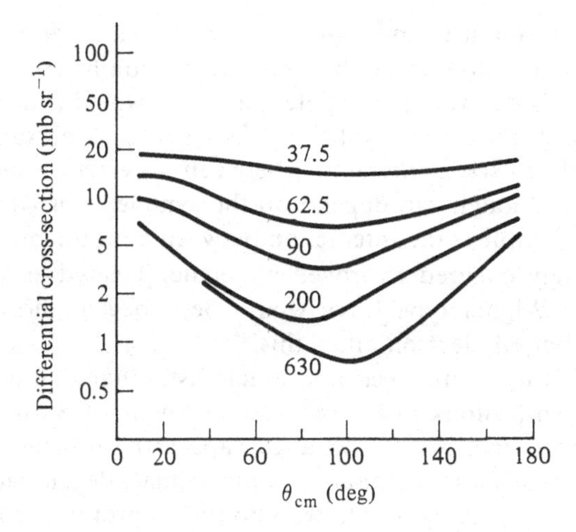

7.7. Differential cross-section for neutron–proton scattering. The laboratory neutron energies in MeV are shown on the curves. (From Burcham, W. E., *Elements of Nuclear Physics*. Longman (1979).)

i.e. the cross-section is proportional to the square of the Fourier transform of the potential. If the potential is of the form $e^{-\mu r}/r$ then $f(q) \propto 1/(q^2+\mu^2)$ (equation 1.1). This predicts that forward angle scattering will be larger than backward, particularly when $q \gg \mu$. The magnitude observed at forward angles is consistent with a potential of the range of ~1 fm. At 300 MeV $k_i \approx 2\ \text{fm}^{-1}$ so $q_{max} \approx 4\ \text{fm}^{-1}$ so forward scattering would be expected to be much larger than backward. The answer is that $\pi^{\pm}$ exchange takes place in addition to π^0 exchange and there is an appreciable probability that the incoming proton charge exchanges and becomes an outgoing neutron, as illustrated in figure 7.8. Such exchange forces give a dependence of the potential energy on the symmetry state of the two nucleons and not just on r.

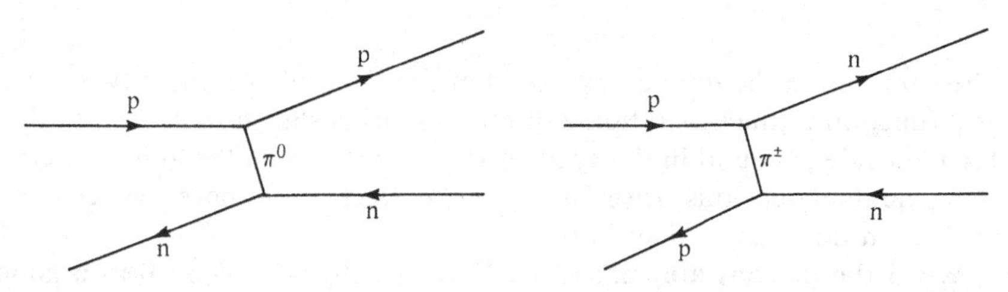

7.8. Representation of π^0 exchange and $\pi^{\pm}$ charge exchange in n–p scattering.

Consider a neutron at $\mathbf{r}_1$ and a proton at $\mathbf{r}_2$ with a separation $\mathbf{r}_1 - \mathbf{r}_2 = \mathbf{r}$. Let the wavefunction describing their relative motion be $\phi(r)$. If the force between n and p is described by a potential $V(r)$, which is independent of the symmetry of ϕ, the force is called a Wigner force. An example of such a force is the electrostatic interaction between an electron and a proton. Forces in general though can depend on the symmetry of $\phi(r)$ as well as on r. A classic example is the interaction between a proton and a hydrogen atom in the singly charged hydrogen molecule. Treated as two separate particles only a Wigner type force would be expected; however, taking account of the bound electron alters this.

The motion of the bound electron is much faster than that of any relative motion of the two protons so the two protons, which we will label A and B, can be treated as being fixed a distance r apart (the adiabatic approximation). When the protons are far apart the approximate degenerate eigenfunctions will be ϕ_A, describing an electron bound to proton A in a hydrogen eigenstate with a bare proton at B, and ϕ_B, likewise. The eigenfunctions ϕ_A and ϕ_B satisfy:

$$\langle \phi_A | T + V_A | \phi_A \rangle = \langle \phi_B | T + V_B | \phi_B \rangle = -\varepsilon$$

where ε is the binding energy of a hydrogen atom. The Hamiltonian H for the H_2^+ molecule is:

$$\begin{aligned} H &= T + V_A + V_B + V_{AB} \\ &= H_e + V_{AB} \end{aligned}$$

where H_e is the electron Hamiltonian, T is the electron's kinetic energy, V_A and V_B its potential energy arising from its Coulomb interaction with proton A and proton B, and V_{AB} is the electrostatic potential energy of the two protons. H is symmetric with respect to interchange of A and B so non-degenerate eigenstates ϕ_s and ϕ_a of H will be either symmetric or antisymmetric under exchange of A and B, i.e.:

$$\phi_s = \frac{1}{\sqrt{2}}(\phi_A + \phi_B)$$

$$\phi_a = \frac{1}{\sqrt{2}}(\phi_A - \phi_B)$$

The variation in the total energy $U(r)$ of the molecule for these two states as a function of the separation r of the protons is shown in figure 7.9. The H_2^+ molecule is bound in the symmetric but unbound in the antisymmetric state; qualitatively this arises because the electron is more between the protons in the ψ_s state than in the ψ_a state.

When the protons are sufficiently far apart that $\langle \phi_A | \phi_B \rangle = 0$ to a good approximation, a simple expression for $U(r)$ can be derived. The electron

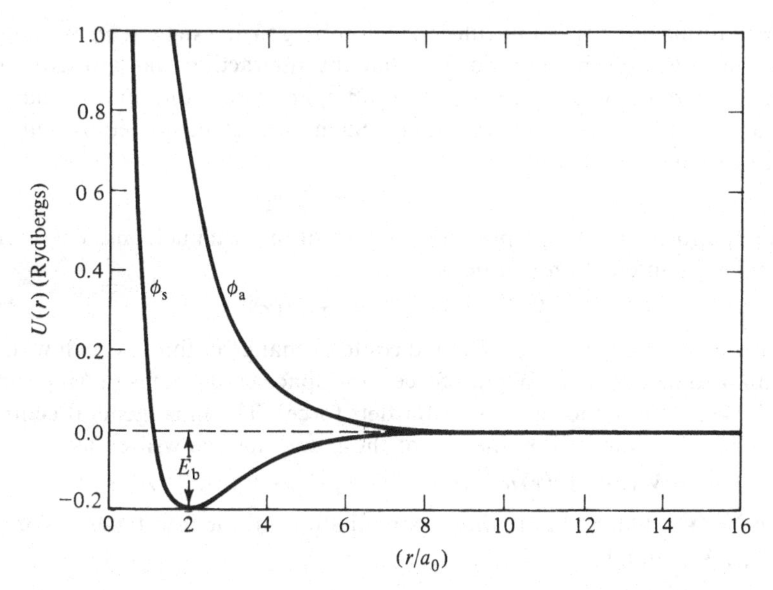

7.9. Total energies $U(r)$ of the lowest symmetric ϕ_s and antisymmetric ϕ_a states of H_2^+ as a function of the interproton distance r. E_b is the binding energy of the H_2^+ molecule. (From Leighton, R. B., *Principles of Modern Physics.* McGraw-Hill (1959).)

energy in the state ϕ_s is then given by:

$$\begin{aligned}\langle\phi_s|H_e|\phi_s\rangle &= \tfrac{1}{2}\langle\phi_A|T+V_A+V_B|\phi_A+\phi_B\rangle \\ &\quad +\tfrac{1}{2}\langle\phi_B|T+V_A+V_B|\phi_A+\phi_B\rangle \\ E_e &= -\varepsilon+\langle\phi_A|V_B|\phi_A\rangle+\langle\phi_A|V_B|\phi_B\rangle\end{aligned}$$

using symmetry. The total energy $U(r)$ of the molecule is $E_e+V_{AB}+\varepsilon$ with $U(r)$ defined as zero at $r=\infty$. The term $\langle\phi_A|V_B|\phi_A\rangle\approx -V_{AB}$ since ϕ_A is spherically symmetric and to a good approximation does not overlap proton B, i.e. $\phi_A(r)\approx 0$. Therefore:

$$\begin{aligned}U(r) &= -J(r) \quad \text{for } \phi_s \\ &= +J(r) \quad \text{for } \phi_a\end{aligned} \tag{7.4}$$

where

$$J(r) = -\int \phi_A^* V_B \phi_B \, d\tau$$

$J(r)$ is called the exchange energy and results from the sharing of the electron between the two protons.

The idea that the binding of a proton and neutron, with the neutron thought of as a proton plus a negatively charged particle, might be analogous to the binding of the H_2^+ molecule led Heisenberg to first suggest an exchange force. Heisenberg's hypothesis was that the interaction was attractive in states antisymmetric with respect to both space (P_M) and spin exchange (P_σ) and repulsive in symmetric states. Such an exchange is equivalent to charge exchange (P_l) with

$$P_l = -P_M P_\sigma = \tfrac{1}{2}(1 + \boldsymbol{\tau}_1 \cdot \boldsymbol{\tau}_2)$$

where $\boldsymbol{\tau}_1$ and $\boldsymbol{\tau}_2$ are the isospin Pauli vectors of the two nucleons. Therefore Heisenberg's interaction can be written as:

$$V_H(r) = -\tfrac{1}{2}(1 + \boldsymbol{\tau}_1 \cdot \boldsymbol{\tau}_2) J(r)$$

$J(r)$ positive. The particle exchanged could exchange nothing, which would give an ordinary force (a Wigner force); the space coordinates (a Majorana force); or the spin coordinates (a Bartlett force). The most general central interaction is a linear combination of these and may be written as:

$$V(r) = V_1(r) + V_2(r)\boldsymbol{\sigma}_1 \cdot \boldsymbol{\sigma}_2 + V_3(r)\boldsymbol{\tau}_1 \cdot \boldsymbol{\tau}_2 + V_4(r)\boldsymbol{\sigma}_1 \cdot \boldsymbol{\sigma}_2\, \boldsymbol{\tau}_1 \cdot \boldsymbol{\tau}_2$$

where V_1, V_2, V_3 and V_4 are linear combinations of the four types of forces V_W, V_H, V_M and V_B.

7.4.1 Meson exchange

In the theory of scalar meson exchange the interaction potential of a neutron with a proton by the exchange of a π^+ is given by:

$$-\left[\frac{g_s \tau_A^+}{\sqrt{2}} \frac{1}{r} \exp(-\mu r)\right] \cdot \left(\frac{g_s}{\sqrt{2}} \tau_B^-\right)$$

where g_s is the meson–nucleon coupling constant, $g_s \tau_A^+/\sqrt{2}$ represents the process of a proton emitting a π^+, $g_s \tau_B^-/\sqrt{2}$ that of a neutron absorbing a π^+ and $-(1/r)\exp(-\mu r)$ the probability amplitude of the π^+. (In terms of $J(r)$ (equation 7.4) above, $\phi_A \to -(1/r)\exp(-\mu r)$ and $\int V_B \phi_B \, d\tau \to -g_s^2/2$.) In addition there is the exchange of a π^- and of π^0, which are necessary to give charge independence and the total interaction potential is given by:

$$V_s(r) = -g_s^2(\tfrac{1}{2}\tau_A^+ \tau_B^- + \tfrac{1}{2}\tau_A^- \tau_B^+ + \tau_A^0 \tau_B^0) \frac{1}{r} \exp(-\mu r)$$

$$= -g_s^2 \boldsymbol{\tau}_A \cdot \boldsymbol{\tau}_B \frac{1}{r} \exp(-\mu r)$$

The exchange operator $\boldsymbol{\tau}_A \cdot \boldsymbol{\tau}_B$ depends on the charge exchange symmetry of the neutron–proton wavefunction and commutes with the total isospin $\mathbf{T} = \tfrac{1}{2}(\boldsymbol{\tau}_A + \boldsymbol{\tau}_B)$. The operator $\boldsymbol{\tau}_A \cdot \boldsymbol{\tau}_B = 1$ if $T = 1$, and $= -3$ if $T = 0$, so the potential $V_s(r)$ is attractive in $T = 1$ states and repulsive in $T = 0$ states. As the deuteron ($T = 0$) is bound this potential is not correct.

The pion is not a scalar but a pseudoscalar particle and the potential would be expected to be different, as mentioned above. The wavefunction for a scalar neutral meson in the presence of a source of strength g_s satisfies:

$$\nabla^2\phi - \mu^2\phi = 4\pi g_s\delta(x)$$

For a pseudoscalar neutral meson the wavefunction satisfies:

$$\nabla^2\phi - \mu^2\phi = 4\pi g_s\delta(x)\boldsymbol{\sigma}\cdot\nabla$$

The operator $\boldsymbol{\sigma}\cdot\mathbf{p}(=-\mathrm{i}\hbar\boldsymbol{\sigma}\cdot\nabla)$ is a pseudoscalar quantity. The interaction energy V_p of a point nucleon (B) in the field (ϕ_A) due to a nucleon of strength g_s at A is

$$V_p = g_s^2\boldsymbol{\sigma}_B\cdot\nabla_B\,\phi_A$$
$$= g_s^2(\boldsymbol{\sigma}_B\cdot\nabla)(\boldsymbol{\sigma}_A\cdot\nabla)\frac{e^{-\mu r}}{r}$$

Performing the differentiations and generalising to both neutral and charged mesons, which introduces $\boldsymbol{\tau}_A\cdot\boldsymbol{\tau}_B$ as above, gives ($r>0$) the one-pion exchange potential (OPEP):

$$V_P = g_s^2\left(\tfrac{1}{3}\boldsymbol{\sigma}_A\cdot\boldsymbol{\sigma}_B + S_{AB}\left[\tfrac{1}{3}+\frac{1}{\mu r}+\frac{1}{(\mu r)^2}\right]\right)\boldsymbol{\tau}_A\cdot\boldsymbol{\tau}_B\frac{\mu^2 e^{-\mu r}}{r}$$

where $S_{AB} = 3(\boldsymbol{\sigma}_A\cdot\mathbf{r})(\boldsymbol{\sigma}_B\cdot\mathbf{r})/r^2 - \boldsymbol{\sigma}_A\cdot\boldsymbol{\sigma}_B$.

Pseudoscalar meson exchange therefore predicts a tensor force and is spin dependent. Moreover since $\boldsymbol{\sigma}_A\cdot\boldsymbol{\sigma}_B = 1$ if $S=1$ and $=-3$ if $S=0$, V_P is attractive in the $S=1$, $T=0$ deuteron ground state. It does also account for the long-range part of the nucleon-nucleon potential; however, V_P has no repulsive core. The constant g_s is not dimensionless but a dimensionless coupling constant f^2 can be defined by:

$$g_s^2 = \frac{f^2}{4\pi}\hbar c$$

where f^2 is now analogous to the fine structure constant $\alpha = e^2/4\pi\varepsilon_0\hbar c = 1/137$. Its value from pion-nucleon scattering experiments is 0.08. Second-order effects are proportional to f^4 and correspond to two-meson exchange. Their range is inversely proportional to the total mass exchanged and is therefore shorter than one-meson exchange and hence not important for the 1.4 fm range part of the nucleon potential. However exchange of heavy vector mesons (1^- mesons such as the ω and ρ meson) could explain the repulsive core and also account for the existence of a spin-orbit interaction.

7.5 The nuclear force within nuclei

Both the behaviour of the binding energy and the size of nuclei imply that the nuclear force saturates, i.e. a nucleon only interacts with a limited number, but not all, of the other nucleons within the nucleus. Initially it was thought that a mixture of an ordinary (Wigner) force and an exchange force, in particular Majorana exchange which accounts for the strong binding of the α-particle, was the explanation of saturation but it is only part of the reason.

7.5.1 Saturation of the nuclear force

The Pauli exclusion principle causes the average nucleon kinetic energy to increase with increasing density and this effect can be estimated using the Fermi gas model. The total kinetic energy $\langle T \rangle$ is given by:

$$\langle T \rangle \propto \int_0^{\varepsilon_F} \frac{\hbar^2 k^2}{2m} k^2 \, \mathrm{d}k \Omega \propto \Omega k_F^5 \propto A\rho^{2/3}$$

where $\rho = A/\Omega$. The total potential energy $\langle V \rangle$ is given by:

$$\langle V \rangle = \frac{-A(A-1)}{2} \int P(r_i) P(r_j) V(r_{ij}) \, \mathrm{d}r_i^3 \, \mathrm{d}r_j^3$$

where $P(r) = 1/\Omega$ is the probability of a nucleon being at r. The interaction $V(r_{ij})$ is just a function of r_{ij}, i.e. a Wigner force. If the range of V is $\ll R$, the nuclear radius, then, neglecting points closer to the surface than the range, $\int \mathrm{d}^3 r_i \, \mathrm{d}r_j^3 = \int \mathrm{d}r_{ij}^3 \, \mathrm{d}^3 R$ so:

$$\begin{aligned} \langle V \rangle &\approx -\frac{A(A-1)}{2} \int \frac{1}{\Omega} \frac{1}{\Omega} V(r_{ij}) \, \mathrm{d}^3 r_{ij} \, \mathrm{d}R^3 \\ &\approx -\frac{A^2}{\Omega} \frac{1}{2} \int V(r_{ij}) \, \mathrm{d}^3 r_{ij} \\ &= -A\rho \bar{V} \end{aligned} \tag{7.5}$$

where $\bar{V} = \frac{1}{2} \int V(r_{ij}) \, \mathrm{d}r_{ij}^3$ is the average interaction potential. (The factor $\frac{1}{2}$ is because each nucleon has been counted twice in the integral.) The average nucleon binding energy $\bar{B} = -\langle E \rangle / A$ is given by:

$$\bar{B} = -\left(\frac{\langle T \rangle}{A} + \frac{\langle V \rangle}{A} \right)$$

$$\therefore \quad \bar{B} = c_1 \rho - c_2 \rho^{2/3}$$

where c_1 and c_2 are independent of A. Therefore the average binding energy would increase with increasing density so the Pauli principle alone is insufficient to give saturation. Additional elements that are significant for saturation are the tensor and exchange forces and the repulsive core.

The tensor force in second-order perturbation theory can couple S states below the Fermi surface to unoccupied D states above and then couple these back down to the S states. The additional binding energy B_T is given by:

$$B_{\mathrm{T}} \propto \sum_{f} \frac{\langle i|V_{\mathrm{T}}|f\rangle\langle f|V_{\mathrm{T}}|i\rangle}{E_i - E_f}$$

where V_{T} is the attractive tensor potential, E_i is below and E_f above the Fermi surface. As the density increases ε_{F} increases as $\rho^{2/3}$ so $E_i - E_f$ increases. If $\langle i|V_{\mathrm{T}}|f\rangle$ remains constant then there will be a consequent decrease in B_{T}. So a tensor force could be significant for saturation.

The effect of an exchange force can be easily seen if the nuclear force is approximated by a Wigner plus Majorana force (a Serber force). Let ε be the fraction of nucleons within the range of the force in an antisymmetric spatial state. Then using equation 7.5 above:

$$\begin{aligned}\langle V\rangle &= -A\rho\bar{V}_{\mathrm{W}} - A\rho(1-\varepsilon)\bar{V}_{\mathrm{M}} + A\rho\varepsilon\bar{V}_{\mathrm{M}} \\ &= -A\rho[\bar{V}_{\mathrm{W}} + (1-2\varepsilon)\bar{V}_{\mathrm{M}}]\end{aligned}$$

Now ε increases with increasing ρ. This is because the wavefunction of relative motion in an antisymmetric state is α sin kr, which for small r is $\alpha\, kr$, so the probability is $\alpha\, k^2r^2$. The mean k is $\alpha\, k_{\mathrm{F}}$ so the probability of being within the range of the nuclear force is $\alpha\, k_{\mathrm{F}}^2 \alpha\, \rho^{2/3}$. Therefore ε increases as ρ increases making $\langle V\rangle$ less negative.

The repulsive core of the nucleon-nucleon potential will obviously give rise to saturation; however, it is the least important. The resultant total binding energy per nucleon as a function of nucleon separation is shown in figure 7.10. It is relatively weak with a minimum corresponding to the nucleons being ~2.4 fm apart. Its weakness means that the nucleons behave rather like a gas within a container, which gives rise to the Fermi gas model of a nucleus.

7.5.2 The average nuclear potential well

Let the wavefunction ψ of the nucleus be given by:

$$\psi = \text{antisym} \prod_i \phi_i(r_i)$$

where $H_i\phi_i = [T_i + U_i(r_i)]\phi_i = E_i\phi_i$ with

$$T_i = \frac{p_i^2}{2m}$$

Neglect the spin-orbit interaction and take:

$$\begin{aligned}U_i(r_i) &= -V_i \quad \text{for } r < R \\ &= 0 \quad \text{for } r > R\end{aligned}$$

so it is not assumed that all particles have the same well depth. The total Hamiltonian H, assuming only pair forces, is given in terms of the average potential, $U_i(r_i)$, felt by the ith particle through its interactions with all the

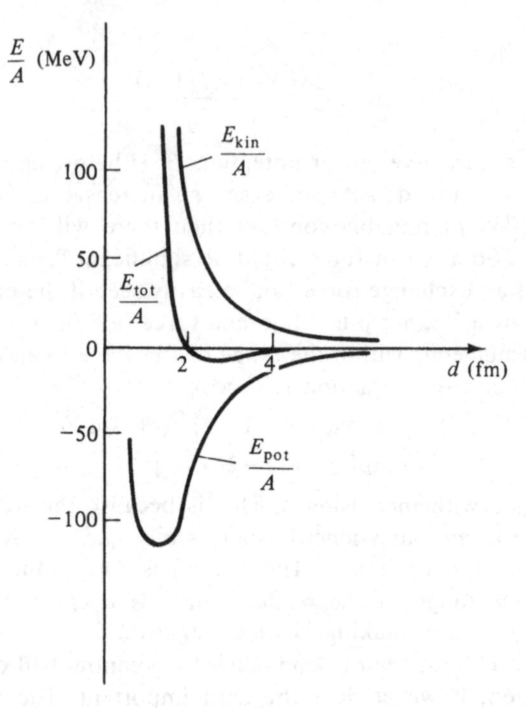

7.10. A plot of the kinetic energy, potential energy and total energy per nucleon as a function of nucleon separation. (From Ring, P. and Shuck, P., *The Nuclear Many-Body Problem.* Springer-Verlag (1980).)

other nucleons by:

$$H = \sum_i T_i + \tfrac{1}{2} \sum_i U_i(r_i)$$

where the factor of $\frac{1}{2}$ arises to avoid counting the potential energies of pairs of nucleons twice. The energy, E, of the nucleus is given by:

$$E = \langle H \rangle = \sum_i \langle T_i \rangle + \tfrac{1}{2} \sum_i \langle U_i \rangle$$

The potential well, $U_i(r_i)$, is a square well and using the Fermi gas model approximation yields:

$$\sum_i \langle T_i \rangle = \tfrac{3}{5} T_F A \quad \text{and} \quad \langle U_i \rangle = -V_i$$

where T_F is the kinetic energy at the Fermi level:

$$V_{AV} \equiv +\frac{1}{A} \sum_i \langle U_i \rangle$$

$$= -\frac{1}{A} \sum_i V_i$$

so

$$E=\tfrac{3}{5}T_F A+\tfrac{1}{2}V_{AV}A$$

Consider the binding energy per nucleon, B, and the nucleon separation energy, S, in this model:

$$B=-\frac{E}{A}=-\tfrac{3}{5}T_F-\tfrac{1}{2}V_{AV} \tag{7.6}$$

$$S=-T_F-V_F=AB(A,Z)-(A-1)B(A-1,Z) \tag{7.7}$$

The nuclear force saturates so $B(A, Z)$ is independent of A, i.e. $B(A, Z) = B$, so:

$$S=AB-(A-1)B=B$$

i.e. for a saturating nuclear force the separation energy equals the binding energy per nucleon. The magnitude of B is given by the coefficient of the volume term in the semi-empirical mass formula (~16 MeV) as no surface, Coulomb or n-p asymmetry effects are being considered in this model. Combining equations 7.6 and 7.7 gives:

$$S=-\tfrac{1}{5}T_F-V_{AV}+V_F \tag{7.8}$$

So if the potential wells, V_i, are all the same then $V_{AV}=V_F$ and S is negative, which is totally wrong. Likewise the total binding energy $(-E)$ will be incorrect.

The answer is to assume the potential wells are momentum dependent. Let:

$$V_i=V_0-\frac{p_i^2}{p_F^2}V_1$$

then $V_{AV}=-(V_0-\tfrac{3}{5}V_1)$ and $V_F=-(V_0-V_1)$.

Substituting in equations 7.6 and 7.7 gives:

$$V_1=\tfrac{1}{2}(5S+T_F) \quad \text{and} \quad V_0=\tfrac{1}{2}(7S+3T_F)$$

Introduce an effective mass m^* through:

$$E_i=\frac{p_i^2}{2m}-V_i=\frac{p_i^2}{2m^*}-V_0$$

Substituting for V_i, V_1 and V_0 yields:

$$\frac{m}{m^*}=1+\frac{V_1}{T_F}=\frac{3}{2}+\frac{5}{2}\frac{S}{T_F}$$

The kinetic energy T_F depends only on the nuclear density, ρ, and for $R=1.2A^{1/3}$, $T_F=33$ MeV. Taking $S=B=16$ MeV makes $m/m^*=2.7$, i.e. a nucleon has a smaller effective mass when bound in a nucleus. This change is analogous to the change in the effective mass of an electron when bound in a solid.

The momentum dependence of the potential well ensures the saturation of the nuclear force. This is in agreement with the discussion above which showed that saturation occurs because the nucleon–nucleon potential is weaker for higher densities. These densities correspond to smaller separations and hence higher relative momenta. Therefore V_i is smaller for higher momenta.

7.5.3 Why independent-particle motion?

The shell model assumes that it is a good approximation to take the total wavefunction as the product of independent-particle eigenfunctions of the average potential well. A pair of nucleons interact and cannot be closer than the repulsive core allows, so it is not immediately obvious why independent-particle wavefunctions are a good approximation. The reason lies in the effect of the Pauli principle.

Consider two nucleons within a hypothetical nucleus with $N = Z$ very large and with no Coulomb interactions; such a medium is called nuclear matter. Let their momenta ($\hbar = 1$) be $\mathbf{k}_\alpha$ and $-\mathbf{k}_\alpha$, so their centre of mass momentum is zero. If the nucleons were free, elastic scattering to states $\mathbf{k}_\beta$ and $-\mathbf{k}_\beta$ ($k_\beta = k_\alpha$) would be possible. However, in nuclear matter such states are occupied so the effect of the nucleon–nucleon interaction can only be to modify the wavefunction for the pair over a certain separation, as asymptotically (outside the range of interaction) they must remain in the states k_α and $-k_\alpha$. Only Fourier components with $k > k_F$ are possible as all states up to k_F are occupied so the distortion only extends over a distance of order $1/k_F$. The repulsive core has such components but the attractive part of the potential is not deep enough to give a shape that contains such components with an appreciable amplitude, so the distortion extends $\sim 1/k_F$ from the repulsive core; this separation is called the healing distance h. The Fermi gas model (chapter 2, p. 28) gives $k_F = \frac{1}{2}(9\pi)^{1/3}/r_0$ so for $r_0 = 1.2$ fm, $1/k_F \approx 0.8$ fm. The repulsive core separation is ~ 0.4 fm so the healing distance is estimated as $h \approx 1.2$ fm. The average separation d of nucleons within a nucleus is given by $d \approx 2r_0 \approx 2.4$ fm, so $h \approx \frac{1}{2}d$.

The consequences of this rapid healing are that if another nucleon should be within h of either of the pair of nucleons it is very likely that the pair will be greater than h apart, which means that their wavefunctions will have healed to independent-particle wavefunctions. Therefore as a first approximation only the interactions between pairs are important in determining the average potential that nucleons move in. (For an accurate value, though, three or more body interactions are important.) Furthermore, the nucleons' wavefunctions at the average separation distance d are very closely those of independent particles, which provides the reason for the success of the shell model.

7.5.4 The residual interaction between nucleons

The residual interaction v between nucleons is the difference between the actual two-nucleon potential V_α experienced by a nucleon in a state α and the average potential. From the above discussion the residual interaction v is relatively weak and the matrix elements of v, $\langle\alpha|v|\beta\rangle$, between states α and β are only appreciable for unoccupied states close in energy, i.e. v mixes states near the Fermi level. In the shell model, for example, two valence nucleons could be in the configurations $(s_{\frac{1}{2}}\, d_{\frac{5}{2}})_2$ or $(d_{\frac{3}{2}}\, d_{\frac{5}{2}})_2$, or more generally in $|1\rangle = |j_1 j_2 J\rangle$ or $|2\rangle = |j_3 j_4 J\rangle$. The residual interaction v would then be specified by its matrix elements $v_{ik} = \langle i|v|k\rangle$, $i, k = 1, 2$. The operator v is called a two-body operator because it can change the state of two nucleons; cf. the E1 operator which is a one-body operator as a photon only interacts with one nucleon. To calculate these matrix elements from the free two-nucleon potential is a formidable task and depends on the configuration space (e.g. sd or fp shell), and although there are theoretical values for them, such as those of Kuo and Brown, their accuracy is uncertain as the magnitude of higher-order corrections is uncertain.

An alternative approach is to treat the matrix elements as parameters and fit them to observed level spectra and nuclear masses which can be expressed in terms of the matrix elements. Another way is to try and use physical ideas to simplify the evaluation of the residual interaction v. This potential depends on the separation:

$$|\mathbf{r}_i - \mathbf{r}_j| = (r_i^2 + r_j^2 - 2r_i r_j \cos\theta_{ij})^{1/2}$$

where θ_{ij} is the angle between $\mathbf{r}_i$ and $\mathbf{r}_j$. The potential can therefore be expanded as:

$$\begin{aligned} v(|\mathbf{r}_i - \mathbf{r}_j|) &= \sum_i v_l(r_i r_j) P_l(\cos\theta_{ij}) \\ &= \sum_l v_l(r_i r_j) \sum_m Y_{lm}(\theta_i \phi_i) Y^*_{lm}(\theta_j \phi_j) \end{aligned} \qquad (7.9)$$

This is a multipole expansion of v and the angular dependence of v has separated into a sum of products of functions of $(\theta_i \phi_i)$ and functions of $(\theta_j \phi_j)$. The radial dependence though is not in general separable. If the interactions are assumed to take place only on the surface of the nucleus then $v_l(r_i r_j) \Rightarrow v_l(R_0)$ and v would be separable. The justification for this approximation is that the kinetic energy of a nucleon is less at the surface so an interaction between a pair of nucleons on the surface will be at a lower relative momentum, for which the attractive potential is stronger (see above), than within the nucleus; the difference therefore giving rise to a residual interaction acting on the surface. If instead of assuming $r_i = r_j = R_0$ one assumes:

$$v_l(r_i r_j) = \chi r_i^l r_j^l$$

where χ is a constant, then $v(|\mathbf{r}_i-\mathbf{r}_j|)=\chi\sum_{lm} M_{lm}(i)M^*_{lm}(j)$ where $M_{lm}(i)$ is the multipole operator defined by:

$$M_{lm}(i)=r_i^l Y_{lm}(\theta_i\phi_i)$$

This separable approximation is very useful in describing collective 1p-1h states (see chapter 8). For low-lying states the quadrupole-quadrupole component is the most important correction to a spherical field. As most nuclei do have a quadrupole deformation this form for the residual interaction is as expected. The higher l components correspond to small θ_{ij} (see equation 7.9) and hence to smaller range. So the quadrupole-quadrupole term is the most important effectively long-range component to the residual interaction at low excitations.

The important short-range component to the residual interaction is the pairing interaction. As mentioned before, if two identical nucleons with spin j have their angular momenta coupled up to $J=0$ then the overlap of their wavefunctions is a maximum. Hence the expectation value of a short-range interaction, i.e. $\langle jmj-mJ=0|V|jmj-mJ=0\rangle=\langle m-m|V|m-m\rangle$, is a maximum. Making the approximation that $\langle j^2JM|V|j^2J'M'\rangle=\delta_{JJ'}\delta_{MM'}\delta_{J0}$, i.e. the interaction only occurs in the $J=0$ state, means that $\langle m-m|V|m-m\rangle=$ a constant for all m, m', i.e. all matrix elements of the pairing interaction are equal. Using this approximation the occurrence of a pairing gap will be explained in the next chapter.

7.6 Summary

The nuclear force between nucleons is to a good approximation both charge symmetric and charge independent. It has a short range of about 1.4 fm and a repulsive core (~0.4 fm). There are also spin, tensor, spin-orbit and exchange components. All of these can qualitatively be understood in terms of meson-exchange theory, which can quantitatively account for the 1.4 fm range part of the nuclear force.

Within nuclei, the Pauli exclusion principle, together with the tensor and exchange forces, can account for the saturation of the nuclear force in nuclei. It is seen that this also implies that the effective one-body potential is momentum dependent.

Consideration of the effect of the Pauli exclusion principle on the wavefunctions of nucleons in nuclear matter shows that independent-particle wavefunctions provide a very good approximation. This provides the reason for the success of the shell model. The residual interaction between nucleons in nuclei is relatively weak and its relation to the free nucleon-nucleon force is complicated. However, the important short-range component is the pairing interaction and the long-range part may usefully be approximated in terms of a separable multipole-multipole force.

7.7 Questions

1. Derive the expression for d given on p. 221 and estimate d if the scattering length a of the neutron 'mirror' is 10 fm.
2. Assuming a simple square well potential of depth V_0 and radius R for the potential between the neutron and proton in a deuteron, show that the matching condition $\tan KR = -K/\beta$ gives the condition $V_0 R^2$ approximately constant and that the value of the constant is ~1 MeV barn.
3.* The nucleon-nucleon force is usually described as: (a) strong, (b) short-range, (c) charge symmetric, (d) charge-independent, (e) containing both 'ordinary' and 'exchange' terms, (f) spin-dependent, (g) non-central, (h) saturating in nuclei. Explain the meaning of these statements and discuss briefly the experimental evidence for them.
4. Discuss why ^{8}He is stable against nucleon emission but ^{5}He is not.
5.* The lifetimes of the charged and neutral pions are respectively 2.6×10^{-8} s and 8×10^{-17} s. Why, in spite of the difference between these two numbers, is it useful to regard the pions as members of an isospin triplet? Are the pion lifetimes relevant to nuclear forces?
6.* What is meant by the statement that strong interactions conserve isospin? Why does isospin conservation imply the charge independence of the nucleon-nucleon force, while pion-nucleon interactions are not generally charge-independent? Are pion-pion interactions charge-independent? The reaction $d + d \rightarrow {}^4He + \pi^0$ has never been observed. Why is this? Can the reaction occur at all?
7.* In pion-nucleon scattering there is a $T = 3/2$ resonance seen at a centre of mass energy of 1232 MeV. At this energy the total cross-sections for π^+ on proton and π^- on proton are in the ratio 3:1. Deduce the branching ratio for the decay of the $T = 3/2$ resonance formed in π^+n to π^0p and π^+n.
8. The potential energy operator V acting on the wavefunction ψ is $V(\mathbf{r})\psi(\mathbf{r})$ for a local potential and $\int V(\mathbf{r}, \mathbf{r}')\psi(\mathbf{r}')\, d^3r'$ for a non-local potential. Show that such a non-local potential is equivalent to a momentum dependent potential by showing that

$$\int V(\mathbf{r}, \mathbf{r}')\psi(\mathbf{r}')\, d^3r' = \int U(\mathbf{r}, \mathbf{p})\phi(\mathbf{p}) \exp(i\mathbf{p} \cdot \mathbf{r}/\hbar)\, d^3p$$

where $\psi(\mathbf{r}) = (2\pi)^{-3/2} \int \exp(i\mathbf{p} \cdot \mathbf{r}/\hbar)\phi(\mathbf{p})\, d^3p$ and the momentum dependent potential $U(\mathbf{r}, \mathbf{p}) = (2\pi)^{-3/2} \int \exp(i\mathbf{p} \cdot \mathbf{x}/\hbar) V(\mathbf{r}, \mathbf{r} + \mathbf{x})\, d^3x$.

8 Deformed nuclei and collective motion

The discussion in chapter 2 of the change in energy, both kinetic and potential, when a spherical nucleus is deformed, showed that many nuclei would be expected to have quite significant quadrupole moments, in agreement with experiment. Moreover, a deformed shell model description was shown by Nilsson to provide a good quantitative description of the shape and spin of many nuclear ground states. However, what was not clear was what aspect of the Nilsson model accounted for nuclei being mainly prolate (cigar-shaped), as can be seen clearly in figure 2.15, and Lemmer and Weisskopf's argument that it is because of the shape of the nuclear potential is first presented.

After discussing the ground-state deformation of nuclei, a microscopic description of collective motion in terms of an independent-particle model is developed in the rest of this chapter. The ground state of the nucleus is generally described by a deformed intrinsic wavefunction. If the deformation is significant, then a variational approach to the motion of deformed nuclei, which avoids the problem of redundant variables inherent in the collective model, is shown to give rise to a rotational spectrum of excited states. The cranking-model expression for the moment of inertia is then derived and the influence of the pairing residual interaction is discussed.

In the shell model, if the residual interaction is diagonalised, then collective features are reproduced. In an approximate treatment of the residual interaction, the schematic model, collective vibrational states are found to correspond in the shell model to a coherent superposition of one-particle one-hole states. Both rotational and vibrational excited states in nuclei are therefore shown to be described in terms of independent-particle models.

8.1 The ground-state deformation of nuclei

Consider first a simple model of A identical nucleons moving independently in an anisotropic oscillator potential well. The total energy, $E(A, \delta)$, is given by, using equations 2.12 and 2.13:

$$E(A,\delta)=\tfrac{3}{4}\sum_i\left[(N_i+\tfrac{3}{2})-\tfrac{1}{3}\delta(3n_{z_i}-N_i)+\tfrac{1}{9}\delta^2(N_i+\tfrac{3}{2})\right]\hbar\omega_0 \tag{8.1}$$

where the sum is over all the nucleons. The energy varies with δ and by differentiating $E(A, \delta)$ with respect to δ the minimum value is at the deformation:

$$\bar{\delta}=\frac{3}{2}\frac{\sum_i(3n_{z_i}-N_i)}{\sum_i(N_i+\frac{3}{2})} \tag{8.2}$$

The energy $E(A, \bar{\delta})$ is lower than the energy at zero deformation by an amount $\frac{1}{9}\bar{\delta}^2E(A, 0)$.

Equation (8.2) shows that only particles in unfilled shells can polarise the nuclear shape since $\sum_i(3n_{z_i}-N_i)$ is zero for a filled shell. This conclusion is not altered by introducing an effective mass to simulate the momentum dependence of the potential well, as this only alters ω_0 in equation (8.1) and therefore leaves equation (8.2) unchanged.

There are minimum values of $E(A, \delta)$ for both oblate and prolate deformations and figure 8.1 shows a plot of the difference in energies $D(A)$ given

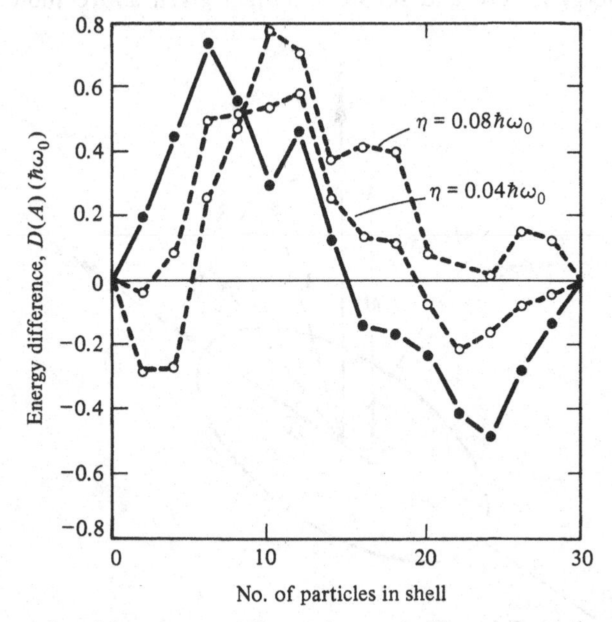

8.1. Difference in minimum energy for prolate and oblate deformations as a function of the number of particles in the $N=4$ shell. The solid curve is for the oscillator potential while the broken curves show the effect of steepening the oscillator walls. (From Lemmer, R. H. and Weisskopf, V. F., *Nucl. Phys.* **25** (1961) 624.)

by:

$$D(A)=E(A,\bar{\delta}_0)-E(A,\bar{\delta}_p)=\tfrac{1}{9}(\bar{\delta}_\rho^2-\bar{\delta}_0^2)E(A,0)$$

as a function of the number of particles in the $N=4$ shell. It can be seen that prolate shapes ($\bar{\delta}>0$) are preferred in the first half of the shell, oblate ($\bar{\delta}<0$) in the second half. However, the oblate are less favoured than the prolate because of the term $\alpha\,\delta^2$ in equation 8.1, which favours a spherical shape, as the magnitude of this term increases with mass number and so is larger in the second half of the shell.

The actual nuclear potential well has steeper sides than an oscillator well and Lemmer and Weisskopf simulated this by adding a perturbing potential ηr^4. The effect of this perturbation is to favour prolate shapes in the second half of the shell and oblate in the first half. The total effect is shown in figure 8.2 for $\eta=0.04\hbar\omega_0$ and $\eta=0.08\hbar\omega_0$; the latter makes a prolate shape preferred for most of the $N=4$ shell and is also a value which gives a reasonable nuclear potential well shape.

This argument has neglected the spin–orbit interaction. However, as this splits energy levels symmetrically, the average spin–orbit contribution to the total energy is zero and so the argument given above should still be

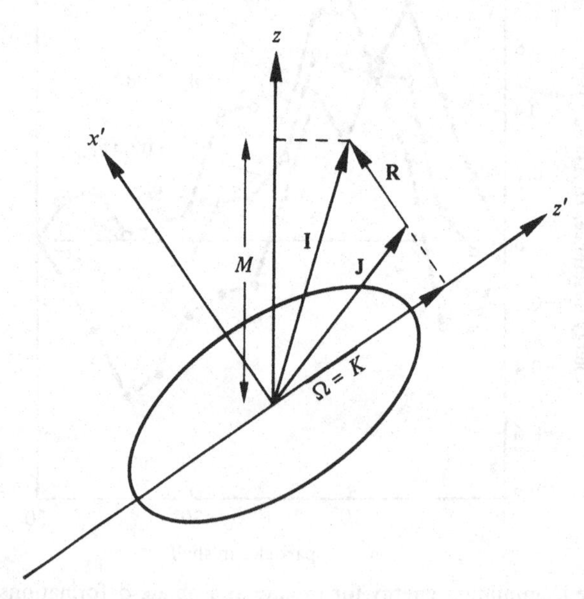

8.2. Diagram showing the body-fixed axes x' and z' for an axially-symmetric deformed nucleus and space-fixed axis z. **I** is the total, **R** the rotational and **J** the intrinsic angular momentum. The z' components of **I** and **J** are K and Ω, respectively, and the z component of **I** is M (in units of $\hbar$).

qualitatively correct, i.e. nuclei are mainly prolate because of the shape of the nuclear potential well, which is like a deformed oscillator potential with steepened sides.

8.2 The collective model

8.2.1 Symmetries of the wavefunction for a deformed nucleus

In the collective model the wavefunction for a deformed nucleus in the adiabatic approximation, as discussed in chapter 2, can be written as:

$$\psi = \phi D(\boldsymbol{\theta})$$

where ϕ is the intrinsic and D the rotational wavefunction. The wavefunction $D(\boldsymbol{\theta})$ is an eigenfunction of $\mathbf{I}^2$, $\mathbf{I}_z$ and $\mathbf{I}_{z'}$ with eigenvalues $I(I+1)$, M and K, respectively (see figure 8.2). The intrinsic wavefunction ϕ is not an eigenstate of $\mathbf{J}^2$, since the nucleons move in a non-spherical potential, but for nuclei with axial symmetry $\langle J_{z'}\rangle = \Omega = K$ is a constant of the motion.

The wavefunction ψ must be invariant with respect to a reversal of the z' axis (as it must be a single-valued function of x, y and z) and in particular to a rotation R_2 about y' of 180°. If the intrinsic Hamiltonian is invariant with respect to R_2 then $\phi_{\bar{K}} \equiv R_2\phi_K$ has the same energy as ϕ_K. Furthermore, the rotation R_2 affects ϕ and D simultaneously and:

$$R_2 D^I_{MK} = (-1)^{I+K} D^I_{M-K}$$

so for $\psi = R_2\psi$:

$$\psi(IKM) = \sqrt{\frac{(2I+1)}{16\pi^2}}\,(\phi_K D^I_{MK} + (-1)^{I+K}\phi_{\bar{K}} D^I_{M-K})$$

where $R_2^2\phi_K = +\phi_K$ for even-A and $= -\phi_K$ for odd-A nuclei.

For $K=0$ a rotation R_1 about x' of 180° is equivalent to R_2, i.e. $R_2\phi_0 = R_1\phi_0$ where:

$$R_1\phi_0 = r\phi_0; \qquad R_1^2\phi_0 = r^2\phi_0 = \phi_0 \quad \text{so } r = \pm 1$$

and r is called the signature of the state. So for $\psi(I0M)$ to be invariant under R_2:

$$r(-1)^I = 1$$

The rotational spectrum of an even–even nucleus therefore contains states with only even or only odd values of I. Experimentally an $r=+1$ state lies lowest, e.g. ^{238}U (figure 2.29). The parity operator operates only on the intrinsic wavefunction so all states in a band have the same parity π. For $K=0$ bands, π and r may be different.

8.2.2 $K=\frac{1}{2}$ bands

As mentioned in chapter 2, the coupling term $H_{\text{coup}}=-\hbar^2(I'_+J'_- + I'_-J'_+)/2\mathcal{I}$, where $I'_+ = I_{x'}+\text{i}\,I_{y'}$, etc., in the Hamiltonian (chapter 2, p. 62) cannot be neglected for $K=\frac{1}{2}$ bands as H_{coup} mixes the $K=\pm\frac{1}{2}$ states. The wavefunction for $K=\frac{1}{2}$ is given by

$$\psi(I\tfrac{1}{2}M)=\sqrt{\frac{2I+1}{16\pi^2}}(\phi_{1/2}D^I_{M1/2}+(-1)^{I+1/2}\phi_{\overline{1/2}}D^I_{M-1/2})$$

The intrinsic wavefunction $\phi_{1/2}$ is not an eigenfunction of $\mathbf{J}^2$ but can be expanded in terms of eigenfunctions of $\mathbf{J}^2$:

$$\phi_{1/2}=\sum_J C_J\phi_{J,1/2}$$

and

$$\phi_{\overline{1/2}}\equiv R_2\phi_{1/2}=\sum_J C_J(-1)^{J+1/2}\phi_{J,-1/2}$$

The effect of J'_- on $\phi_{1/2}$ is:

$$J'_-\phi_{1/2}=\sum_J C_J(J+\tfrac{1}{2})\phi_{J,-1/2}$$

and, noting that $\mathbf{I}\times\mathbf{I}=-\text{i}\mathbf{I}$ so I'_+ behaves like a lowering operator,

$$I'_+D^I_{M1/2}=(I+\tfrac{1}{2})D^I_{M-1/2}$$

with similar results for J'_+ and I'_-. The energy shift $\Delta E(K=\frac{1}{2})$ due to H_{coup} is therefore given by:

$$\Delta E(K=\tfrac{1}{2})=-\frac{\hbar^2}{2\mathcal{I}}\langle I\tfrac{1}{2}M|I'_+J'_- + I'_-J'_+|I\tfrac{1}{2}M\rangle$$

$$=-\frac{\hbar^2}{2\mathcal{I}}(-1)^{I+1/2}(I+\tfrac{1}{2})\sum_J|C_J|^2(-1)^{J+1/2}(J+\tfrac{1}{2})$$

$$=\frac{\hbar^2}{2\mathcal{I}}a(-1)^{I+1/2}(I+\tfrac{1}{2})$$

where $a\equiv-\sum_J|C_J|^2(-1)^{J+1/2}(J+\frac{1}{2})$ is called the decoupling parameter. The rotational energy $E_{K=1/2}(I)$ is therefore:

$$E_{K=1/2}(I)=E^0_{K=1/2}+\frac{\hbar^2}{2\mathcal{I}}[I(I+1)+a(-1)^{I+1/2}(I+\tfrac{1}{2})] \quad (8.3)$$

Figure 8.3 shows two examples of $K=\frac{1}{2}$ bands and how well equation 8.3 accounts for them.

Although the wavefunction $\psi(IKM)$ provides a good description of many rotational bands, it involves redundant variables, i.e. instead of $3A$ coordinates it has $3A+3$. A more satisfactory procedure for generating a wavefunction for a deformed nucleus, which avoids this problem, was proposed by Peierls and Yoccoz and involves a variational approach.

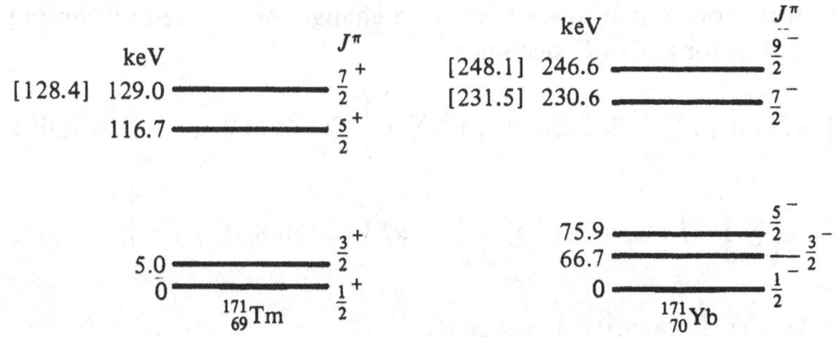

keV J^π

[128.4] 129.0 $\frac{7}{2}^+$

116.7 $\frac{5}{2}^+$

5.0 $\frac{3}{2}^+$

0 $\frac{1}{2}^+$

$^{171}_{69}$Tm

keV J^π

[248.1] 246.6 $\frac{9}{2}^-$

[231.5] 230.6 $\frac{7}{2}^-$

75.9 $\frac{5}{2}^-$

66.7 $\frac{3}{2}^-$

0 $\frac{1}{2}^-$

$^{171}_{70}$Yb

8.3. Two examples of $K=\frac{1}{2}$ bands. The energies predicted by equation 8.3 for the higher levels drawn are shown in square brackets.

8.3 Variational approach to collective motion

Assume that a nucleus can be described by a wavefunction ϕ made up from single-particle eigenfunctions u_i. The best wavefunction of this form is determined by the variational principle as the one that minimises the total energy E given by:

$$E=\langle\phi|H|\phi\rangle/\langle\phi\,|\,\phi\rangle$$

where $H=\sum_i T_i+\frac{1}{2}\sum_{ij}^{i\neq j} V_{ij}$ is the total Hamiltonian for the nucleus, $T_i=p_i^2/2m$ is the kinetic energy operator, $V_i=\sum_{j\neq i} V(r_ir_j)$ is the potential energy operator for a single nucleon and $V(r_ir_j)$ is the two-nucleon interaction. Varying ϕ by an infinitesimal amount causes a change:

$$\delta E=(\langle\phi\,|\,\phi\rangle\delta\langle\phi|H|\phi\rangle-\langle\phi|H|\phi\rangle\delta\langle\phi\,|\,\phi\rangle)/\langle\phi\,|\,\phi\rangle^2$$

which equals zero if

$$\langle\phi\,|\,\phi\rangle\delta\langle\phi|H|\phi\rangle=\langle\phi|H|\phi\rangle\delta\langle\phi\,|\,\phi\rangle$$

i.e. $$\langle\phi\,|\,\phi\rangle\{\langle\phi|H|\delta\phi\rangle+\langle\delta\phi|H|\phi\rangle\}=\langle\phi|H|\phi\rangle\{\langle\phi\,|\,\delta\phi\rangle+\langle\delta\phi\,|\,\phi\rangle\}$$

This is true for arbitrary $|\delta\phi\rangle$ so replacing $|\delta\phi\rangle$ by $\mathrm{i}\,|\,\delta\phi\rangle$ gives

$$\langle\phi\,|\,\phi\rangle\mathrm{i}\{\langle\phi|H|\delta\phi\rangle-\langle\delta\phi|H|\phi\rangle\}=\langle\phi|H|\phi\rangle\mathrm{i}\{\langle\phi\,|\,\delta\phi\rangle-\langle\delta\phi\,|\,\phi\rangle\}$$

Combining these equations gives the conditions for zero δE:

$$\langle\phi\,|\,\phi\rangle\langle\delta\phi|H|\phi\rangle=\langle\phi|H|\phi\rangle\langle\delta\phi\,|\,\phi\rangle$$

or

$$\langle\phi\,|\,\phi\rangle\langle\phi|H|\delta\phi\rangle=\langle\phi|H|\phi\rangle\langle\phi\,|\,\delta\phi\rangle$$

For simplicity consider ϕ given by a simple product $\phi=\prod_i u_i$ (Hartree theory), where $u_i\equiv u_i(r_i)$. (If ϕ is antisymmetrised by taking a Slater determinant of single-particle eigenfunctions u_i then minimising E gives the Hartree–Fock equations.) Then for a change $\langle\delta\phi|$ caused by varying u_i^* the condition for zero δE becomes:

$$\prod_i \int u_i^* u_i \, \mathrm{d}r_i^3 \left[\sum_i \int \delta u_i^* T_i u_i \, \mathrm{d}^3 r_i + \tfrac{1}{2} \sum_{\substack{ij \\ i\neq j}} \iint \delta(u_i^* u_j^*) V(r_i r_j) u_i u_j \, \mathrm{d}^3 r_i \, \mathrm{d}^3 r_j\right]$$

$$= \left[\sum_i \int u_i^* T_i u_i \, \mathrm{d}^3 r_i + \tfrac{1}{2} \sum_{\substack{ij \\ i\neq j}} \iint u_i^* u_j^* V(r_i r_j) u_i u_j \, \mathrm{d}^3 r_i \, \mathrm{d}^3 r_j\right]$$

$$\times \prod_{j\neq i} \int u_j^* u_j \, \mathrm{d}^3 r_j \int \delta u_i^* u_i \, \mathrm{d}^3 r_i$$

This is satisfied (n.b. the effect of $\sum_{ij}^{i\neq j}$) if:

$$T_i u_i + \sum_{j\neq i} \int u_j^* V(r_i r_j) u_i u_j \, \mathrm{d}^3 r_j = \alpha_i u_i$$

or

$$[T_i + U(r_i)] u_i = \alpha_i u_i$$

where $U(r_i) = \sum_{j\neq i} \int V(r_i r_j) u_j^* u_j \, \mathrm{d}^3 r_j$ is the self-consistent single-particle potential and α_i is the single-particle energy. Therefore for a very short-range interaction the lowest energy and best wavefunction is obtained if the single-particle potential has the same density distribution as that of the nucleons that give rise to it.

From an initial set of single-particle eigenfunctions a single-particle potential $U(r_i)$ can be calculated which in turn can be used to calculate a better set of single-particle wavefunctions. These then can be used to generate a better potential and this procedure can be iterated until self-consistency is achieved. In general $U(r_i)$ will be non-spherical and an approximate form is given by the Nilsson potential.

Consider first an even–even deformed nucleus. Let the potential well U in which the nucleons move be ellipsoidal in shape with axial symmetry. Any orientation (Ω) of the well in space generates a wavefunction $\phi(\Omega)$ of the same energy, where $\phi(\Omega)$ is a wavefunction made up from single-particle eigenfunctions and is a function of Ω and of all the nucleons' coordinates. This degeneracy is removed by writing:

$$\psi = \int f(\Omega) \phi(\Omega) \, \mathrm{d}\Omega \tag{8.4}$$

This wavefunction has no redundant variables as the variable Ω is integrated over. Minimising the expectation value of the energy with respect to variations in $f(\Omega)$ will also give a better wavefunction and a lower energy.

The energy is given by $E=\langle\psi|H|\psi\rangle/\langle\psi|\psi\rangle$, where H is the total Hamiltonian given above, and for a minimum the change δE brought about by an infinitesimal change in ψ must be zero. An invariance of E with respect to an infinitesimal change in ψ caused by an infinitesimal rotation corresponds to ψ being an eigenstate of angular momentum (see the discussion in chapter 1, p. 10). So minimising the energy is equivalent to choosing $f(\Omega)$ such as to make ψ an angular momentum eigenstate. This means that $f(\Omega)$ projects out of ψ functions of good angular momentum.

To see more formally the equivalence of projection to the variational principle, consider a wavefunction $\phi(r)\equiv\phi(r_1, r_2, r_3, \ldots)$ which is a solution of the Hamiltonian for the nucleus determined by the variational method described above. Just as a wavefunction for a nucleus can not depend on the orientation (Ω above) of a potential well, as it would not then be rotationally invariant, it also can not depend on the position of the centre of the well. This dependence is avoided by considering a wavefunction ψ of the same form as that of equation 8.4 above given by:

$$\psi=\int f(x)\phi(r-x)\,\mathrm{d}x$$

where x represents a displacement of the centre of the potential well, and where the function $\phi(r-x)\equiv\phi(r_1-x, r_2-x, r_3-x, \ldots)$ is a solution of the Hamiltonian H for the nucleus with the same energy as $\phi(r)$. The function $f(x)$ is determined by the variational principle:

$$\delta E=\delta(\langle\psi|H|\psi\rangle/\langle\psi|\psi\rangle)=0$$

Let $f'\equiv f(x')$, $\phi'\equiv\phi(r-x')$ etc., then varying $f(x)$ gives

$$\delta E=\left\{\int f'^*\phi'^*\phi f\,\mathrm{d}x\,\mathrm{d}x'\,\mathrm{d}^3r\int f'^*\phi'^*H\phi\delta f\,\mathrm{d}x\,\mathrm{d}x'\,\mathrm{d}^3r\right.$$
$$\left.-\int f'^*\phi'^*H\phi f\,\mathrm{d}x\,\mathrm{d}x'\,\mathrm{d}^3r\int f'^*\phi'^*\phi\delta f\,\mathrm{d}x\,\mathrm{d}x'\,\mathrm{d}^3r\right\}\Big/\langle\psi|\psi\rangle^2$$

Therefore $\delta E=0$ when $\delta f=cf$, where c is a constant. This is satisfied if $f(x)=\exp(cx)$. Furthermore the integrands involved in the expression for E can only depend physically on $x-x'$, giving the additional condition on f that $f^*(x')f(x)$ be just a function of $x-x'$. Hence c is purely imaginary and $f(x)=\exp(ikx)$. The function $f(x)$, which minimises the energy, is therefore a momentum eigenstate and projects out of ψ states of good momentum.

Likewise the function $f(\Omega)$, which minimises the energy E_{lm} of a state of angular momentum l and l_z component m, is the angular momentum eigenstate $Y_{lm}(\theta, \phi)$ where (θ, ϕ) specify the orientation Ω. For an even-even nucleus only even l are possible since ϕ is of even parity, so that equation 8.4 is identically zero for l odd. The energy E_{lm} is given by:

$$E_{lm} = \frac{\int \langle\phi^*|H|\phi'\rangle Y'^*_{lm} Y_{lm} \, d\Omega \, d\Omega'}{\int \langle\phi^*|\phi'\rangle Y'^*_{lm} Y_{lm} \, d\Omega \, d\Omega'}$$

where

$$\begin{aligned} |\phi'\rangle &\equiv \phi(\theta'\phi') \\ &= R(\Omega' - \Omega)\phi(\theta, \phi) \end{aligned}$$

$Y'^*_{lm} = Y^*_{lm}(\theta'\phi')$ etc. and H is the Hamiltonian. If the matrix elements are expressed as:

$$\langle\phi^*|H|\phi'\rangle = h(\Theta) = \sum h_l P_l(\cos \Theta) \tag{8.5}$$

and

$$\langle\phi^*|\phi'\rangle = n(\Theta) = \sum n_l P_l(\cos \Theta) \tag{8.6}$$

where Θ is the angle between Ω and Ω' then the addition theorem for spherical harmonics gives

$$E_{lm} = h_l / n_l$$

If the well has a large anisotropy the overlap between wavefunctions ϕ of different orientations is small unless $\Omega' \approx \Omega$, i.e. Θ is small. The overlap integrals involve integrating over all the nucleons' coordinates so the anisotropy does not have to be very large for the integrals to be sharply peaked about $\Theta = 0$.

The functions h_l are given by:

$$h_l = \frac{2l+1}{2} \int h(\Theta) P_l(\cos \Theta) \, d \cos \Theta$$

For Θ small, $P_l(\cos \Theta)$ can be expanded to give:

$$h_l \approx \frac{2l+1}{2} \int h(\Theta)[1 - \tfrac{1}{4}l(l+1)\Theta^2] \sin \Theta \, d\Theta$$

Likewise

$$n_l \approx \frac{2l+1}{2} \int n(\Theta)[1 - \tfrac{1}{4}l(l+1)\Theta^2] \sin \Theta \, d\Theta$$

Then

$$E_{lm} = \frac{H_0 - \frac{1}{4}l(l+1)H_2}{N_0 - \frac{1}{4}l(l+1)N_2}$$

where

$$H_n = \int h(\Theta)\Theta^n \sin \Theta \, d\Theta$$

and

$$N_n = \int n(\Theta)\Theta^n \sin\Theta \, d\Theta$$

For large anisotropy $n(\Theta)$ and $h(\Theta)$ are only large for small Θ so $H_2 \ll H_0$ and $N_2 \ll N_0$ so

$$E_{lm} \approx \frac{H_0}{N_0} + \tfrac{1}{4}l(l+1)\left(\frac{N_2 H_0}{N_0^2} - \frac{H_2}{N_0}\right)$$

This is of the form of a rotational band and also gives a value for the moment of inertia. As discussed above, the function $f(\Omega) = Y_{lm}(\Omega)$ can also be viewed as projecting out of the deformed wavefunction $\phi(\Omega)$ the wavefunction ϕ_{lm} of good angular momentum. It is clear that all members of the rotational band will have the same parity as that of $\phi(\Omega)$.

The treatment for nuclei with $K \neq 0$, which includes all odd-A nuclei, is similar to that for even–even nuclei. For axial symmetry the expression for ψ becomes:

$$\psi^I_M = \int D^{I*}_{MK}(\Omega)\phi_K(\Omega) \, d\Omega \tag{8.7}$$

This is like equation 8.4 but the values of I are not limited to just even or just odd values when $\phi_K(\Omega)$ has a definite signature, as they are when $K = 0$. If $\phi_K(\Omega)$ has a definite signature then an expansion of $\phi_K(\Omega)$ in eigenfunctions of $\mathbf{I}^2$ and $\mathbf{I}_{z'}$ has the form:

$$\phi_K(\Omega) = \sum_{I \geq K} c_I(\phi_{IK} \pm (-1)^{I+K}\phi_{I-K})$$

so

$$R_1\phi_K(\Omega) = r\phi_K(\Omega) = \pm(-1)^{-K}\phi_K(\Omega)$$

since $R_1\phi_{IK} = (-1)^I\phi_{I-K}$. Therefore the signature r is ± 1 for even-A and $\pm$i for odd-A nuclei. The values of I are $I \geq K$ and are only limited to even or to odd values when $K = 0$.

The energy E_{IM} involves overlap integrals like $h(\Theta)$ and $n(\Theta)$ above, equations 8.5 and 8.6. If the angle between Ω and Ω' is given by the Euler angles (α, β, γ) then the functions analogous to h_l and n_l are:

$$h^I_{KK} = \int h(\beta) d^I_{KK}(\beta) \, d\cos\beta$$

and

$$n^I_{KK} = \int n(\beta) d^I_{KK}(\beta) \, d\cos\beta$$

where the reduced rotation matrix $d^I_{KM}(\beta)$ is related to the rotation matrix D^I_{KM} by $D^I_{KM}(\alpha\beta\gamma) = \exp(-\mathrm{i}K\alpha)d^I_{KM}(\beta)\exp(-\mathrm{i}M\gamma)$ (appendix p. 265). For large anisotropies $h(\beta)$ and $n(\beta)$ are sharply peaked about $\beta = 0$ so it is useful to expand $d^I_{KK}(\beta)$ about $\beta = 0$ (for $K = \frac{1}{2}$ it is necessary to also consider the expansion about $\beta = \pi$):

$$d^I_{KK}(\beta) = 1 - \tfrac{1}{4}\beta^2[J(J+1) - K^2] + \cdots$$

Substituting this in the expressions for h^I_{KK} and n^I_{KK} gives:

$$E_{IM} \approx \frac{H_0}{N_0} + \tfrac{1}{4}[J(J+1) - K^2]\left(\frac{N_2 H_0}{N_0^2} - \frac{H_2}{N_0}\right)$$

where H_n and N_n are as defined above.

Rotational spectra are therefore predicted for both odd-A and even-A nuclei, if the potential well in which the nucleons move is significantly deformed. The expression for the moment of inertia is complicated to evaluate and requires the wavefunction given by the superposition of intrinsic wavefunctions $\phi(\Omega)$ (equation 8.4 or 8.7) to be a good approximation to the true wavefunction for an accurate value to be obtained. A somewhat simpler expression for the moment of inertia was deduced by Inglis using a semi-classical argument called the cranking model.

8.4 The cranking model for the moment of inertia

The rotational model describes the nucleus as a deformed intrinsic state which rotates about an axis perpendicular to its symmetry axis, the z axis. The cranking model considers the energy associated with this motion.

Consider a single-particle deformed potential, V, of fixed shape which rotates in space. The time-dependent Schrödinger equation is:

$$\left[\frac{-\hbar^2}{2m}\nabla^2_{r\theta\phi} + V(r\theta\phi t)\right]\psi(r\theta\phi t) = \mathrm{i}\hbar\frac{\partial\psi(r\theta\phi t)}{\partial t}$$

where θ is the angle between r and the x axis, ϕ is the angle of rotation about the x axis and

$$V(r\theta\phi t) = V(r\theta\phi - \omega t)$$

corresponding to a rotation with angular frequency ω.

Let

$$\alpha = \phi - \omega t \quad \text{then} \quad \left.\frac{\partial}{\partial\phi}\right)_{r\theta t} = \left.\frac{\partial}{\partial\alpha}\right)_{r\theta t}$$

so

$$\text{l.h.s.} \Rightarrow \left[\frac{-\hbar^2}{2m}\nabla^2_{r\theta\alpha} + V(r\theta\alpha)\right]\psi(r\theta\alpha t)$$

The r.h.s. is $\mathrm{i}\hbar\partial\psi/\partial t)_{r\theta\phi}$.

Now

$$\delta\psi = \left.\frac{\partial\psi}{\partial\alpha}\right)_{r\theta t}\delta\alpha + \left.\frac{\partial\psi}{\partial t}\right)_{r\theta\alpha}\delta t$$

so

$$\left.\frac{\partial\psi}{\partial t}\right)_{r\theta\phi} = \left.\frac{\partial\psi}{\partial\alpha}\right)_{r\theta t}\left.\frac{\partial\alpha}{\partial t}\right)_{r\theta\phi} + \left.\frac{\partial\psi}{\partial t}\right)_{r\theta\alpha}$$

$$= -\omega\left.\frac{\partial\psi}{\partial\alpha}\right)_{r\theta t} + \left.\frac{\partial\psi}{\partial t}\right)_{r\theta\alpha}$$

so

$$\left[\frac{-\hbar^2}{2m}\nabla^2_{r\theta\alpha} + V(r\theta\alpha)\right]\psi(r\theta\alpha t) = -\mathrm{i}\hbar\omega\left.\frac{\partial\psi}{\partial\alpha}\right)_{r\theta t} + \mathrm{i}\hbar\left.\frac{\partial\psi}{\partial t}\right)_{r\theta\alpha}$$

Let $\psi = \phi(r\theta\alpha)\exp(-\mathrm{i}\varepsilon t/\hbar)$. Noting that $\omega l_x = (\hbar/\mathrm{i})\omega\partial/\partial\alpha$ then

$$\left[-\frac{\hbar^2}{2m}\nabla^2_{r\theta\alpha} + V(r\theta\alpha) - \omega l_x\right]\phi(r\theta\alpha) = \varepsilon\phi(r\theta\alpha) \tag{8.8}$$

This is the equation that $\phi(r\theta\alpha)$ satisfies but to obtain the energy the expectation value of H is required:

$$\langle\psi|H|\psi\rangle = \varepsilon_\omega$$

Now

$$H\psi = \mathrm{i}\hbar\left.\frac{\partial\psi}{\partial t}\right)_{r\theta\phi} = \omega l_x\psi + \mathrm{i}\hbar\left.\frac{\partial\psi}{\partial t}\right)_{r\theta\alpha}$$

The operators on the r.h.s. are in terms of $(r\theta\alpha t)$ so putting $\psi = \phi(r\theta\alpha)\exp(-\mathrm{i}\varepsilon t/\hbar)$ gives:

$$\varepsilon_\omega = \langle\psi|H|\psi\rangle = \langle\psi|\omega l_x|\psi\rangle + \varepsilon$$

or

$$\varepsilon_\omega = \varepsilon + \omega\langle\phi|l_x|\phi\rangle \tag{8.9}$$

To evaluate ε_ω assume that ωl_x is a small perturbation and let ϕ_n be the eigenfunctions in the static deformed potential with no Coriolis forces, that is when $\omega = 0$, with ϕ_0 being the lowest state. Then to first order in ω equation 8.8 has the solution:

$$\phi = \phi_0 + \sum_{n\neq 0}\frac{\langle n|\omega l_x|0\rangle}{\varepsilon_n - \varepsilon_0}\phi_n \tag{8.10}$$

where ε_n are the eigenvalues in the static deformed potential. Introducing equation 8.10 into 8.9 we obtain to second order in ω:

$$\varepsilon_\omega = \varepsilon + 2\omega^2 \sum_n \frac{|\langle 0|l_x|n\rangle|^2}{\varepsilon_n - \varepsilon_0}$$

Now ε is also a function of ω as can be seen from equation 8.8. Considering $-\omega l_x$ as a perturbation it contributes to the energy ε only in second order:

$$\varepsilon = \varepsilon_0 - \omega^2 \sum_n \frac{|\langle 0|l_x|n\rangle|^2}{\varepsilon_n - \varepsilon_0}$$

so

$$\varepsilon_\omega = \varepsilon_0 + \omega^2 \sum_n \frac{|\langle 0|l_x|n\rangle|^2}{\varepsilon_n - \varepsilon_0}$$

$$= \varepsilon_0 + \tfrac{1}{2}\mathscr{I}\omega^2 \quad \text{with } \mathscr{I} = 2\sum_n \frac{|\langle 0|l_x|n\rangle|^2}{\varepsilon_n - \varepsilon_0} \tag{8.11}$$

which is Inglis's cranking formula for the moment of inertia. The operator $l_x = \frac{1}{2}(l_+ - l_-)$ links m with $m \pm 1$ states so for configurations involving closed shells $\langle l_x \rangle = 0$ as all m states are occupied. Therefore only valence nucleons contribute to the moment of inertia. The above derivation assumes intrinsic spin zero. Including intrinsic spin the Coriolis term becomes $-\omega j_x$.

If the residual interactions between nucleons are neglected then upon rotation all nucleons would rotate and the moment of inertia would be expected to equal $\mathscr{I}_{\text{rigid}}$. The cranking formula does give $\mathscr{I}_{\text{rigid}}$, which is surprising as only the valence nucleons contribute. To see how this is possible consider the expression for the mass of a nucleus:

$$M = 2\sum_n \frac{|\langle 0|p_z|n\rangle|^2}{\varepsilon_n - \varepsilon_0}$$

obtained in an analogous way. Just as for the rotational energy, the translational energy only involves states near the Fermi surface (the valence nucleons). How these can give rise to the whole nucleus moving is illustrated in figure 8.4*a*. The movement of the nucleus corresponds to a simple displacement in momentum space of the entire Fermi sphere. This may be achieved by moving all the states, or by taking a few from one side of the sphere and putting them on the other; the latter only involves states near the Fermi surface. For rotation a similar diagram may be drawn (figure 8.4*b*) and a similar equivalence is possible. Although only a few states are involved very near the Fermi surface, because the particles are identical every particle participates in these few states and a rigid rotation of the system can be completely equivalent to motion in which only a small fraction of the particles appear to be changing their state.

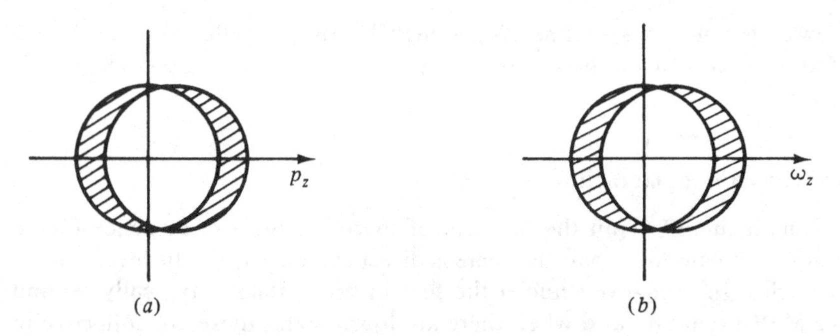

8.4. Illustration of the effect of (a) translation and (b) rotation on the distribution of nucleons near the Fermi surface.

A typical moment of inertia is $\sim\frac{1}{3}\mathscr{I}_{\text{rigid}}$ and this reduction is caused by the residual interaction. Using first-order perturbation theory gives surprisingly no change in the moment of inertia. However a residual interaction for which perturbation theory does not converge can give an effect. In particular, the attractive pairing residual interaction gives a lowering of the moment of inertia by lowering the energy of the ground state of nuclei. This introduces a pairing gap and increases the difference $(\varepsilon_n - \varepsilon_0)$ in the cranking-model expression for $\mathscr{I}$ (equation 8.11). Calculating this effect gives good agreement with experiment.

The increase in the moment of inertia seen at very high spin in some nuclei (chapter 2, p. 70) to values close to $\mathscr{I}_{\text{rigid}}$ is thought to be due to a decrease in pairing caused by the increase in the Coriolis force at very high spin. The pairing is in states near the Fermi surface. The breaking of one pair gives rise to the phenomenon of backbending (chapter 2, p. 64), and after a few pairs have been broken the effect of pairing can be lost. Before discussing the pairing interaction in more detail, a brief description of the effect of including the Coriolis term, $-\omega j_x$, in the Nilsson model is given.

8.4.1 The cranked Nilsson model

To predict the excited states of deformed nuclei in the Nilsson model at high angular momentum, it is necessary to add the Coriolis term $-\omega j_x$ to the Nilsson Hamiltonian to give what is called the cranked Nilsson Hamiltonian. The result is that the energy levels depend not only on deformation but also on the rotational angular momentum I. The relation between I and the angular frequency ω is $\omega = (E_I - E_{I-2})/2\hbar$ in an even-even nucleus (chapter 2, equation 2.15, p. 65). The effect of the Coriolis term is to break the two-fold degeneracy of the Nilsson levels, in particular for levels with high j and small Ω. For large frequencies levels from higher major shells are brought down to near the Fermi surface. Large gaps between levels (new magic numbers) can develop at high I which will affect the calculation of shell effects in the description of the shape of these states (chapter 5, p. 146).

8.5 The pairing interaction

Although the effect on the moment of inertia is indirect evidence for an energy gap caused by pairing, there is direct evidence from the level spectra of nuclei. In even–even nuclei the first intrinsic state is typically around 1–2 MeV excitation and when there are lower states these are collective in character, while in odd–even nuclei there are often low-lying intrinsic excited states. There is also evidence from the odd–even effect seen in the binding energy of nuclei. The pairing interaction also tends to keep nuclei spherical and delays the onset of significant deformation as more nucleons are added to a closed shell. This is because the gain in binding energy upon deformation ($b_r^2/4a_r$ in the Rainwater model) must be greater than the loss in binding energy caused by the consequent decrease in pairing.

The discussion of the pairing interaction in chapter 2 explained that a short-range attractive interaction would be most effective between two nucleons in a $J=0$ state and that the interaction would be expected to increase with increasing orbital angular momentum l. In general in a nucleus there are many different configurations near the Fermi surface in which two identical nucleons can couple to $J=0$ and this enhances the effect of the pairing interaction and gives rise to an energy gap. This gap is analogous to the energy gap in superconductors caused by the pairing of electrons. To see how this comes about a very simplified model of this pairing interaction will be analysed.

8.5.1 The energy gap

Consider a nucleus with two identical nucleons outside a closed shell. Let there be n essentially degenerate 0^+ states, in which the two nucleons could be, satisfying:

$$H_0\psi_i = \varepsilon\psi_i \qquad i=1,\ldots,n$$

where H_0 is the shell model Hamiltonian neglecting residual interaction.

The pairing interaction, V_p, has matrix elements $\langle\psi_i|V_p|\psi_j\rangle$ between these states and these are taken to be all the same, i.e.

$$\langle\psi_i|V_p|\psi_j\rangle = -V \qquad i,j=1,\ldots,n$$

which is the approximation discussed in chapter 7, p. 238.

The eigenfunctions ψ for the total Hamiltonian $H=H_0+V_p$ satisfy:

$$(H_0+V_p)\psi = E\psi$$

Substituting $\psi = \sum_i a_i \psi_i$ leads to the matrix equation:

$$\begin{pmatrix} \varepsilon - V & -V & -V & -V & \cdots & a_1 \\ -V & \varepsilon - V & -V & -V & \cdots & a_2 \\ -V & -V & \varepsilon - V & \cdots & & a_3 \\ \vdots & \vdots & \vdots & & & \vdots \end{pmatrix} = E \begin{pmatrix} a_1 \\ a_2 \\ a_3 \\ \vdots \end{pmatrix}$$

and the problem reduces to diagonalising this matrix.

For a non-trivial solution the determinant:

$$\begin{vmatrix} \varepsilon - V - E & -V & -V & \cdots \\ -V & \varepsilon - V - E & -V & \cdots \\ -V & -V & \varepsilon - V - E & \cdots \\ \vdots & \vdots & \vdots & \end{vmatrix} = 0$$

This has the solution:

$$(-1)^n (E - \varepsilon)^{n-1} (E - \varepsilon + nV) = 0$$

which means that $n-1$ states have the unperturbed energy $E = \varepsilon$ and one state is more bound and has $E = \varepsilon - nV$. If the energy gap generated is called 2Δ then Δ corresponds to the $\delta(A)$ term in the semi-empirical mass formula. The pairing energy Δ is therefore given by:

$$\Delta = \tfrac{1}{2} nV \equiv \frac{1}{2} \left(\frac{n}{V}\right) V^2 \approx \tfrac{1}{2} \rho_{\mathrm{F}} V^2 \tag{8.12}$$

where ρ_{F} is the density of levels at the Fermi surface. This approximation assumes that the pairing interaction mixes states within an energy interval of V about the Fermi level and hence $\rho_{\mathrm{F}} \approx n/V$. Equation 8.12 predicts a larger pairing energy gap, the higher the level density at the Fermi surface.

The wavefunction for the depressed 0^+ ground state in the above model is given by:

$$\psi_{\mathrm{gs}} = \sqrt{\frac{1}{n}} (\psi_1 + \psi_2 + \cdots + \psi_n)$$

i.e. a coherent superposition of many shell model 0^+ configurations. The ground state is therefore a collective rather than a pure shell model state. Independent experimental evidence for such a wavefunction is provided by the observation of enhanced two-proton $\sigma_{2\mathrm{p}}$ and two-neutron $\sigma_{2\mathrm{n}}$ transfer cross-sections. In the zero-range approximation these cross-sections are proportional to the square of a matrix element M of the form:

$$M = \int \exp(\mathrm{i}\mathbf{q} \cdot \mathbf{r}) \psi_{2\mathrm{n}}(r) \, \mathrm{d}^3 r$$

where q is the momentum transfer and $\psi_{2\mathrm{n}}$ is the wavefunction of the two nucleons. If $\psi_{2\mathrm{n}}$ is a coherent superposition of the form above then M will be of the order of:

$$M \approx \sqrt{n} \int \exp(\mathrm{i}\mathbf{q} \cdot \mathbf{r}) \phi_{2n}(r) \, \mathrm{d}^3 r$$

where ϕ_{2n} is a typical 2n configuration. Hence the cross-section will be approximately n times that of the reaction to a pure 2n state.

The pairing correlations also cause the occupancy of levels to fall smoothly through the Fermi level to zero, rather than abruptly. A particle-hole excitation of the ground state therefore becomes a two part-particle part-hole excitation called a two quasi-particle state.

8.6 Shell model description of collective states

For nuclei with not too many nucleons in unfilled shells e.g. s–d shell nuclei, it is possible to diagonalise the residual interaction in a shell model basis using either two-body matrix elements, such as those of Kuo and Brown, derived from the two-nucleon interaction, or elements obtained by fitting to known energy levels. The technique is like that described above for diagonalising the pairing interaction. An example is the description of ^{19}F in terms of seven nucleons outside a ^{12}C core, rather than just three nucleons outside an ^{16}O core. The low-lying states of ^{19}F are: ground state $\frac{1}{2}^+$; 110 keV $\frac{1}{2}^-$; 197 keV $\frac{5}{2}^+$; 1346 keV $\frac{5}{2}^-$; 1459 keV $\frac{3}{2}^-$; 1554 keV $\frac{3}{2}^+$. The seven nucleons are allowed to occupy the $1p_{1/2}$, $1d_{5/2}$ and $2s_{1/2}$ levels and many different configurations can contribute to one state. For example, $\{(\pi p_{1/2})^1(\nu p_{1/2})^2(\pi d_{5/2})^2(\nu s_{1/2})^2\}$ is a configuration which could be part of the $J^\pi = \frac{1}{2}^-$ state at 110 keV excitation in ^{19}F and is, referred to ^{16}O, an example of a (4p–1h) configuration. Diagonalising the residual interaction in such a basis gives a good description of the low-lying states of ^{19}F and, in particular, the occurrence of the low-lying states of opposite parity.

Rotational features are also accounted for in such shell model calculations, e.g. rotational bands in ^{20}Ne, reflecting that in these nuclei there are deformed intrinsic nuclear wavefunctions. Vibrational levels are also predicted. The method has proved very successful in describing many energy levels and, if core polarisation is allowed for by the use of effective charges, many γ-decay, β-decay and magnetic moment data. However, it is limited by the size of matrix required to describe the nuclear states (e.g. for $J^\pi = 4^+$, $T = 2$ states in ^{52}Cr described by (2p, 1f) shell model configurations, there are over three million states).

To overcome this problem various truncation schemes have been used. One technique is to include only configurations required by an approximate form of the residual interaction, such as the pairing plus quadrupole force. Another involves assuming the residual interaction pairs off nucleons in both s-pairs ($l=0$) and in d-pairs ($l=2$). These pairs act rather like bosons and several level spectra, of vibrational, rotational and transitional form, can be fit if the interactions between these bosons are taken into account. This model is called the interacting boson approximation (IBA). Relating the strengths of these interactions, however, to the nucleon–nucleon interaction is a difficult problem. A further technique is to assume states in nuclei can be described in terms of the motion of clusters of nucleons e.g. $^{19}F = {}^{15}N + \alpha$.

8.6.1 The cluster model

Buck, Dover and Vary proposed in 1975 a type of cluster model which accounts quantitatively for many features of both light and heavy nuclei. In this model, states in a nucleus are assumed to correspond to two sub-nuclei orbiting each other in a deep potential well. Hartree–Fock variational calculations for light nuclei, such as ^{24}Mg, indicate that there are indeed states which look like two clusters, providing a justification for this assumption. The shape of the potential well has a Gaussian form which is the shape generated by means of a double-folding calculation. Solving the Schrödinger equation for the relative motion of the two nuclei, the cluster and the core nucleus, in this well gives rise to a rotational band of states.

The allowed values of the principal quantum number N and angular momentum L for these states are related to the single particle quantum numbers n_j and l_j of the cluster nucleons (assuming the cluster nucleus is in its ground state) by having a constant value of

$$2N + L = \sum_{j=1}^{n_c} (2n_j + l_j)$$

where n_c is the number of cluster nucleons. To satisfy the Pauli exclusion principle the cluster nucleons must occupy levels above the Fermi surface of the core nucleus.

As an example of this model consider the ground state rotational band in ^{20}Ne described as $\alpha + {}^{16}O$. In the shell model the cluster nucleons are all in the 2s1d shell and as $n_j = 1$ for the 2s and $n_j = 0$ for the 1d level then $2N + L = 8$ giving rise to states with $J^{\pi} = 0^+, 2^+, 4^+, 6^+$ and 8^+. While in a harmonic oscillator shaped well these states would be degenerate, in a Gaussian shaped potential well the degeneracy is lifted and there is an $L(L+1)$ energy dependence in good agreement with experiment. Such an energy dependence is like that found in a molecular rotational band; however, in this model the energy of excited states in the band is not purely rotational but is a balance between the kinetic and potential energies of the clusters.

Other examples are the 0^+ and 1^- rotational bands in ^{16}O, with bandheads at $E_x = 6.05$ and 9.63 MeV, respectively, which may be described as $\alpha + {}^{12}$C cluster states. In the shell model the cluster nucleons are either all in the sd shell or three of them are and one is in the fp shell giving $2N+L=8$ or 9. The 0^+ band therefore has 0^+, 2^+, 4^+, 6^+ and 8^+ and the 1^- band 1^-, 3^-, 5^-, 7^- and 9^- levels, in agreement with observation. Their excitation energies are also reasonably well reproduced by a Gaussian shaped potential well, but with the parameters of the well dependent on the parity of the band, reflecting the effects of space-exchange on the potential well.

Several other observables are also well described by this model. For example in ^{24}Mg a ^{12}C$+^{12}$C cluster model reproduces the interband and intraband transitions of all the known positive- and negative-parity bands, and it may be capable of describing the 'quasi-molecular' resonances observed in the ^{12}C$+^{12}$C reaction at high energies. This cluster model has also recently been successfully applied to a description of alpha decays across the periodic table and to a description of exotic decays from heavy nuclei such as $^{223}\mathrm{Ra} \rightarrow {}^{209}\mathrm{Pb} + {}^{14}\mathrm{C}$ and $^{232}\mathrm{U} \rightarrow {}^{208}\mathrm{Pb} + {}^{24}\mathrm{Ne}$. For the exotic decays all the available lifetime data can be reproduced to within factors of 0.1 to 0.5, consistent with this factor being the preformation probability, i.e. the simple cluster-core configuration is a quite significant fraction of the ground state of the heavy nucleus.

8.6.2 Vibrational states and the schematic model

Near closed shells the low-lying collective states in even–even nuclei are 2^+ vibrational states. These correspond to a coherent superposition of 2^+ configurations, mainly of particles within the same shell ($0\hbar\omega$ excitations). The collective 2^+ state is depressed relative to the pure 2^+ configurations just like the 0^+ ground state is depressed relative to the pure 0^+ configurations. Further from closed shells nuclei are generally deformed and the collective 2^+ state lies at lower excitation and is a rotational state. The octupole vibrations (3^-) correspond to a coherent superposition of 3^- states and these are also depressed. Figure 8.5 shows the position of the lowest 2^+ and 3^- states in various even–even nuclei. Near closed shells the first 2^+ levels are at a relatively high excitation reflecting the low density of states when only a few nucleons from a closed shell. The effect of closed shells is much less pronounced for the lowest 3^- levels. This is because the 3^- configurations involve promoting a nucleon to the next major shell ($1\hbar\omega$ excitations) to give the negative parity; the number of such configurations is therefore not so affected by whether the occupied shell is full or not.

This also applies to the 1^- giant dipole state. Both the 1^- and 3^- states are superpositions of particle–hole (p–h) configurations and are vibrational states in a collective model description. There is also a 2^+ giant quadrupole

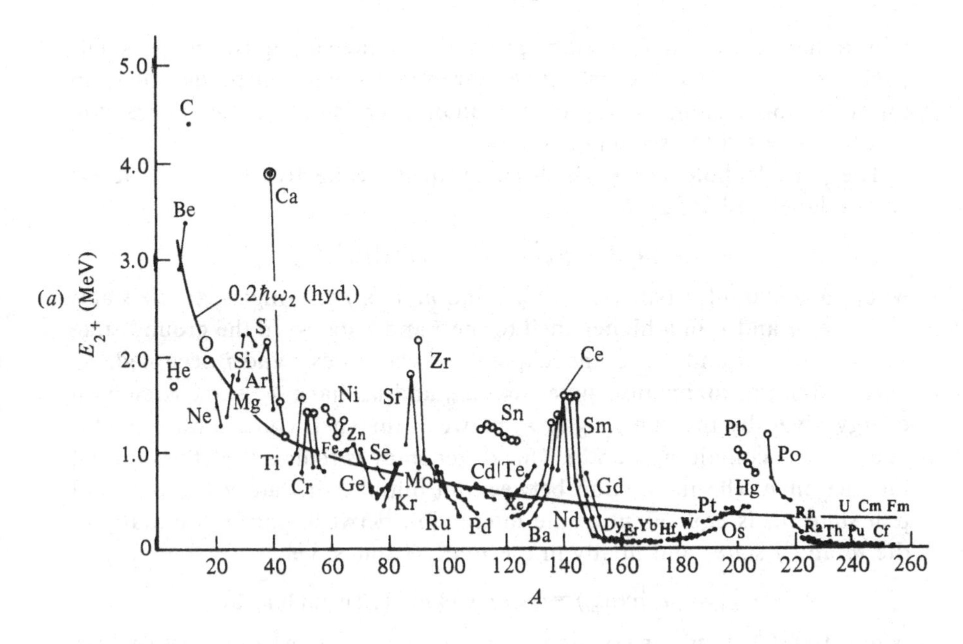

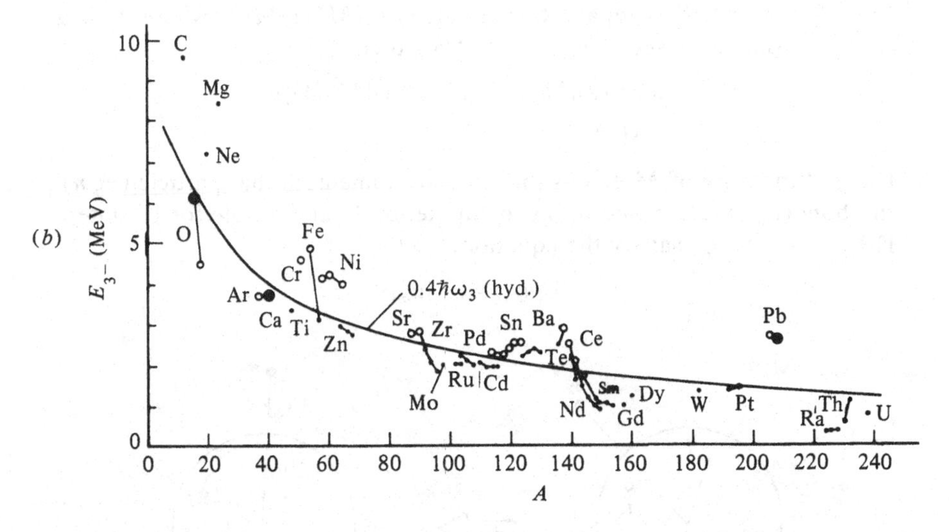

8.5. Excitation energies of (a) the first excited 2^+ state and (b) the collective 3^- state in even–even nuclei. (From Nathan, D. and Nilsson, S. G. in *α-, β- and γ-ray Spectroscopy* (ed. Siegbahn, K.), North-Holland, p. 601 (1965).)

vibrational (p–h) state formed by promoting a nucleon up two major shells ($2\hbar\omega$ excitations). These collective vibrational states can be described in the shell model using the approximation of a separable residual interaction in what is called the schematic model.

The particle–hole states which make up the collective 1^-, 2^+, 3^- levels can be labelled by:

$$|u_{mi}\rangle = |p_m h_i\rangle; \qquad |u_{nj}\rangle = |p_n h_j\rangle$$

where p, h stand for particle or hole, and m, i, n, j are single-particle states with the m and n in a higher shell to the i and j states. In the ground state $|0\rangle$ the states m and n are unoccupied and the states i and j occupied.

As a first approximation states like u_{mi} and u_{nj} have the same excitation energy given by the energy gap ε_g between the shell containing m and n and the one containing i and j. This degeneracy is removed by the residual interaction v. The interaction between an initial p–h state u_{mi} and a final p–h state u_{nj} is equivalent to the interaction between particles initially in the m and j states and finally in the n and i states, i.e.:

$$v_{njmi} = \langle u_{nj}|v|u_{mi}\rangle \equiv \langle p_n(2)p_i(1)|v(1,2)|p_m(1)p_j(2)\rangle$$

where 1 and 2 stand for positions $\mathbf{r}_1$ and $\mathbf{r}_2$, i.e. $|h_i\rangle \equiv \langle p_i|$ and $\langle h_j| \equiv |p_j\rangle$, see figure 8.6. If $v(1,2)$ is separable, i.e. $v(1,2) = \lambda M^*(1)M(2)$ where M is a multipole operator (see chapter 7, p. 238), then:

$$\begin{aligned} v_{njmi} &= \lambda\langle p_n(2)|M(2)|p_j(2)\rangle\langle p_i(1)|M^*(1)|p_m(1)\rangle \\ &= \lambda M_{nj} M^*_{mi} \end{aligned}$$

The multipolarity of M equals the angular momentum that particle (m, n) and hole (i, j) are coupled to, i.e. octupole for 3^- and dipole for 1^- states. The p–h states u_{mi} satisfy the equation:

$$H_0|u_{mi}\rangle = \varepsilon_{mi}|u_{mi}\rangle$$

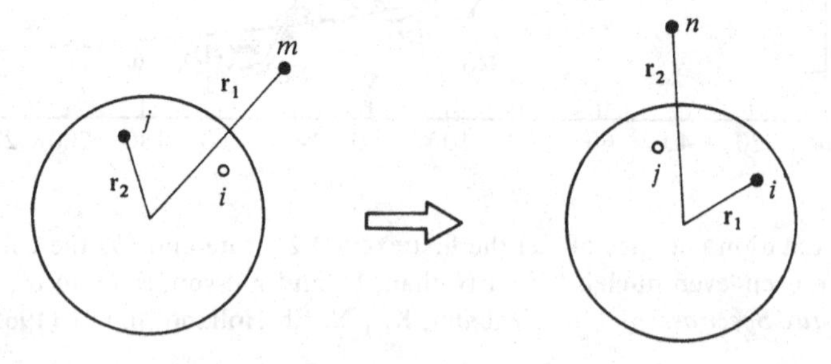

8.6. Illustration of particles initially in the m and j states with a hole in the i state, and finally in the n and i states with a hole in the j state.

where H_0 is the shell model Hamiltonian with no residual interaction. The eigenfunctions ψ of the total Hamiltonian satisfy:

$$(H_0 + v)|\psi\rangle = \varepsilon|\psi\rangle$$

Substituting $|\psi\rangle = \sum_{mi} c_{mi}|u_{mi}\rangle$ and multiplying by $\langle u_{nj}|$ and integrating gives:

$$\sum_{mi} c_{mi} v_{njmi} = (\varepsilon - \varepsilon_{nj})c_{nj}$$

$$\therefore \quad \lambda M_{nj} \sum_{mi} c_{mi} M^*_{mi} = (\varepsilon - \varepsilon_{nj})c_{nj} \tag{8.13}$$

The sum $\sum_{mi} c_{mi} M^*_{mi}$ equals some constant X so:

$$c_{nj} = \frac{\lambda X}{\varepsilon - \varepsilon_{nj}} M_{nj} \tag{8.14}$$

$$\therefore \quad (\lambda X)^2 \sum_{nj} \frac{|M_{nj}|^2}{(\varepsilon - \varepsilon_{nj})^2} = 1 \tag{8.15}$$

since $\sum_{nj} c^*_{nj} c_{nj} = 1$. Rearranging and summing equation 8.13 gives:

$$\sum_{nj} \frac{\lambda M_{nj}}{\varepsilon - \varepsilon_{nj}} M^*_{nj} \sum_{mi} c_{mi} M^*_{mi} = \sum_{nj} c_{nj} M^*_{nj}$$

$$\therefore \quad \frac{1}{\lambda} = \sum_{nj} \frac{|M_{nj}|^2}{\varepsilon - \varepsilon_{nj}}$$

This can be solved graphically (see figure 8.7). It is seen that all eigenvalues

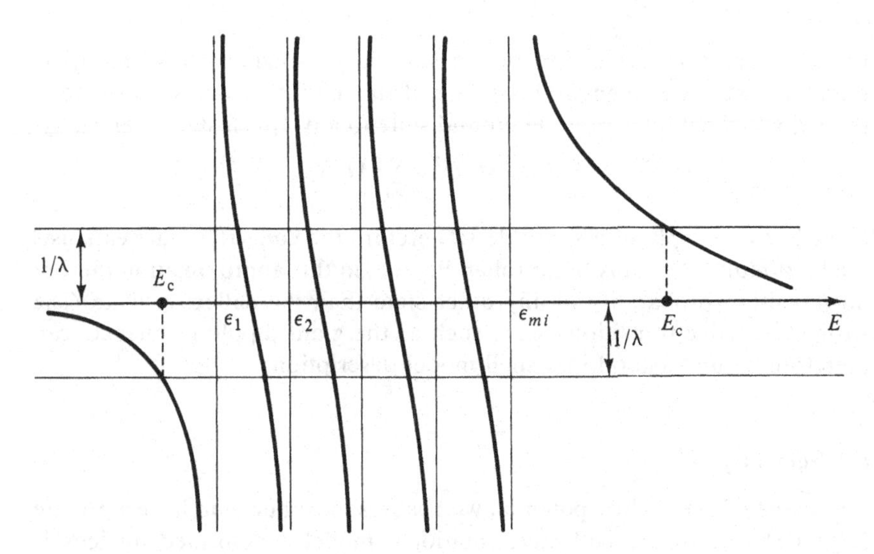

8.7. Graphical solution of the eigenvalue equation for the excitation energies E in the schematic model. E_c is the energy of the collective state. (From Ring, P. and Shuck, P., *The Nuclear Many-Body Problem.* Springer-Verlag (1980).)

ε lie between the values of ε_{nj} except one which is either higher ($1/\lambda$ positive) or lower ($1/\lambda$ negative). In general the residual interaction is attractive (λ negative) for $T=0$ states and repulsive for $T=1$ states; examples are the 3^-, $T=0$ and 1^-, $T=1$ collective states in even-even nuclei.

The one state whose eigenvalue is pushed up or down is formed from a coherent superposition of all the p-h states. If all the p-h states are degenerate and have $\varepsilon_{mi}=\varepsilon_g$ then the coefficients c_{mi} in the collective state $|u_c\rangle$ are given from equations 8.14 and 8.15 above by:

$$c_{mi}=\left(\sum_{mi}|M_{mi}|^2\right)^{-1/2} M_{mi} \tag{8.16}$$

and

$$|u_c\rangle=\sum_{mi} c_{mi}|u_{mi}\rangle$$

and the energy of the collective state ε_c by:

$$\varepsilon_c=\varepsilon_g+\lambda\sum_{mi}|M_{mi}|^2$$

This collective state has a very enhanced electromagnetic transition probability to the ground state $|0\rangle$. The square of the transition matrix element is:

$$|\langle u_c|M|0\rangle|^2=\sum_{mi}|M_{mi}|^2$$

i.e. a coherent superposition of the squares of single-particle multipole transition matrix elements. The square of the matrix elements for the total transition probability from the ground state to any 1p-1h state is given by:

$$\sum_{mi}|\langle u_{mi}|M|0\rangle|^2=\sum_{mi}|M_{mi}|^2$$

(This is an example of a sum rule.) Therefore the collective state exhausts the total multipole transition probability, i.e. in this approximation there is no transition probability to any other state than the collective state. The strong collective transitions seen, such as the giant dipole resonance, can therefore be understood in a shell model description.

8.7 Summary

The shape of the nuclear potential well causes most nuclei to have a prolate (cigar) shape. In the collective rotational model a deformed nucleus is described by the product of an intrinsic and a rotational wavefunction. While this accounts for many features, it involves redundant variables. This problem is avoided by the variational method of Peierls and Yoccoz, which is equivalent to projecting states of good angular momentum from a deformed intrinsic wavefunction. These states form a rotational band if the nucleus is significantly deformed.

Calculation of the effective moment of inertia is complicated in this method but a semi-classical argument gives a somewhat simpler expression: the Inglis cranking formula. Using this formula the effect of pairing on the moment of inertia is seen to be significant and can explain the reduction of observed values from the rigid body value. The energy gap caused by the pairing interaction, analogous to that in a superconductor, is calculated and it is seen that the ground state is a collective superposition of many configurations.

Collective rotational states in nuclei derive in an independent-particle model description from deformed intrinsic wavefunctions. In a spherical shell model basis these contain many configurations. Vibrational states correspond to a coherent superposition of many particle–hole configurations and using the schematic model their collective character can be understood. In summary, many collective as well as single-particle phenomena can be well described by an independent-particle model of the nucleus.

8.8 Questions

1. A ground state rotational band is observed in ${}^{20}_{10}\mathrm{Ne}$ whose J^{π} sequence is 0^+, 2^+, 4^+, 6^+ and 8^+. Explain why no state with higher spin is seen in this band. The states up to the 6^+ are at 0 MeV (0^+), 1.63 MeV (2^+), 4.25 MeV (4^+) and 8.78 MeV (6^+) excitation in ${}^{20}_{10}\mathrm{Ne}$. Comment on the fact that the following states are found at 0.11 MeV ($\frac{1}{2}^-$), 1.35 MeV ($\frac{5}{2}^-$), 1.46 MeV ($\frac{3}{2}^-$), 4.00 MeV ($\frac{7}{2}^-$), 4.03 MeV ($\frac{9}{2}^-$), 7.17 MeV ($\frac{11}{2}^-$) and 8.29 MeV ($\frac{13}{2}^-$) excitation in ${}^{19}_{9}\mathrm{F}$.
2. Discuss the similarity in the low-lying spectra of ${}^{18}_{8}\mathrm{O}$ and ${}^{24}_{10}\mathrm{Ne}$ [${}^{18}_{8}\mathrm{O}$: 0 MeV (0^+), 1.48 MeV (2^+), 3.56 MeV (4^+), 3.63 MeV (0^+), 3.92 MeV (2^+), ${}^{24}_{10}\mathrm{Ne}$: 0 MeV (0^+), 1.98 MeV (2^+), 3.87 MeV (2^+), 3.96 MeV (4^+), 4.76 MeV (0^+)].
3. Show that when the $N=4$ levels in an anisotropic oscillator potential well are filled with two particles the equilibrium deformation $\bar{\delta}_{\mathrm{p}}=0.14$ and $\bar{\delta}_0=-0.07$.
4. Derive the expression for the mass M of a nucleus:
$$M=2\sum_n [|\langle 0|p_z|n\rangle|^2/(\varepsilon_n-\varepsilon_0)]$$
5. By taking ϕ to be an antisymmetrised product wavefunction $\phi=$ (Antisym)$\prod_i u_i$, show that the result of applying the variational principle to determine ϕ is that the u_i must satisfy the Hartree–Fock

equations:

$$T_i u_i(r_i) + \sum_{j \neq i} \int u_j^*(r_j) V(r_i r_j) u_i(r_i) u_j(r_j) \, \mathrm{d}^3 r_j$$

$$- \sum_{j \neq i} \int u_j^*(r_j) V(r_i r_j) u_i(r_j) u_j(r_i) \, \mathrm{d}^3 r_j = \alpha_i u_i(r_i)$$

6. Derive equation 8.16 for the coefficients c_{mi} in the collective state $|u_c\rangle$ when all the p–h states are degenerate.
7. Describe the shell and collective models of the nucleus. Give examples of phenomena that illustrate each and discuss the collective model in terms of an independent-particle model of the nucleus.

Appendix Rotations

Rotations may be described using the Euler angles (α, β, γ) in the following way. If a set of axes x, y, z are rotated by (α, β, γ) then the new axes x', y', z' are obtained by first rotating γ about the original z-axis, then by β about the original y-axis and finally by α about the original z-axis. Such a rotation is described by the rotation operator:

$$D(\alpha, \beta, \gamma) = \exp(-\mathrm{i}\alpha J_z) \exp(-\mathrm{i}\beta J_y) \exp(-\mathrm{i}\gamma J_z)$$

The effect of a rotation upon a wavefunction with angular momentum quantum numbers IM is described by the rotation matrix:

$$\langle IM'|D(\alpha, \beta, \gamma)|IM\rangle = D^I_{M'M}(\alpha, \beta, \gamma)$$

Because the functions $|IM\rangle$ are chosen to be eigenfunctions of J_z then the rotation matrix can be expressed as:

$$D^I_{M'M}(\alpha, \beta, \gamma) = \exp[-\mathrm{i}(\alpha M' + \gamma M)] d^I_{M'M}(\beta)$$

where $d^I_{M'M}(\beta)$ is called the reduced rotation matrix.

If the effect of a rotation (α, β, γ) on the spherical harmonic $Y_{lm}(\theta, \phi)$ is to give $Y_{lm}(\theta', \phi')$, i.e. to alter (θ, ϕ) to (θ', ϕ'), then:

$$Y_{lm}(\theta', \phi') = \sum_{m'} D^l_{m'm}(\alpha, \beta, \gamma) Y_{lm'}(\theta, \phi)$$

which follows from the definition of the rotation matrix given above. Spherical harmonics are examples of spherical tensors. A general spherical tensor T_k of rank k is a quantity with $2k+1$ components T_{kq} which transform under rotations according to:

$$T'_{kq} = \sum_p T_{kp} D^k_{pq}(\alpha, \beta, \gamma)$$

A spherical tensor of rank zero is a scalar and of rank one is a vector. The tensor component to the nuclear force discussed in chapter 7 is a spherical tensor of rank two.

The wavefunctions $|\alpha_1 j_1 m_1\rangle$ and $|\alpha_2 j_2 m_2\rangle$, where α_1 and α_2 represent other quantum numbers describing the states, are spherical tensors of rank j_1 and j_2. Their product, which represents the coupling of the two angular momenta, can be expressed in terms of wavefunctions $|\beta JM\rangle$, where $|j_1 - j_2| \leq J \leq j_1 + j_2$, by:

$$|\alpha j_1 j_2 m_1 m_2\rangle \equiv |\alpha_1 j_1 m_1\rangle |\alpha_2 j_2 m_2\rangle$$
$$= \sum_{JM} |\beta JM\rangle\langle JM|j_1 j_2 m_1 m_2\rangle$$

The coefficients $\langle JM|j_1 j_2 m_1 m_2\rangle$ are called Wigner or Clebsch-Gordon coefficients and they vanish unless $M = m_1 + m_2$. For example, the coupling of a $j = \frac{1}{2}$ $m = \frac{1}{2}$ state with a $j = 1$ $m = 0$ state can give $J = \frac{1}{2}$ or $J = \frac{3}{2}$, and their product is given in terms of $J = \frac{1}{2}$ and $J = \frac{3}{2}$ functions by:

$$|\tfrac{1}{2}1\tfrac{1}{2}0\rangle = |\tfrac{1}{2}\tfrac{1}{2}\rangle\langle\tfrac{1}{2}\tfrac{1}{2}|\tfrac{1}{2}1\tfrac{1}{2}0\rangle + |\tfrac{3}{2}\tfrac{1}{2}\rangle\langle\tfrac{3}{2}\tfrac{1}{2}|\tfrac{1}{2}1\tfrac{1}{2}0\rangle$$
$$= \sqrt{\tfrac{1}{3}}|\tfrac{1}{2}\tfrac{1}{2}\rangle + \sqrt{\tfrac{2}{3}}|\tfrac{3}{2}\tfrac{1}{2}\rangle$$

Spherical tensors can be operators as well as wavefunctions so the product of a spherical tensor operator T_{kq} and a wavefunction $|\alpha jm\rangle$ can be expressed in terms of wavefunctions $|\beta KQ\rangle$ by:

$$T_{kq}|\alpha jm\rangle = \sum_{KQ} |\beta KQ\rangle\langle KQ|kjqm\rangle$$

so the matrix element $\langle\alpha' j' m'|T_{kq}|\alpha jm\rangle$ factorizes into two parts:

$$\langle\alpha' j' m'|T_{kq}|\alpha jm\rangle = \langle\alpha' j' m'|\beta j' m'\rangle\langle j' m'|kjqm\rangle$$

i.e. only the part of $T_{kq}|\alpha jm\rangle$ which transforms as a function with angular momentum j' can contribute to the matrix element which must be a scalar. Since the inner product of the wavefunctions $\langle\alpha' j' m'|\beta j' m'\rangle$ is independent of m', the matrix element of the tensor operator T_{kq} can be written as:

$$\langle\alpha' j' m'|T_{kq}|\alpha jm\rangle = (-)^{2k}\langle\alpha' j'\| T_k \|\alpha j\rangle\langle j' m'|kjqm\rangle$$

This relation is known as the Wigner-Eckart theorem and $\langle\alpha' j'\| T_k \|\alpha j\rangle$ is called the reduced matrix element. All the dependence of the matrix element on the orientations of the wavefunctions, i.e. on the m' and m values, is contained in the Clebsch-Gordon coefficient.

An example of the use of the Wigner-Eckart theorem is in the derivation of the isobaric multiplet mass equation. The Coulomb interaction between the nucleons in a nucleus is given by:

$$H_c = (e^2/4\pi\varepsilon_0) \sum_{i>j} (\tfrac{1}{2} - t_{z_i})(\tfrac{1}{2} - t_{z_j})(1/r_{ij})$$

This may be rewritten as:

$$H_c = (e^2/4\pi\varepsilon_0) \sum_{i>j} (1/r_{ij})\{[\tfrac{1}{4} + \mathbf{t}_i \cdot \mathbf{t}_j/3]$$
$$-[(t_{z_i} + t_{z_j})/2] + [t_{z_i} t_{z_j} - \mathbf{t}_i \cdot \mathbf{t}_j/3]\}$$

where the first term is a scalar, the second a vector and the third a second rank tensor in isospin space. Also contributing to the mass differences of members of an isobaric multiplet are the neutron-proton mass difference, which can be expressed as the sum of isoscalar $(M_n + M_p)A/2$ and isovector $(M_n - M_p)T_z$ terms, and the spin-orbit interaction, which can be expressed like the Coulomb interaction in terms of an isoscalar, an isovector and a second rank isotensor. The resulting total perturbation H' can be written therefore in terms of rank zero, rank one and rank two isotensors, i.e. $H' = H^{(0)} + H^{(1)} + H^{(2)}$, with the m component of $H^{(1)}$ and $H^{(2)}$ zero.

Using the Wigner-Eckart theorem the expectation value of H' is given by:

$$\begin{aligned}\langle \alpha T T_z | H' | \alpha T T_z \rangle = {} & \langle \alpha T \| H^{(0)} \| \alpha T \rangle + \{ T_z / \sqrt{T(T+1)} \} \langle \alpha T \| H^{(1)} \| \alpha T \rangle \\ & + \{(3T_z^2 - T(T+1)) / \sqrt{(2T-1)T(T+1)(2T+3)}\} \\ & \times \langle \alpha T \| H^{(2)} \| \alpha T \rangle\end{aligned}$$

i.e. $\langle \alpha T T_z | H' | \alpha T T_z \rangle = a + bT_z + cT_z^2$ which is the isobaric multiplet mass equation.

Answers to selected questions

1.2	5.1 fm
1.6	0.21 fm; 7.1×10^{-25} s
1.7	^{17}F: 11.07 MeV; ^{17}O: 10.97 MeV
2.1	^{31}Na
2.2	−15.67 MeV; −4.14 MeV; −0.60 MeV
2.6	−25.0 b
2.8	0.277
2.12	$2\mathscr{I}/\hbar^2 \approx 170\ \text{MeV}^{-1}$
3.2	3.236 MeV
3.7	2.64×10^{-13} s; 222 μm
3.11	3.4×10^{-7}
3.15	99.84%
4.2	$\sim10^{-13}$ s
4.5	~100 s
4.7	3.3×10^{-15}; 1.0 μm/s
4.9	5.9×10^{-20} s
5.2	2.4 : 7.8 : 24.5
5.5	^{235}U: 2.1 kg; ^{239}Pu: 3.15 kg
6.1	10 148 keV
6.2	2.95×10^{25}
6.8	14.3 b; 2^+
7.1	~1 m

Bibliography

Undergraduate texts

Bowler, M. G. (1973). *Nuclear Physics.* Pergamon Press.
Brink, D. M. (1965). *Nuclear Forces.* Pergamon Press.
Burcham, W. E. (1979). *Elements of Nuclear Physics.* Longman.
Cohen, B. L. (1971). *Concepts of Nuclear Physics.* McGraw-Hill.
Cottingham, W. N. and Greenwood, D. A. (1986). *An Introduction to Nuclear Physics.* Cambridge University Press.
Enge, H. A. (1966). *Introduction to Nuclear Physics.* Addison-Wesley.
Feynman, R. P. (1965). *Lectures in Physics* Vol. III. Addison-Wesley.
Frauenfelder, H. and Henley, E. M. (1974). *Subatomic Physics.* Prentice-Hall.
Gasiorowicz, S. (1974). *Quantum Physics.* Wiley.
Jones, G. A. (1987). *Properties of Nuclei* (2nd edn). Oxford University Press.
Krane, H. S. (1987). *Introductory Nuclear Physics.* Wiley.
Meyerhof, W. E. (1967). *Elements of Nuclear Physics.* McGraw-Hill.
Pearson, J. M. (1986). *Nuclear Physics, Energy and Matter.* Adam Hilger.
Perkins, D. H. (1987). *Introduction to High Energy Physics* (3rd edn). Addison-Wesley.
Satchler, G. R. (1980). *Introduction to Nuclear Reactions.* Wiley.
Segrè, E. (1977). *Nuclei and Particles* (2nd edn). Benjamin.

Graduate texts

Bass, R. (1980). *Nuclear Reactions with Heavy Ions.* Springer-Verlag.
Blatt, J. M. and Weisskopf, V. F. (1952). *Theoretical Nuclear Physics.* Wiley.
Bock, R. (editor) (1980). *Heavy Ion Collisions.* North-Holland.
Bohr, A. and Mottelson, B. R. (1969, 1975). *Nuclear Structure* Vols. 1 and 2. Benjamin.
Bromley, D. A. (editor) (1984). *Treatise on Heavy-Ion Science.* Plenum Press.
Cerny, J. (editor) (1974). *Nuclear Spectroscopy and Reactions.* Academic Press.
De Benedetti, S. (1964). *Nuclear Interactions.* Wiley.
De Shalit, A. and Feshbach, H. (1974). *Theoretical Nuclear Physics* Vol. 1. Wiley.
Elton, L. R. B. (1959). *Introductory Nuclear Theory*, Pitman.
England, J. B. A. (1974). *Techniques in Nuclear Structure Physics* Vols. 1 and 2. Macmillan.

Hamilton, W. D. (editor) (1975). *The Electromagnetic Interaction in Nuclear Spectroscopy.* North-Holland.
Irvine, J. M. (1975). *Heavy Nuclei, Superheavy Nuclei and Neutron Stars.* Oxford University Press.
Lawson, R. D. (1980). *Theory of the Nuclear Shell Model.* Oxford University Press.
Marmier, P. and Sheldon, E. (1969). *Physics of Nuclei and Particles.* Academic Press.
Mathews, J. and Walker, R. L. (1970). *Mathematical Methods for Physicists* (2nd edn). Benjamin.
Migdal, A. B. (1977). *Qualitative Methods in Quantum Theory.* Benjamin.
Preston, M. A. and Braduri, R. K. (1975). *Structure of the Nucleus.* Addison-Wesley.
Ragnarsson, I. (in press). *Shapes and Shells in Nuclear Structure.* Cambridge University Press.
Rho, M. and Wilkinson, D. H. (editors) (1979). *Mesons in Nuclei.* North-Holland.
Ring, P. and Schuck, P. (1980). *The Nuclear Many-Body Problem.* Springer-Verlag.
Rowe, D. J. (1970). *Nuclear Collective Motion.* Methuen.
Satchler, G. R. (1983). *Direct Nuclear Reactions.* Oxford University Press.
Siegbahn, K. (1966). *α-, β- and γ-ray Spectroscopy* Vols. 1 and 2. North-Holland.
Wilkinson, D. H. (editor) (1969). *Isospin in Nuclear Physics.* North-Holland.
Wu, C. S. and Moszkowski, S. A. (1966). *Beta Decay.* Wiley.

Index